Graduate Texts in Mathematics

53

Yu. I. Manin

A Course in Mathematical Logic

Translated from the Russian by
Neal Koblitz

Springer Science+Business Media, LLC

Yu. I. Manin
V. A. Steklov Mathematical
Institute of the Academy of Sciences
Moscow V-333
Ul. Vavilova 42
USSR

Neal Koblitz
Department of Mathematics
Harvard University
Cambridge, Massachusetts 02138
USA

AMS Subject Classifications: 02-01, 02Bxx, 02Fxx

Library of Congress Cataloging in Publication Data

Manin, IU I
A course in mathematical logic.
(Graduate texts in mathematics; 53)
Includes index.
1. Logic, Symbolic and mathematical. I. Title.
II. Series.
QA9.M296 511'.3 77-1838

Originally published by Springer-Verlag New York, Inc. in 1977
MyCopy version of the original edition 1977

9 8 7 6 5 4 3 2 1

DOI 10.1007/978-1-4757-4385-2
www.springer.com/mycopy

To my son

Preface

1. This book is above all addressed to mathematicians. It is intended to be a textbook of mathematical logic on a sophisticated level, presenting the reader with several of the most significant discoveries of the last ten or fifteen years. These include: the independence of the continuum hypothesis, the Diophantine nature of enumerable sets, the impossibility of finding an algorithmic solution for one or two old problems.

All the necessary preliminary material, including predicate logic and the fundamentals of recursive function theory, is presented systematically and with complete proofs. We only assume that the reader is familiar with "naive" set theoretic arguments.

In this book mathematical logic is presented both as a part of mathematics and as the result of its self-perception. Thus, the substance of the book consists of difficult proofs of subtle theorems, and the spirit of the book consists of attempts to explain what these theorems say about the mathematical way of thought.

Foundational problems are for the most part passed over in silence. Most likely, logic is capable of justifying mathematics to no greater extent than biology is capable of justifying life.

2. The first two chapters are devoted to predicate logic. The presentation here is fairly standard, except that semantics occupies a very dominant position, truth is introduced before deducibility, and models of speech in formal languages precede the systematic study of syntax.

The material in the last four sections of Chapter II is not completely traditional. In the first place, we use Smullyan's method to prove Tarski's theorem on the undefinability of truth in arithmetic, long before the introduction of recursive functions. Later, in the seventh chapter, one of the proofs of the incompleteness theorem is based on Tarski's theorem. In the second place, a large section is devoted to the logic of quantum mechanics and to a proof of von Neumann's theorem on the absence of "hidden variables" in the quantum mechanical picture of the world.

The first two chapters together may be considered as a short course in logic apart from the rest of the book. Since the predicate logic has received the widest dissemination outside the realm of professional mathematics, the author has not resisted the temptation to pursue certain aspects of its relation to linguistics, psychology, and common sense. This is all discussed in a series of digressions, which, unfortunately, too often end up trying to explain "the exact meaning of a proverb" (E. Baratynskiĭ [1]). This series of digressions ends with the second chapter.

The third and fourth chapters are optional. They are devoted to complete proofs of the theorems of Gödel and Cohen on the independence of the continuum hypothesis. Cohen forcing is presented in terms of Boolean-valued models; Gödel's constructible sets are introduced as a subclass of von Neumann's universe. The number of omitted formal deductions does not exceed the accepted norm; due respects are paid to syntactic difficulties. This ends the first part of the book: "Provability."

The reader may skip the third and fourth chapters, and proceed immediately to the fifth. Here we present elements of the theory of recursive functions and enumerable sets, formulate Church's thesis, and discuss the notion of algorithmic undecidability.

The basic content of the sixth chapter is a recent result on the Diophantine nature of enumerable sets. We then use this result to prove the existence of versal families, the existence of undecidable enumerable sets, and, in the seventh chapter, Gödel's incompleteness theorem (as based on the definability of provability via an arithmetic formula). Although it is possible to disagree with this method of development, it has several advantages over earlier treatments. In this version the main technical effort is concentrated on proving the basic fact that all enumerable sets are Diophantine, and not on the more specialized and weaker results concerning the set of recursive descriptions or the Gödel numbers of proofs.

[1] Nineteenth century Russian poet (translator's note). The full poem is:

> We diligently observe the world,
> We diligently observe people,
> And we hope to understand their deepest meaning.
> But what is the fruit of long years of study?
> What do the sharp eyes finally detect?
> What does the haughty mind finally learn
> At the height of all experience and thought,
> What?—the exact meaning of an old proverb.
>
> 1828

The last section of the sixth chapter stands somewhat apart from the rest. It contains an introduction to the Kolmogorov theory of complexity, which is of considerable general mathematical interest.

The fifth and sixth chapters are independent of the earlier chapters, and together make up a short course in recursive function theory. They form the second part of the book: "Computability."

The third part of the book, "Provability and Computability," relies heavily on the first and second parts. It also consists of two chapters. All of the seventh chapter is devoted to Gödel's incompleteness theorem. The theorem appears later in the text than is customary because of the belief that this central result can only be understood in its true light after a solid grounding both in formal mathematics and in the theory of computability. Hurried expositions, where the proof that provability is definable is entirely omitted and the mathematical content of the theorem is reduced to some version of the "liar paradox," can only create a distorted impression of this remarkable discovery. The proof is considered from several points of view. We pay special attention to properties which do not depend on the choice of Gödel numbering. Separate sections are devoted to Feferman's recent theorem on Gödel formulas as axioms, and to the old but very beautiful result of Gödel on the length of proofs.

The eighth and final chapter is, in a way, removed from the theme of the book. In it we prove Higman's theorem on groups defined by enumerable sets of generators and relations. The study of recursive structures, especially in group theory, has attracted continual attention in recent years, and it seems worthwhile to give an example of a result which is remarkable for its beauty and completeness.

3. This book was written for very personal reasons. After several years or decades of working in mathematics, there almost inevitably arises the need to stand back and look at this research from the side. The study of logic is, to a certain extent, capable of fulfilling this need.

Formal mathematics has more than a slight touch of self-caricature. Its structure parodies the most characteristic, if not the most important, features of our science. The professional topologist or analyst experiences a strange feeling when he recognizes the familiar pattern glaring out at him in stark relief.

This book uses material arrived at through the efforts of many mathematicians. Several of the results and methods have not appeared in monograph form; their sources are given in the text. The author's point of view has formed under the influence of the ideas of Hilbert, Gödel, Cohen, and especially John von Neumann, with his deep interest in the external world, his open-mindedness and spontaneity of thought.

Various parts of the manuscript have been discussed with Yu. V. Matijasevič, G. V. Čudnovskiĭ, and S. G. Gindikin. I am deeply grateful to all of these colleagues for their criticism.

W. D. Goldfarb of Harvard University very kindly agreed to proofread the entire manuscript. For his detailed corrections and laborious rewriting of part of Chapter IV, I owe a special debt of gratitude.

I wish to thank Neal Koblitz for his meticulous translation.

Yu. I. Manin

Moscow, September 1974

Interdependence of Chapters

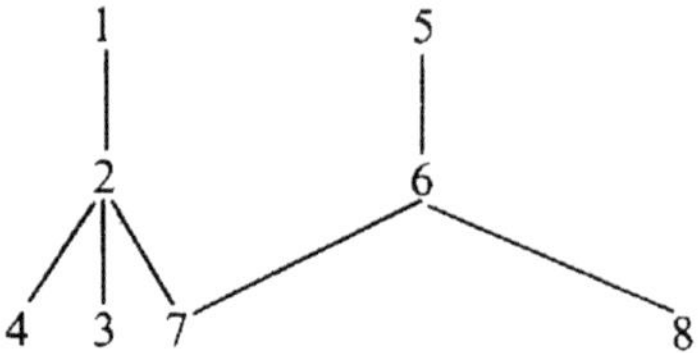

Contents

Part I

PROVABILITY

[1] "A metaphorical compound word or phrase used especially in Old English and Old Norse poetry, e.g., 'swan-road' for 'ocean' "—Webster's New Collegiate Dictionary (translator's note).

Part III

PROVABILITY AND COMPUTABILITY

I

PROVABILITY

CHAPTER I

Introduction to formal languages

Gelegentlich ergreifen wir die Feder
Und schreiben Zeichen auf ein weisses Blatt,
Die sagen dies und das, es kennt sie jeder,
Es ist ein Spiel, das seine Regeln hat.

H. Hesse, "Buchstaben"

We now and then take pen in hand
And make some marks on empty paper.
Just what they say, all understand.
It is a game with rules that matter.

H. Hesse, "Alphabet"
(translated by Prof. Richard S. Ellis)

1 General information

1.1. Let A be any abstract set. We call A an *alphabet*. Finite sequences of elements of A are called *expressions* in A. Finite sequences of expressions are called *texts*.

We shall speak of a *language with alphabet A* if certain expressions and texts are distinguished (as being "correctly composed," "meaningful," etc.). Thus, in the Latin alphabet A we may distinguish English word forms and grammatically correct English sentences. The resulting set of expressions and texts is a working approximation to the intuitive notion of the "English language."

The language Algol 60 consists of distinguished expressions and texts in the alphabet $\{\text{Latin letters}\} \cup \{\text{digits}\} \cup \{\text{logical signs}\} \cup \{\text{separators}\}$. *Programs* are among the most important distinguished texts.

In natural languages the set of distinguished expressions and texts usually has unsteady boundaries. The more formal the language, the more rigid these boundaries are.

The rules for forming distinguished expressions and texts make up the *syntax* of the language. The rules which tell how they correspond with reality make up the *semantics* of the language. Syntax and semantics are described in a *metalanguage*.

1.2. "Reality" for the languages of mathematics consists of certain classes of (mathematical) arguments or certain computational processes using (abstract) automata. Corresponding to these designations, the languages are divided into formal and algorithmic languages. (Compare: in natural languages, the declarative versus imperative moods, or—on the level of texts—statement versus command.)

Different formal languages differ from one another, in the first place, by the scope of the formalizable types of arguments—their expressiveness; in the second place, by their orientation toward concrete mathematical theories; and in the third place, by their choice of elementary modes of expression (from which all others are then synthesized) and written forms for them.

In the first part of this book a certain class of formal languages is examined systematically. Algorithmic languages are brought in episodically.

The "language–parole" dichotomy, which goes back to Humboldt and Saussure, is as relevant to formal languages as to natural languages. In §3 of this chapter we give models of "speech" in two concrete languages, based on set theory and arithmetic, respectively; because, as many believe, habits of speech must precede the study of grammar.

The language of set theory is among the richest in expressive means, despite its extreme economy. In principle, a formal text can be written in this language corresponding to almost any segment of modern mathematics—topology, functional analysis, algebra, or logic.

The language of arithmetic is one of the poorest, but its expressive possibilities are sufficient for describing all of elementary arithmetic, and also for demonstrating the effects of self-reference à la Gödel and Tarski.

1.3. As a means of communication, discovery, and codification, no formal language can compete with the mixture of mathematical argot and formulas which is common to every working mathematician.

However, because they are so rigidly normalized, formal texts can themselves serve as an object for mathematical investigation. The results of this investigation are themselves *theorems of mathematics*. They arouse great interest (and strong emotions) because they can be interpreted as *theorems about mathematics*. But it is precisely the possibility of these and still broader interpretations that determines the general philosophical and human value of mathematical logic.

1.4. We have agreed that the expressions and texts of a language are elements of certain abstract sets. In order to work with these elements, we must somehow fix them materially. In the modern European tradition (as opposed to the ancient Babylonian tradition, or the latest American tradition, using computer memory), the following notation is customary. The elements of the alphabet are indicated by certain symbols on paper (letters of different kinds of type, digits, additional signs, and also combinations of these). An expression in an alphabet A is written in the form of a sequence of symbols, read from left to right, with hyphens when necessary. A text is written as a sequence of written expressions, with spaces or punctuation marks between them.

1.5. If written down, most of the interesting expressions and texts in a formal language either would be physically extremely long, or else would be psychologically difficult to decipher and learn in an acceptable amount of time, or both.

They are therefore replaced by "abbreviated notation" (which can sometimes turn out to be physically longer). The expression "$xxxxxx$" can be briefly written "$x \cdots x$ (six times)" or "x^6." The expression "$\forall z(z \in x \Leftrightarrow z \in y)$" can be briefly written "$x = y$." Abbreviated notation can also be a way of denoting any expression of a definite type, not only a single such expression; (any expression $101010 \cdots 10$ can be briefly written "the sequence of length $2n$ with ones in odd places and zeros in even places" or "the binary expansion of $\frac{2}{3}(4^n - 1)$.")

Ever since our tradition started, with Vieta, Descartes, and Leibniz, abbreviated notation has served as an inexhaustible source of inspiration and errors. There is no sense in, or possibility of, trying to systematize its devices; they bear the indelible imprint of the fashion and spirit of the times, the artistry and pedantry of the authors. The symbols Σ, $\int$, $\in$ are classical models worthy of imitation. Frege's notation, now forgotten, for "P and Q" (actually "not [if P, then not Q]," whence the asymmetry):

shows what should be avoided. In any case, abbreviated notation permeates mathematics.

The reader should become used to the trinity

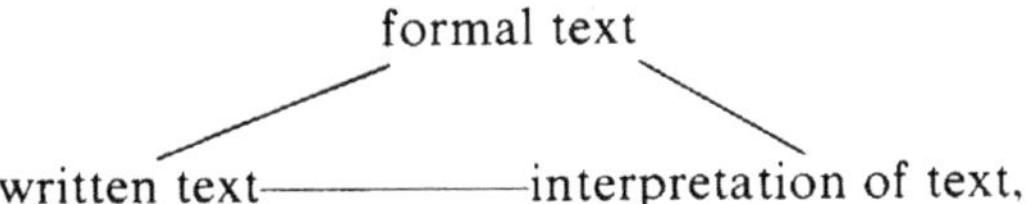

which replaces the unconscious identification of a statement with its form and its sense, as one of the first priorities in his study of logic.

2 First order languages

In this section we describe the most important class of formal languages $\mathcal{L}_1$—the first order languages—and give two concrete representatives of this class: the Zermelo–Fraenkel language of set theory $\mathrm{L_1Set}$, and the Peano language of arithmetic $\mathrm{L_1Ar}$. Another name for $\mathcal{L}_1$ is *predicate languages*.

2.1. The alphabet of any language in the class $\mathcal{L}_1$ is divided into six disjoint subsets. The following table lists the generic name for the elements in each subset, the standard notation for these elements in the general case, the special notation used in this book for the languages $\mathrm{L_1Set}$ and $\mathrm{L_1Ar}$. We then describe the rules for forming distinguished expressions and briefly discuss semantics.

The distinguished expressions of any language L in the class $\mathcal{L}_1$ are divided into two types: *terms* and *formulas*. Both types are defined recursively.

2.2. **Definition.** *Terms* are the elements of the least subset of the expressions of the language which satisfies the two conditions:

(a) Variables and constants are (atomic) terms.
(b) If f is an operation of degree r and $t_1, \ldots, t_r$ are terms, then $f(t_1, \ldots, t_r)$ is a term.

In (a) we identify an element with a sequence of length one. The alphabet does not include commas, which are part of our abbreviated notation: $f(t_1, t_2, t_3)$ means the same as $f(t_1t_2t_3)$. In §1 of Chapter II we

Language Alphabets

Subsets of the Alphabet	Names and Notation		
	General	in $\mathrm{L_1Set}$	in $\mathrm{L_1Ar}$
connectives and quantifiers	$\Leftrightarrow$ (equivalent); $\Rightarrow$ (implies); $\vee$ (inclusive or); $\wedge$ (and); $\neg$ (not); $\forall$ (universal quantifier); $\exists$ (existential quantifier)		
variables	$x, y, z, u, v, \ldots$ with indices		
constants	$c \cdots$ with indices	$\varnothing$ (empty set)	$\bar{0}$ (zero); $\bar{1}$ (one)
operations of degree $1, 2, 3, \ldots$	$f, g, \ldots$ with indices	none	$+$ (addition, degree 2); $\cdot$ (multiplication, degree 2)
relations (predicates) of degree $1, 2, 3, \ldots$	$p, q, \ldots$ with indices	$\in$ (is an element of, degree 2); $=$ (equals, degree 2)	$=$ (equality, degree 2)
parentheses	((left parenthesis);)(right parenthesis)		

explain how a sequence of terms can be uniquely deciphered despite the absence of commas.

If two sets of expressions in the language satisfy conditions (a) and (b), then the intersection of the two sets also satisfies these conditions. Therefore the definition of the set of terms is correct.

2.3. **Definition.** *Formulas* are the elements of the least subset of the expressions of the language which satisfies the two conditions:

(a) If p is a relation of degree r and $t_1, \ldots, t_r$ are terms, then $p(t_1, \ldots, t_r)$ is an (atomic) formula.
(b) If P and Q are formulas (abbreviated notation!), and x is a variable, then the expressions

$$(P)\Leftrightarrow(Q), \quad (P)\Rightarrow(Q), \quad (P)\vee(Q), \quad (P)\wedge(Q),$$
$$\neg(P), \quad \forall x(P), \quad \exists x(P)$$

are formulas.

It is clear from the definitions that any term is obtained from atomic terms in a finite number of steps, each of which consists in "applying an operation symbol" to the earlier terms. The same is true for formulas. In Chapter II, §1 we make this remark more precise.

The following initial interpretations of terms and formulas are given for the purpose of orientation and belong to the so-called "standard models" (see Chapter II, §2 for the precise definitions).

2.4. EXAMPLES AND INTERPRETATIONS

(a) The terms stand for (are notation for) the objects of the theory. Atomic terms stand for indeterminate objects (variables) or concrete objects (constants). The term $f(t_1, \ldots, t_r)$ is the notation for the object obtained by applying the operation denoted by f to the objects denoted by $t_1, \ldots, t_r$. Here are some examples from L_1Ar:

$\bar{0}$ denotes zero;

$\bar{1}$ denotes one;

$+(\bar{1}, \bar{1})$ denotes two ($1 + 1 = 2$ in the usual notation);

$+\left(\bar{1} + (\bar{1}, \bar{1})\right)$ denotes three;

$\cdot\left(+(\bar{1}, \bar{1}) + (\bar{1}, \bar{1})\right)$ denotes four ($2 \times 2 = 4$).

Since this normalized notation is different from what we are used to in arithmetic, in L_1Ar we shall usually write simply $t_1 + t_2$ instead of $+(t_1, t_2)$ and $t_1 \cdot t_2$ instead of $\cdot(t_1, t_2)$. This convention may be considered as another use of abbreviated notation.

x stands for an indeterminate integer;

$x + \bar{1}$ (or $+(x, \bar{1})$) stands for the next integer.

In the language L_1Set all terms are atomic:

x stands for an indeterminate set;
$\varnothing$ stands for the empty set.

(b) The formulas stand for statements (arguments, propositions, . . .) of the theory. When translated into formal language, a statement may be either true, false, or indeterminate (if it concerns indeterminate objects); see Chapter II for the precise definitions. In the general case the atomic formula $p(t_1, \ldots, t_r)$ has roughly the following meaning: "The ordered r-tuple of objects denoted by $t_1, \ldots, t_r$ has the property denoted by p." Here are some examples of atomic formulas in L_1Ar. Their general structure is $= (t_1, t_2)$, or, in nonnormalized notation, $t_1 = t_2$:

$$\bar{0} = \bar{1}, \qquad x + \bar{1} = y.$$

Here are some examples of formulas which are not atomic:

$$\neg(\bar{0} = \bar{1}),$$
$$(x = \bar{0}) \Leftrightarrow (x + \bar{1} = \bar{1}),$$
$$\forall x\big((x = \bar{0}) \vee \big(\neg(x \cdot x = \bar{0})\big)\big).$$

Some atomic formulas in L_1Set:

$$y \in x \qquad (y \text{ is an element of } x),$$

and also $\varnothing \in y$, $x \in \varnothing$, etc. Of course, normalized notation must have the form $\in(xy)$, and so on.

Some nonatomic formulas:

$$\exists x\big(\forall y(\neg(y \in x))\big): \qquad \text{there exists an } x \text{ of which no } y \text{ is an element.}$$

Informally this means: "The empty set exists." We once again recall that an informal interpretation presupposes some standard interpretive system, which will be introduced explicitly in Chapter II.

$$\forall y(y \in z \Rightarrow y \in x): \qquad z \text{ is a subset of } x.$$

This is an example of a very useful type of abbreviated notation: four parentheses are omitted in the formula on the left. We shall not specify precisely when parentheses may be omitted; in any case, it must be possible to reinsert them in a way that is unique or is clear from the context without any special effort.

We again emphasize: the abbreviated notation for formulas are only material designations. Abbreviated notation is chosen for the most part with psychological goals in mind: speed of reading (possibly with a loss in formal uniqueness), tendency to encourage useful associations and discourage harmful ones, suitability to the habits of the author and reader, and so on. The mathematical objects in the theory of formal languages are the formulas themselves, and not any particular designations.

Digression: names

On several occasions we have said that a certain object (a sign on paper, an element of an alphabet as an abstract set, etc.) is a notation for, or denotes, another element. A convenient general term for this relationship is naming.

The letter x is the name of an element of the alphabet; when it appears in a formula, it becomes the name of a set or a number; the notation $x \in y$ is the name of an expression in the alphabet A, and this expression, in turn, is the name of an assertion about indeterminate sets; and so on.

When we form words, we often identify the names of objects with the objects themselves: we say "the variable x," "the formula P," "the set z." This can sometimes be dangerous. The following passage from Rosser's book *Logic for Mathematicians* points up certain hidden pitfalls:

> The gist of the matter is that, if we have a statement such as "3 is greater than $\frac{9}{12}$" about the rational number $\frac{9}{12}$ and containing a name "$\frac{9}{12}$" of this rational number, one can replace this name by any other name of the same rational number, for instance, "$\frac{3}{4}$." If we have a statement such as "3 divides the denominator of '$\frac{9}{12}$'" about a name of a rational number and containing a name of this name, one can replace this name of the name by some other name of the same name, but not in general by the name of some other name, if it is a name of some other name of the same rational number.

Rosser adds that "failure to observe such distinctions carefully can seldom lead to confusion in logic and still more seldom in mathematics." However, these distinctions play a significant role in philosophy and in mathematical practice.

"A rose by any other name would smell as sweet"—this is true because roses exist outside of us and smell in and of themselves. But, for example, it seems that Hilbert spaces only "exist" insofar as we talk about them, and the choice of terminology here makes a difference. The word "space" for the set of equivalence classes of square integrable functions was at the same time a codeword for an entire circle of intuitive ideas concerning "real" spaces. This word helped organize the concept and led it in the right direction.

A successfully chosen name is a bridge between scientific knowledge and common sense, between new experience and old habits. The conceptual foundation of any science consists of a complicated network of names of things, names of ideas, and names of names. It evolves itself, and its projection on reality changes.

3 Beginners' course in translation

3.1. We recall that the formulas in L_1Set stand for statements about sets; the formulas in L_1Ar stand for statements about natural numbers; these formulas contain names of sets and numbers, which may be indeterminate.

In this section we give the first basic examples of two-way translation "argot$\Leftrightarrow$formal language." One of our purposes will be to indicate the great expressive possibilities in L_1Set and L_1Ar, despite the extremely limited modes of expression.

As in the case of natural languages, this translation cannot be given by rigid rules, is not uniquely determined, and is a creative process. Compare Hesse's quatrain with its translation in the epigraph to this book: the most important aim of translation is to "understand . . . just what they say."

Before reading further, the reader should look through the Appendix to Chapter II: "*The von Neumann Universe*." The semantics implicit in L_1Set relates to this universe, and not to arbitrary "Cantor" sets.

A more complete picture of the meaning of the formulas can be obtained from §2 of Chapter II.

Translation from L_1Set *to argot.*

3.2. $\forall x(\neg(x \in \varnothing))$: "for all (sets) x it is false that x is an element of (the set) $\varnothing$" (or "$\varnothing$ is the empty set").

The second assertion is only equivalent to the first in the von Neumann universe, where the elements of sets can only be sets, and not real numbers, chairs, or atoms.

3.3. $\forall z(z \in x \Leftrightarrow z \in y) \Leftrightarrow x = y$: "if for all z it is true that z is an element of x if and only if z is an element of y, then it is true that x coincides with y; and conversely," or "a set is uniquely determined by its elements."

In the expression 3.3 at least six parentheses have been omitted; and the subformulas $z \in x$, $z \in y$, $x = y$ have not been normalized according to the rules of $\mathfrak{L}_1$.

3.4. $\forall u\, \forall v\, \exists x\, \forall z(z \in x \Leftrightarrow (z = u \vee z = v))$: "for any two sets u, v there exists a third set x such that u and v are its only elements."

This is one of the axioms of Zermelo–Fraenkel. The set x is called the "unordered pair of sets u, v" and is denoted $\{u, v\}$ in the Appendix.

3.5. $\forall y\, \forall z(((z \in y \wedge y \in x) \Rightarrow z \in x) \wedge (y \in x \Rightarrow \neg(y \in y)))$: "the set x is partially ordered by the relation $\in$ between its elements."

We mechanically copied the condition $y \in x \Rightarrow \neg(y \in y)$ from the definition of partial ordering. This condition is automatically fulfilled in the von Neumann universe, where no set is an element of itself.

A useful exercise would be to write the following formulas:

"x is totally ordered by the relation $\in$";
"x is linearly ordered by the relation $\in$";
"x is an ordinal."

3.6. $\forall x(y \in z)$: The literal translation "for all x it is true that y is an element of z" sounds a little strange. The formula $\forall x\, \exists x(y \in z)$, which agrees with the rules for constructing formulas, looks even worse. It would be possible to make the rules somewhat more complicated, in order to rule out such formulas, but in general they cause no harm. In Chapter II we shall see that, from the point of view of "truth" or "deducibility," such a formula is equivalent to the formula $y \in z$. It is in this way that they must be understood.

Translation from argot to $\mathrm{L_1Set}$.

We choose several basic constructions having general mathematical significance and show how they are realized in the von Neumann universe, which only contains sets obtained from $\varnothing$ by the process of "collecting into a set," and in which all relations must be constructed from $\in$.

3.7. "*x is the direct product $y \times z$.*"

This means that the elements of x are the ordered pairs of elements of y and z, respectively. The definition of an unordered pair is obvious: the formula

$$\forall u(u \in x \Leftrightarrow (u = y_1 \vee u = z_1))$$

"means," or may be briefly written in the form, $x = \{y_1, z_1\}$ (compare 3.4). The ordered pair y_1 and z_1 is introduced using a device of Kuratowski and Wiener: this is the set x_1 whose elements are the unordered pairs $\{y_1, y_1\}$ and $\{y_1, z_1\}$.

We thus arrive at the formula

$$\exists y_2\, \exists z_2(\text{“}x_1 = \{y_2, z_2\}\text{”} \wedge \text{“}y_2 = \{y_1, y_1\}\text{”} \wedge \text{“}z_2 = \{y_1, z_1\}\text{”}),$$

which will be abbreviated

$$x_1 = \langle y_1, z_1 \rangle$$

and will be read: "x_1 is the ordered pair with first element y_1 and second element z_1." The abbreviated notation for the subformulas is in quotes; we shall later omit the quotation marks.

Finally, the statement "$x = y \times z$" may be written in the form:

$$\forall x_1(x_1 \in x \Leftrightarrow \exists y_1\, \exists z_1(y_1 \in y \wedge z_1 \in z \wedge \text{“}x_1 = \langle y_1, z_1 \rangle\text{”})).$$

In order to remind the reader for the last time of the liberties taken in abbreviated notation, we write this same formula adhering to all the canons of $\mathfrak{L}_1$:

$$\forall x_1 \Big[(\in (x_1 x))$$

$$\Leftrightarrow \Big[\exists y_1 \Big(\exists z_1 \Big(((\in(y_1 y)) \wedge (\in(z_1 z))) \wedge \Big(\exists y_2 \Big(\exists z_2 \Big(\Big(\Big(\forall u ((\in(u x_1))$$

$$\Leftrightarrow ((=(u y_2)) \vee (=(u z_2))))) \Big) \wedge \Big(\forall u ((\in (u y_2))$$

$$\Leftrightarrow (=(u y_1))))) \Big) \wedge \Big(\forall u ((\in (u z_2)) \Leftrightarrow ((=(u y_1)) \vee (=(u z_1))))) \Big) \Big) \Big) \Big) \Big) \Big) \Big] \Big]$$

Exercise: Find the open parenthesis corresponding to the fifth closed parenthesis from the end. In §1 of Chapter II we give an algorithm for solving such problems.

3.8. "*f is a mapping from the set u to the set v.*"

First of all, mappings, or functions, are identified with their graphs; otherwise, we would not be able to consider them as elements of the universe. The following formula successively imposes three conditions on f: f is a subset of $u \times v$; the projection of f onto u coincides with all of u; and, each element of u corresponds to exactly one element of v:

$$\forall z (z \in f \Rightarrow (\exists u_1 \ \exists v_1 (u_1 \in u \wedge v_1 \in v \wedge \text{“} z = \langle u_1, v_1 \rangle \text{”})))$$

$$\wedge \forall u_1 (u_1 \in u \Rightarrow \exists v_1 \ \exists z (v_1 \in v \wedge \text{“} z = \langle u_1, v_1 \rangle \text{”} \wedge z \in f))$$

$$\wedge \forall u_1 \ \forall v_1 \ \forall v_2 (\exists z_1 \ \exists z_2 (z_1 \in f \wedge z_2 \in f \wedge \text{“} z_1 = \langle u_1, v_1 \rangle \text{”} \wedge \text{“} z_2 = \langle u_1, v_2 \rangle \text{”})$$

$$\Rightarrow v_1 = v_2).$$

Exercise: Write the formula "f is the projection of $y \times z$ onto z."

3.9. "*x is a finite set.*"

Finiteness is far from being a primitive concept. Here is Dedekind's definition: "there does not exist a one-to-one mapping f of the set x onto a proper subset." The formula:

$$\neg \exists f \big(\text{“}f \text{ is a mapping from } x \text{ to } x\text{”} \wedge \forall u_1 \ \forall u_2 \ \forall v_1 \ \forall v_2 ((\text{“}\langle u_1, v_1 \rangle \in f\text{”}$$

$$\wedge \text{“}\langle u_2, v_2 \rangle \in f\text{”} \wedge \neg (u_1 = u_2)) \Rightarrow \neg (v_1 = v_2)) \wedge \exists v_1 (v_1 \in x \wedge \neg \exists u_1$$

$$(\text{“}\langle u_1, v_1 \rangle \in f\text{”}))) \big).$$

The abbreviation "$\langle u_1, v_1\rangle \in f$" means, of course, $\exists y$("$y = \langle u_1, v_1\rangle$"$\wedge y \in f$).

3.10. "*x is a nonnegative integer.*"

The natural numbers are represented in the von Neumann universe by the finite ordinals, so that the required formula has the form:

$$\text{"}x\text{ is totally ordered by the relation } \in\text{"}\wedge\text{"}x\text{ is finite."}$$

EXERCISE: Figure out how to write the formulas "$x + y = z$" and "$x \cdot y = z$," where x, y, z are integers $\geqslant 0$.

After this it is possible in the usual way to write the formulas "x is an integer," "x is a rational number," "x is a real number" (following Cantor or Dedekind), etc., and then construct a formal version of analysis. The written statements will have acceptable length only if we periodically extend the language L_1Set (see §8 of Chapter II). For example, in L_1Set we are not allowed to write term-names for the numbers 1, 2, 3, . . . ($\varnothing$ is the name for 0), although we may construct the formulas "x is the finite ordinal containing 1 element," "x is the finite ordinal containing 2 elements," etc. If we use such roundabout methods of expression, the simplest numerical identities become incredibly long; but, of course, in logic we are mainly concerned with the theoretical possibility of writing them.

3.11. "*x is a topological space.*"

In the formula we must give the topology of x explicitly. We define the topology, for example, in terms of the set y of all open subsets of x. We first write that y consists of subsets of x and contains x and the empty set:

$$P_1: \quad \forall z(z \in y \Rightarrow \forall u(u \in z \Rightarrow u \in x)) \wedge x \in y \wedge \varnothing \in y.$$

The intersection w of any two elements u, v in y is open, i.e., belongs to y:

$$P_2: \quad \forall u\, \forall v\, \forall w\big((u \in y \wedge v \in y \wedge \forall z((z \in u \wedge z \in v) \Leftrightarrow z \in w)) \Rightarrow w \in y\big).$$

It is harder to write "the union of any set of open subsets is open." We first write:

$$P_3: \quad \forall u(u \in z \Leftrightarrow \forall v(v \in u \Rightarrow v \in y)),$$

that is, "z is the set of all subsets of y." Then:

$$P_4: \quad \forall u\, \forall w\big((u \in z \wedge \forall v_1(v_1 \in w \Leftrightarrow \exists v(v \in u \wedge v_1 \in v))) \Rightarrow w \in y\big).$$

This means (taking into account P_3, which defines z): "If u is any subset of y, i.e., a set of open subsets of x, then the union w of all these subsets belongs to y, i.e., is open." Now the final formula may be written as follows:

$$P_1 \wedge P_2 \wedge \forall z(P_3 \Rightarrow P_4).$$

The following comments on this formula will be reflected in precise definitions in Chapter II, §§1 and 2. The letters x, y have the same meaning in all the P_i, while z plays different roles: in P_1 it is a subset of x, and in P_3 and P_4 it is the set of subsets of x. We are allowed to do this because, as soon as we "bind" z by the quantifier $\forall$, say in P_1, z no longer stands for an (indeterminate) individual set, and becomes a temporary designation for "any set." Where the "scope of action" of $\forall$ ended, z can be given a new meaning. In order to "free" z for later use, $\forall z$ was also put before $P_3 \Rightarrow P_4$.

Translation from argot to $\mathrm{L_1Ar}$.

3.12. "$x < y$" : $\exists z\big(y = (x + z) + \bar{1}\big)$. Recall that the variables are names for nonnegative integers.

3.13. "*x is a divisor of y*" : $\exists z(y = x \cdot z)$.

3.14. "*x is a prime number*": "$\bar{1} < x$"$\wedge$("*y is a divisor of x*"$\Rightarrow(y = \bar{1} \vee y = x)$).

3.15. "*Fermat's big theorem*": $\forall x_1\, \forall x_2\, \forall x_3\, \forall u$("$\bar{2} < u$"$\wedge$"$x_1^u + x_2^u = x_3^u$"$\Rightarrow$"$x_1x_2x_3 = \bar{0}$"). It is not clear how to write the formula $x_1^u + x_2^u = x_3^u$ in $\mathrm{L_1Ar}$. Of course, for any *concrete* $u = 1, 2, 3$ there is a corresponding atomic formula in $\mathrm{L_1Ar}$, but how do we make u into a variable? This is not a trivial problem. In the second part of the book we show how to find an atomic formula $p(x, u, y, z_1, \ldots, z_n)$ such that the assertion that $\exists z_1 \cdots \exists z_n p(x, u, y, z_1, \ldots, z_n)$ in the domain of natural numbers is equivalent to $y = x^u$. Then $x_1^u + x_2^u = x_3^u$ can be translated as follows:

$$\exists y_1\, \exists y_2\, \exists y_3(\text{“}x_1^u = y_1\text{”} \wedge \text{“}x_2^u = y_2\text{”} \wedge \text{“}x_3^u = y_3\text{”} \wedge y_1 + y_2 = y_3).$$

The existence of such a p is a nontrivial number theoretic fact, so that here the very possibility of performing a translation becomes a mathematical problem.

3.16. "*The Riemann hypothesis.*" The Riemann zeta-function $\zeta(s)$ is defined by the series $\sum_{n=1}^{\infty} n^{-s}$ in the halfplane $\operatorname{Re} s > 1$. It can be continued meromorphically onto the entire complex s-plane. The Riemann hypothesis is the assertion that the nontrivial zeros of $\zeta(s)$ lie on the line $\operatorname{Re} s = \frac{1}{2}$. Of course, in this form the Riemann hypothesis cannot be translated into $\mathrm{L_1Ar}$. However, there are several purely arithmetic assertions which are demonstrably equivalent to the Riemann hypothesis. Perhaps the simplest of them is the following.

Let $\mu(n)$ be the Möbius function on the set of integers $\geqslant 1$: it equals 0 if n is divisible by a square, and equals $(-1)^r$, where r is the number of prime divisors of n, if n is square-free. We then have:

$$\text{Riemann hypothesis} \Leftrightarrow \forall \varepsilon > 0\; \exists x\; \forall y \left[y > x \Rightarrow \left(\left| \sum_{n=1}^{y} \mu(n) \right| < y^{1/2+\varepsilon} \right) \right].$$

Only the exponent is not an integer on the right; but ε need only run through numbers of the form $1/z$, z an integer $\geqslant 1$, and then we can raise the inequality to the $(2z)$th power. The formula

$$\left(\sum_{n=1}^{y} \mu(n)\right)^{2z} < y^{z+2}$$

can then be translated into $\mathrm{L_1Ar}$, although not completely trivially. The necessary techniques will be developed in the second part of the book.

The last two examples were given in order to show the complexity that is possible in problems which can be stated in $\mathrm{L_1Ar}$, despite the apparent simplicity of the modes of expression and the semantics of the language.

We conclude this section with some remarks concerning higher order languages.

3.17. *Higher order languages*. Let L be any first order language. Its modes of expression are limited in principle by one important consideration: we are not allowed to speak of *arbitrary* properties of objects of the theory, that is, arbitrary subsets of the set of all objects. Syntactically, this is reflected in the prohibition against forming expressions such as $\forall p(p(x))$, where p is a relation of degree 1; relations must stand for fixed rather than variable properties.

Of course, certain properties can be defined using nonatomic formulas. For example, in $\mathrm{L_1Ar}$ instead of "x is even" we may write $\exists y\big(x = (\bar{1}+\bar{1})\cdot y\big)$. However, there is a continuum of subsets of the integers but only a countable set of definable properties (see §2 of Chapter II), so there are automatically properties which cannot be defined by formulas. Thus, it is impossible to replace the forbidden expression $\forall p(p(x))$ by a sequence of expressions $P_1(x), P_2(x), P_3(x), \ldots$.

Languages in which quantifiers may be applied to properties and/or functions (and also, possibly, to properties of properties, and so on) are called higher order languages. One such language—$\mathrm{L_2Real}$—will be considered in Chapter III for the purpose of illustrating a simplified version of Cohen forcing.

On the other hand, the same extension of expressive possibilities can be obtained without leaving $\mathfrak{L}_1$. In fact, in the first order language $\mathrm{L_1Set}$ we may quantify over all subsets of any set, over all subsets of the set of subsets, and so on. Informally this means we are speaking of all properties, all properties of properties, . . . (with transfinite extension). In addition, any higher order language with a "standard interpretation" in some type of structured sets can be translated into $\mathrm{L_1Set}$ so as to preserve the meanings and truth values in this standard interpretation. (An apparent exception is the languages for describing Gödel–Bernays classes and "large" categories; but it seems, based on our present understanding of paradoxes, that no higher order languages can be constructed from such a language.)

The attentive reader will notice the contrast between the *possibility* of writing a formula in L_1Set in which $\forall$ is applied to all subsets (informally, to all properties) of finite ordinals (informally, of integers), and the *impossibility* of writing a formula in L_1Set which would define any concrete subset in the continuum of undefinable subsets. (There are fewer such subsets in L_1Set than in L_1Ar, but still a continuum.) We shall examine these problems more closely in Chapter II when we discuss "Skolem's paradox."

Let us summarize. Almost all the basic logical and set theoretic principles used in the day to day work of the mathematician are contained in the first-order languages and, in particular, in L_1Set. Hence, those languages will be the subject of study in the first and third parts of the book. But concrete oriented languages can be formed in other ways, with various degrees of deviation from the rules of $\mathfrak{L}_1$. In addition to L_2Real, examples of such languages examined in Chapter II include SELF (Smullyan's language for self-description) and SAr, which is a language of arithmetic convenient for proving Tarski's theorem on the undefinability of truth.

Digression: syntax

1. The most important feature that most artificial languages have in common is the ability to encompass a rich spectrum of modes of expression starting with a small finite number of generating principles.

In each concrete case the choice of these principles (including the alphabet and syntax) is based on a compromise between two extremes. Economical use of modes of expression leads to unified notation and simplified mechanical analysis of the text. But then the texts become much longer and farther removed from natural language texts. Enriching the modes of expression brings the artificial texts closer to the natural language texts, but complicates the syntax and the formal analysis. (Compare machine languages with such programming languages as Algol, Fortran, Cobol, etc.)

We now give several examples based on our material.

2. *Dialects of* $\mathfrak{L}_1$

(a) Without changing the logic in $\mathfrak{L}_1$, it is possible to discard parentheses and either of the two quantifiers from the alphabet, and to replace all the connectives by one, namely $\downarrow$ (conjunction of negations). (In addition, constants could be declared to be functions of degree 0, and functions could be interpreted as relations.)

This is accomplished by the following change in the definitions. If $t_1, \ldots, t_r$ are terms, f is an operation of degree r, and p is a relation of degree r, then $ft_1 \cdots t_r$ is a term, and $pt_1 \cdots t_r$ is an atomic formula. If P and Q are formulas, then $\downarrow PQ$ and $\forall xP$ are formulas. The content of $\downarrow PQ$ is "not P and not Q," so that we have the following expressions in this dialect:

$$\neg(P): \qquad \downarrow PP$$

$$(P)\wedge(Q): \qquad \downarrow\downarrow PP\downarrow QQ$$

$$(P)\vee(Q): \qquad \downarrow\downarrow PQ\downarrow PQ$$

Clearly, economizing on parentheses and connectives leads to much repetition of the same formula. Nevertheless, it may become simpler to prove theorems about such a language because of the shorter list of syntactic norms.

(b) Bourbaki's language of set theory has an alphabet consisting of the signs $\square$, τ, $\vee$, $\neg$, $=$, $\in$ and the letters. Expressions in this language are not simply sequences of signs in the alphabet, but sequences in which certain elements are paired together by superlinear connectives. For example:

$$\tau \vee \neg \in \square\, A' \in \square\, A''.$$

The main difference between Bourbaki's language and L_1Set is the use of the "Hilbert choice symbol." If, for example, $\in xy$ is the formula "x is an element of y," then

$$\tau \in \square\; y$$

is a term meaning "some element of the set y."

Bourbaki's language is not very convenient and is not widely used. It became known in the popular literature thanks to an example of a very long abbreviated notation for the term "one," which the authors imprudently introduced:

$$\tau_Z\Big((\exists u)(\exists U)\Big(u=(U,\{\varnothing\},Z)\wedge U\subset\{\varnothing\}\times Z\wedge(\forall x)\big((x\in\{\varnothing\})$$
$$\Rightarrow(\exists y)((x,y)\in U)\big)\wedge(\forall x)(\forall y)(\forall y')\Big(\big((x,y\in U\wedge(x,y')\in U\big)$$
$$\Rightarrow(y=y')\big)\wedge(\forall y)\big((y\in Z)\Rightarrow(\exists x)((x,y)\in U)\big)\Big)\Big)\Big).$$

It would take several tens of thousands of symbols to write out this term completely; this seems a little too much for "one."

(c) A way to greatly extend the expressive possibilities of almost any language in $\mathfrak{L}_1$ is to allow "class terms" of the type $\{x|P(x)\}$, meaning "the class of all objects x having the property P." This idea was used by Morse in his language of set theory and by Smullyan in his language of arithmetic; see §10 of Chapter II.

3. *General remarks*. Most natural and artificial languages are characteristically discrete and linear (one-dimensional). On the one hand, our perception of the external world is not felt by us to be either discrete or linear, although these characteristics are observed on the level of physiological mechanisms (coding by impulses in the nervous system). On the other hand, the languages in which we communicate tend to transmit information in a sequence of distinguishable elementary signs. The main reason for this is probably the much greater (theoretically unlimited) uniqueness and reproducibility of information than is possible with other methods of conveyance. Compare with the well-known advantages of digital over analog computers.

The human brain clearly uses both principles. The perception of images as a whole, along with emotions, are more closely connected with nonlinear and nondiscrete processes—perhaps of a wave nature. It is interesting to examine from this point of view the nonlinear fragments in various languages.

In mathematics this includes, first of all, the use of drawings. But this use does not lend itself to formal description, with the exception of the separate and formalized theory of graphs. Graphs are especially popular objects, because they are as close as possible both to their visual image as a whole and to their description using all the rules of set theory. Every time we are able to connect a problem with a graph, it becomes much simpler to discuss it, and large sections of verbal description are replaced by manipulation with pictures.

A less well-known class of examples is the commutative diagrams and spectral sequences of homological algebra. A typical example is the "snake lemma." Here is its precise formulation.

Suppose we are given a commutative diagram of abelian groups and homomorphisms between them (in the box below), in which the rows are exact sequences:

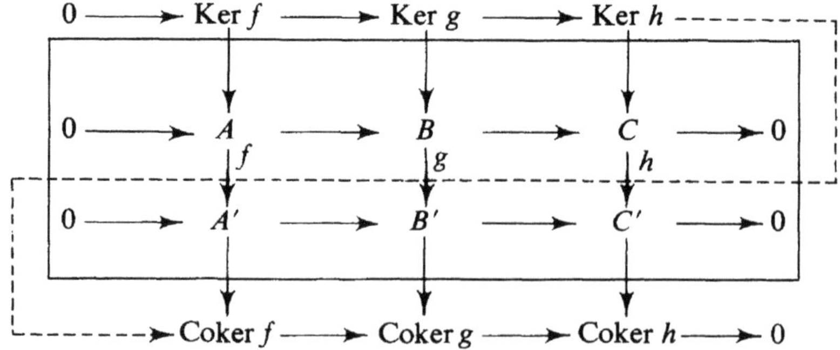

Then the kernels and cokernels of the "vertical" homomorphisms f, g, h form a six-term exact sequence, as shown in the drawing, and the entire diagram of solid arrows is commutative. The "snake" morphism Ker $h \to$

Coker f, which is denoted by the dotted arrow, is the basic object constructed in the lemma.

Of course, it is easy to describe the snake diagram sequentially in a suitable, more or less formal, linear language. However, such a procedure requires an artificial and not uniquely determined breaking up of a clearly two-dimensional picture (as in scanning a television image). Moreover, without having the overall image in mind, it becomes harder to recognize the analogous situation in other contexts and to bring the information together into a single block.

The beginnings of homological algebra saw the enthusiastic recognition of useful classes of diagrams. At first this interest was even exaggerated; see the editor's appendix to the Russian translation of *Homological Algebra* by Cartan and Eilenberg.

There is one striking example of an entire book with an intentional two-dimensional (block) structure: C. H. Lindsey and S. G. van der Meulen, *Informal Introduction to Algol 68* (North-Holland, Amsterdam, 1971). It consists of eight chapters, each of which is divided into seven sections (eight of the 56 sections are empty, to make the system work!). Let (i, j) be the name of the jth section of the ith chapter; then the book can be studied either "row by row" or "column by column" in the (i, j)-matrix, depending on the reader's intentions.

As with all great undertakings, this is the fruit of an attempt to solve what is in all likelihood an insoluble problem, since, as the authors remark, Algol 68 "is quite impossible to describe . . . until it has been described."

CHAPTER II

Truth and deducibility

1 Unique reading lemma

The basic content of this section is Lemma 1.4 and Definitions 1.5 and 1.6. The lemma guarantees that the terms and formulas of any language in $\mathcal{L}_1$ can be deciphered in a unique way, and it serves as a basis for most inductive arguments. (The reader may take the lemma on faith for the time being, provided that he was able independently to verify the last formula in 3.7 of Chapter I. However, the proof of the lemma will be needed in §4 of Chapter VII.) It is important to remember that the theory of any formal language begins by checking that the syntactic rules are free of ambiguity.

We begin with the standard combinatoric definitions, in order to fix the terminology.

1.1. Let A be a set. By a sequence of length n of elements of A we mean a mapping from the set $\{1, \ldots, n\}$ to A. The image of i is called the ith term of the sequence. Corresponding to $n = 0$ we have the empty sequence. Sequences of length 1 will sometimes be identified with elements of A.

A sequence of length n can also be written in the form $a_1, \ldots, a_i, \ldots, a_n$, where a_i is its ith term. The number i is called the index of the term a_i. If $P = (a_1, \ldots, a_n)$ and $Q = (b_1, \ldots, b_m)$ are two sequences, their concatenation PQ is the sequence $(a_1, \ldots, a_n, b_1, \ldots, b_m)$ of length $m + n$ whose ith terms is a_i for $i \leqslant n$ and b_{i-n} for $n + 1 \leqslant i \leqslant n + m$. We similarly define the concatenation of a finite sequence of sequences.

An *occurrence* of the sequence Q in P is any representation of P as a concatenation P_1QP_2. Substituting a sequence R in place of a given occurrence of Q in P amounts to constructing the sequence P_1RP_2.

Let Π^+, Π^- be two disjoint subsets of $\{1, \ldots, n\}$. A map $c : \Pi^+ \to \Pi^-$ is called a *parentheses bijection* if it is bijective and satisfies the conditions:

(a) $c(i) > i$ for all $i \in \Pi^+$;
(b) for every i and j, $j \in [i, c(i)]$ if and only if $c(j) \in [i, c(i)]$.

1.2. **Lemma.** *Given* Π^+ *and* Π^-, *if a parentheses bijection exists, then it is unique.*

This lemma will be applied to expressions in languages in $\mathfrak{L}_1$: Π^+ will consist of the indices of the places in the expression at which "(" occurs, Π^- will consist of the indices of the places at which ")" occurs, and the map c correlates to each left parenthesis the corresponding right parenthesis.

Proof of the lemma. Let the function $\varepsilon : \{1, \ldots, n\} \to \{0, \pm 1\}$ take the value 1 on Π^+, -1 on Π^-, and 0 everywhere else. We claim that for every $i \in \Pi^+$, for any parentheses bijection $c : \Pi^+ \to \Pi^-$, and for any k, $1 \leqslant k \leqslant c(i) - i$, we have the relations:

$$\sum_{j=i}^{c(i)} \varepsilon(j) = 0, \qquad \sum_{j=i}^{c(i)-k} \varepsilon(j) > 0.$$

The lemma follows immediately from these relations, since we obtain the following recipe for determining c from Π^+ and Π^-; $c(i)$ is the least $l > i$ for which $\sum_{j=i}^{l} \varepsilon(j) = 0$.

The first relation holds because the elements of Π^+ and Π^- which appear in the interval $[i, c(i)]$ do so in pairs $(j, c(j))$, and $\varepsilon(j) + \varepsilon(c(j)) = 0$.

To prove the second relation, suppose that for some i and k we have $\sum_{j=i}^{c(i)-k} \varepsilon(j) \leqslant 0$. Since $\varepsilon(i) = 1$, it follows that $\sum_{j=i+1}^{c(i)-k} \varepsilon(j) < 0$. Hence, the number of elements of Π^- in the interval $[i + 1, c(i) - k]$ is strictly greater than the number from Π^+. Let $c(j_0) \in \Pi^-$ be an element in the interval such that $j_0 \notin [i + 1, c(i) - k]$. Then $j_0 \leqslant i$, and in fact, $j_0 < i$, since $c(i)$ is outside the interval. But then only one element of the pair j_0, $c(j_0)$ lies in $[i, c(i)]$, which contradicts the definition of c. □

1.3. Now let A be the alphabet of a language L in $\mathfrak{L}_1$ (see §2 of Chapter I). Finite sequences of elements of A are the expressions in this language. Certain expressions have been distinguished as formulas or terms. We recall that the definitions in §2 of Chapter I imply that:

(a) Any term in L either is a constant, is a variable, or is represented in the form $f(t_1, \ldots, t_r)$, where f is an operation of degree r, and $t_1, \ldots, t_r$ are terms shorter in length.

(b) Any formula in L is represented either in the form $p(t_1, \ldots, t_r)$, where p is a relation of degree r and $t_1, \ldots, t_r$ are terms shorter in length, or in one of the seven forms

$$(P)\Leftrightarrow(Q),\quad (P)\Rightarrow(Q),\quad (P)\vee(Q),\quad (P)\wedge(Q),$$
$$\neg(P),\quad \forall x(P),\quad \exists x(P),$$

where P and Q are formulas shorter in length, and x is a variable.

The following result is then obtained by induction on the length of the expression: *if E is a term or a formula, then there exists a parentheses bijection between the set Π^+ of indices of left parentheses in E and the set Π^- of indices of right parentheses*. In fact, the new parentheses in 1.3(a) and (b) have a natural bijection, while the old ones (which might be contained in the terms $t_1, \ldots, t_r$ or the formulas P, Q) have such a bijection by the induction assumption. In addition, the new parentheses never come between two paired old parentheses.

We can now state the basic result of this section:

1.4. **Unique Reading Lemma.** *Every expression in L is either a term, or a formula, or neither. These alternatives, as well as all of the alternatives listed in* 1.3(a) *and* (b), *are mutually exclusive. Every term* (*resp. formula*) *can be represented in exactly one of the forms in* 1.3(a) (*resp.* 1.3(b)), *and in a unique way.*

In addition, in the course of the proof we show that, *if an expression is the concatenation of a finite sequence of terms, then it is uniquely representable as such a concatenation.*

PROOF. Using induction on the length of the expression E, we describe an informal algorithm for syntactic analysis, which uniquely determines which alternative holds.

(a) If there are no parentheses in E, then E is either a constant term, a variable term, or neither a term nor a formula.

(b) If E contains parentheses, but there is no parentheses bijection between the left and right parentheses, then E is neither a term nor a formula.

(c) Suppose E contains parentheses with a parentheses bijection. Then either E is uniquely represented in one of the nine forms

$$f(E_0)\quad (\text{where } f \text{ is an operation}),$$
$$p(E_0)\quad (\text{where } p \text{ is a relation}),$$
$$(E_1)\Leftrightarrow(E_2),\quad (E_1)\Rightarrow(E_2),\quad (E_1)\vee(E_2),\quad (E_1)\wedge(E_2),$$
$$\neg(E_3),\quad \forall x(E_3),\quad \exists x(E_3),$$

or else E is neither a term nor a formula. Here the pairs of parentheses we have written out are connected by the unique parentheses bijection which is assumed to exist in E; this is what ensures uniqueness. In fact, we obtain the form $f(E_0)$ if and only if the first element of the expression is a function, the second element is "(," and the last element is the ")" which corresponds under the bijection: and similarly for the other forms.

We have thereby reduced the problem to the syntactic analysis of the expressions E_0, E_1, E_2, E_3, which are shorter in length. This almost completes our description of the algorithm, since what remains to be determined about E_1, E_2, E_3 is whether or not they are formulas. However, for E_0 we must determine whether this expression is a concatenation of the right number of terms, and we must ask whether such a representation must be unique.

The answer to the latter question is positive. We have the following recipe for breaking off terms from left to right in a union of terms.

(d) Let E_0 be an expression having a parentheses bijection between its left and right parentheses. If E_0 can be represented in the form tE_0', where t is a term, then this representation is unique. In fact, either E_0 can be uniquely represented in one of the forms

$$xE_0', \quad cE_0', \quad f(E_0'')E_0'$$

(where x is a variable, c is a constant, and f is an operation whose parentheses correspond under the unique parentheses bijection in E_0), or else E_0 cannot be represented in the form tE_0', where t is a term. In the cases $E_0 = xE_0'$ or $E_0 = cE_0'$, this is obviously the only way to break off a term from the left. In the case $E_0 = f(E_0'')E_0'$, the question reduces to whether or not E_0'' is a concatenation of degree (f) terms. By induction on the length of E_0, we may assume that either E_0'' is not such a concatenation, or else it is uniquely representable as a concatenation of terms. The lemma is proved. □

Exercise: State and prove a unique reading lemma for the "parentheses-less" dialect of $\mathfrak{L}_1$ described in 2(a) of "Digression: Syntax" in Chapter I.

Here is the first inductive description of the difference between free and bound occurrences of a variable in terms and formulas. The correctness of the following definitions is ensured by Lemma 1.4.

1.5. **Definition.**

(a) Every occurrence of a variable in an atomic formula or term is free.

(b) Every occurrence of a variable in $\neg(P)$ or in $(P_1) * (P_2)$ (where $*$ is any of the connectives "$\vee$," "$\wedge$," "$\Rightarrow$" or "$\Leftrightarrow$") is free (respectively bound) if and only if the corresponding occurrence in P, P_1, or P_2 is free (respectively bound).

(c) Every occurrence of the variable x in $\forall x(P)$ and $\exists x(P)$ is bound. The occurrences of other variables in $\forall x(P)$ and $\exists x(P)$ are the same as the corresponding occurrences in P.

Suppose the quantifier $\forall$ (or $\exists$) occurs in the formula P. It follows from the definitions that it must be followed in P by a variable and a left parenthesis. The expression which begins with this variable and ends with the corresponding right parenthesis is called the scope of the given (occurrence of the) quantifier.

1.6. **Definition.** Suppose we are given a formula P, a free occurrence of the variable x in P, and a term t. We say that t is free for the given occurrence of x in P if the occurrence does not lie in the scope of any quantifier of the form $\exists y$ or $\forall y$, where y is a variable occurring in t.

In other words, if t is substituted in place of the given occurrence of x, all free occurrences of variables in t remain free in P.

We usually have to substitute a term for each free occurrence of a given variable. It is important to note that this operation takes terms into terms and formulas into formulas (induction on the length). If t is free for each free occurrence of x in P we simply say that t is free for x in P.

1.7. We shall start working with definitions 1.5 and 1.6 in the next section. Here we shall only give some intuitive explanations.

Definition 1.5 allows us to introduce the important class of *closed* formulas. By definition, this consists of formulas without free variables. (They are also called sentences.) The intuitive meaning of the concept of a closed formula is as follows. A closed formula corresponds to an assertion which is completely determined (in particular, regarding truth or falsity); indeterminate objects of the theory are only mentioned in the context "all objects x satisfy the condition . . . " or "there exists an object y with the property . . . ." Conversely, a formula which is not closed, such as $x \in y$ or $\exists x(x \in y)$, may be true or false depending on what sets are being designated by the names x and y (for the first) or by the name y (for the second). Here truth or falsity is understood to mean for a fixed interpretation of the language, as will be explained in §2.

In particular, Definition 1.6 gives the rules of hygiene for changing notation. If we want to call an indeterminate object x by another name y in a given formula, we must be sure that x does not appear in the parts of the formula where this name y is already being used to denote an *arbitrary* indeterminate object (after a quantifier). In other words, y must be free for x. Moreover, if we want to say that x is obtained from certain operations on other indeterminate objects (x = a term containing $y_1, \ldots, y_n$), then the variables $y_1, \ldots, y_n$ must not be bound.

There is a close parallel to these rules in the language of analysis: instead of $\int_1^x f(y)\,dy$ we may confidently write $\int_1^x f(z)\,dz$ but we must not write $\int_1^x f(x)\,dx$; the variable y is bound in the scope of $\int f(y)\,dy$.

2 Interpretation: truth, definability

2.1. Suppose we are given a language L in $\mathfrak{L}_1$ and a set (or class) M. To give an *interpretation* of L in M means to tell how a formula in L can be given a meaning as a statement about the elements of M.

More precisely, an interpretation ϕ of the language L in M consists of a collection of mappings which correlate terms and formulas of the language to elements of M and structures over M (in the sense of Bourbaki). These mappings are divided into *primary* mappings, which actually determine the interpretation, and *secondary* mappings, which are constructed in a natural and unique way from the primary mappings. We shall use the term interpretation to refer to the mappings themselves, and sometimes also to the values they take.

Let us proceed to the systematic definitions. We shall sometimes call the elements of the alphabet of L *symbols*. The notation ϕ for the interpretation will either be included when writing the mappings or omitted, depending on the context.

2.2. *Primary mappings*

(a) An interpretation of the constants is a map from the set of symbols for constants (in the alphabet of L) to M, which takes a symbol c to $\phi(c) \in M$.

(b) An interpretation of the operations is a map from the set of symbols for operations (in the alphabet of L) which takes a symbol f of degree r to a function $\phi(f)$ on $M \times \cdots \times M = M^r$ with values in M.

(c) An interpretation of the relations is a map from the set of symbols for relations (in the alphabet of L) which takes a symbol p of degree r to a subset $\phi(p) \subset M^{r'}$.

Secondary mappings Intuitively, we would like to interpret variables as names for the "generic element" of the set M, which can be given specific values in M. We would like to interpret the term $f(x_1, \ldots, x_r)$ as a function $\phi(f)$ of r arguments which run through values in M, and so on.

In order to give a precise definition, we introduce the *interpretation class* $\overline{M}$:

$$\overline{M} = \text{the set of all maps to } M \text{ from the set of symbols for variables in the alphabet of } L.$$

Thus, every point $\xi \in \overline{M}$ correlates to any variable x a value $\phi(x)(\xi) \in M$, which we shall usually denote simply x^{ξ}. This allows us to consider variables as *functions on* $\overline{M}$ *with values in* M. More generally:

2.3. The interpretation of terms correlates to each term t a function $\phi(t)$ on $\overline{M}$ with values in M. This correspondence is defined inductively by the following compatibilities:

(a) If c is a constant, then $\phi(c)$ is the constant function whose value is defined by the primary mapping.
(b) If x is a variable, then $\phi(x)$ is $\phi(x)(\xi)$ as a function of ξ.
(c) If $t = f(t_1, \ldots, t_r)$, then for all $\xi \in \overline{M}$

$$\phi(t)(\xi) = \phi(f)(\phi(t_1)(\xi), \ldots, \phi(t_r)(\xi)),$$

where the $\phi(t_i)(\xi)$ are defined by the induction assumption, and

$\phi(f) : M^r \to M$ is given by the primary mapping. Instead of $\phi(t)(\xi)$ we shall sometimes write simply t^ξ.

2.4. *Interpretation of atomic formulas.* An interpretation ϕ assigns to every formula P in L a *truth* function $|P|_\phi$. This is a function on the interpretation class $\overline{M}$ which only takes the values 0 ("false") and 1 ("true"). It is defined for atomic formulas as follows:

$$|p(t_1, \ldots, t_r)|_\phi(\xi) = \begin{cases} 1, & \text{if } \langle t_1^\xi, \ldots, t_r^\xi \rangle \in \phi(p), \\ 0, & \text{otherwise.} \end{cases}$$

Intuitively, a statement p about the names $t_1, \ldots, t_r$ for objects in M becomes true if the objects named by $t_1, \ldots, t_r$ satisfy the relation named by p.

2.5. *Interpretation of formulas.* The truth function for nonatomic formulas is defined inductively by means of the following relations (for brevity, we have omitted parentheses and explicit mention of ϕ and ξ):

$$|P \Leftrightarrow Q| = |P|\,|Q| + (1 - |P|)(1 - |Q|):$$

$P \Leftrightarrow Q$ is true when either P and Q are both true or P and Q are both false.

$$|P \Rightarrow Q| = 1 - |P| + |P|\,|Q|:$$

$P \Rightarrow Q$ is only false when P is true and Q is false.

$$|P \vee Q| = \max(|P|, |Q|):$$

$P \vee Q$ is only false when P and Q are both false.

$$|P \wedge Q| = \min(|P|, |Q|):$$

$P \wedge Q$ is only true when P and Q are both true.

$$|\neg P| = 1 - |P|:$$

$\neg P$ is only false when P is true.

Finally, we must describe what happens when quantifiers are introduced. Suppose that $\xi \in \overline{M}$ and x is a variable. By a *variation of ξ along x* we mean any point $\xi' \in \overline{M}$ for which $y^\xi = y^{\xi'}$ whenever y is a variable different from x. Then

$$|\forall x P|(\xi) = \min_{\xi'} |P|(\xi'),$$

$$|\exists x P|(\xi) = \max_{\xi'} |P|(\xi'),$$

where ξ' runs through all variations of ξ along x.

A formula P is called *ϕ-true* if $|P|_\phi(\xi) = 1$ for all $\xi \in \overline{M}$. The interpretation ϕ (or M) is called a *model* for a set of formulas $\mathcal{E}$ if all the elements of $\mathcal{E}$ are ϕ-true.

2.6. Example: Standard Interpretation of L_1Ar. This is the interpretation in the set N of nonnegative integers, in which $\overline{0}$, $\overline{1}$ are interpreted as

0, 1, respectively, and $+$, $\cdot$, $=$ are interpreted as addition, multiplication, and equality, respectively.

2.7. Example: Standard Interpretation of L_1Set. This is the interpretation in the von Neumann universe V, in which $\varnothing$ is interpreted as the empty set, $\in$ is interpreted as the relation "is an element in," and $=$ is interpreted as equality.

All of the examples of translations in Chapter I were based on these standard interpretations. The relationship between those examples and the above definitions is as follows. Let $\Pi(x, y, z)$ be a statement in argot about the indeterminate sets x, y, z in V; and let $P(x, y, z)$ be a translation of Π into the language L_1Set. Then for any point ξ interpreting x, y, z as the names of sets x^ξ, y^ξ, z^ξ in the von Neumann universe, we have:

$$\Pi(x^\xi, y^\xi, z^\xi) \text{ is true} \Leftrightarrow |P(x, y, z)|(\xi) = 1.$$

Thus, every formula expresses, or defines, a property of objects in the interpretation set:

2.8. **Definition.** A set $S \subset M^r$, $r \geqslant 1$, is called ϕ-definable (by the formula P in L with the interpretation ϕ) if there exist variables $x_1, \ldots, x_r$ such that

$$|P|_\phi(\xi) = 1 \Leftrightarrow \langle x_1^\xi, \ldots, x_r^\xi \rangle \in S$$

for all ξ in $\overline{M}$.

One of the most important problems concerning formal languages is to understand the structure of the sets of

$$\phi\text{-true formulas in } L;$$

$$\phi\text{-definable sets in } \bigcup_{r \geqslant 1} M^r.$$

2.9. Example. The sets definable by means of L_1Ar with the standard interpretation constitute the smallest class of sets in $\bigcup_{r \geqslant 1} N^r$ which

(a) contains all sets of the form

$$\{\langle k_1, \ldots, k_r \rangle | F(k_1, \ldots, k_r) = 0\} \subset N^r,$$

where F runs through all polynomials with integral coefficients.

(b) is closed relative to finite intersections, unions, and complements (in the appropriate N^r)

(c) is closed relative to the projections $\overline{\mathrm{pr}}_i : N^r \to N^{r-1}$:

$$\overline{\mathrm{pr}}_i \langle k_1, \ldots, k_r \rangle = \langle k_1, \ldots, k_{i-1}, k_{i+1}, \ldots, k_r \rangle.$$

In fact, sets of type (a) are defined by atomic formulas of the form $t_1^F = t_2^F$, where t_1^F is a term corresponding to the sum of the monomials in F with positive coefficients, and t_2^F corresponds to the sum of the monomials with negative coefficients. Further, if $S_1, S_2 \subset N^r$ are definable by formulas P_1, P_2 (with the same variables), then $S_1 \cap S_2$ is definable by $P_1 \wedge P_2$, $S_1 \cup S_2$ is definable by $P_1 \vee P_2$, and $N^r \setminus S_1$ is definable by $\neg P_1$. Finally, the set $\mathrm{pr}_i(S_1)$ is definable by the formula $\exists x_i(P_1)$. The connectives $\Rightarrow$ and $\Leftrightarrow$ and the quantifier $\forall$ give nothing new, since, without changing the set being defined, we may replace them by combinations of the logical operations already discussed: $\forall x$ may be replaced by $\neg \exists x \neg$, and so on.

This first description of *arithmetical* sets, i.e., L_1Ar-definable sets, will be greatly amplified in the second and third parts of the book. At this point it is not immediately clear how to develop the subtler properties of definability, such as the definability of the set of prime numbers in N (see example 3.14 in Chapter I), the definability of the set of partial fractions in the continued fraction expansion of $\sqrt[3]{2}$, or the definability of the set of pairs

$$\{\langle i, i\text{th digit in the decimal expansion of } \pi\rangle\} \subset N^2.$$

However, as we shall see in §11 and in Chapter VII, the "Gödel numbers of the true formulas of arithmetic" form still a much more complicated set, and this set is not definable.

We now give several simple technical results.

2.10. **Proposition.** *Let P be a formula in L, ϕ an interpretation in M, and $\xi, \xi' \in \overline{M}$. Suppose that x^{ξ} coincides with $x^{\xi'}$ for all variables x occurring freely in P. Then $|P|_{\phi}(\xi) = |P|_{\phi}(\xi')$.*

2.11. **Corollary.** *In any interpretation the closed formulas P have well-defined truth values: $|P|_{\phi}(\xi)$ does not depend on ξ.*

PROOF.

(a) Let t be a term, and suppose that for any variable x in t we have $x^{\xi} = x^{\xi'}$. Then Lemma 1.4 and induction on the length of t give $t^{\xi} = t^{\xi'}$.

(b) Assertion 2.10 holds for atomic formulas P of the form $p(t_1, \ldots, t_r)$. In fact,

$$|P|(\xi) = \begin{cases} 1, & \text{if } \langle t_1^{\xi}, \ldots, t_r^{\xi}\rangle \in \phi(P), \\ 0, & \text{otherwise,} \end{cases}$$

and similarly for $|P|(\xi')$. But if ξ and ξ' coincide on all the variables in P (all of which occur freely), then *a fortiori* they coincide on all the variables in t_i, and, by part (a), we have $t_i^{\xi} = t_i^{\xi'}$, $i = 1, \ldots, r$. Therefore, $|P|(\xi) = |P|(\xi')$.

(c) We now use induction on the total number of connectives and quantifiers in P. If P has the form $\neg Q$ or $Q_1 * Q_2$, then 2.10 for P follows trivially from 2.10 for Q, Q_1, Q_2. Now suppose that P has the form $\forall x(Q)$, and that 2.10 holds for Q. (The case $\exists x(Q)$ can be treated analogously or can be reduced to the case $\forall x$ by replacing $\exists x$ by $\neg \forall x \neg$.) By definition, we have

$$|\forall x Q|(\xi) = \begin{cases} 1, & \text{if } |Q|(\eta) = 1 \text{ for all variations } \eta \text{ of } \xi \text{ along } x, \\ 0, & \text{otherwise;} \end{cases}$$

$$|\forall x Q|(\xi') = \begin{cases} 1, & \text{if } |Q|(\eta') = 1 \text{ for all variations } \eta' \text{ of } \xi' \text{ along } x, \\ 0, & \text{otherwise.} \end{cases}$$

On the right we may let η and η' vary in addition on all variables which do not occur freely in Q. The assertions after the word "if" remain true or false in this wider range of values if they were true or false before, by the induction hypothesis on Q. But then η and η' run through the same values, because ξ and ξ' only differ on variables which do not occur freely in Q, and on x. The proposition is proved. □

The following almost obvious fact is the basis for many phenomena which attest to the inadequacy of formal languages for completely describing intuitive concepts (see "Skolem's paradox" below):

2.12. **Proposition.** *The cardinality of the class of ϕ-definable sets does not exceed*

$$\text{card(alphabet of } L) + \aleph_0.$$

Here and below, by "card(alphabet of L)" we mean the cardinality of the alphabet of L *without the set of variables*.

PROOF. If the language has $\leqslant \aleph_0$ variables, then there are at most

$$\text{card(alphabet of } L) + \aleph_0 \text{ formulas.}$$

If, on the other hand, it has an uncountable set of variables, then we note that every definable set can be defined by a formula whose variables belong to a fixed countable subset of the variables which is chosen once and for all. □

2.13. **Corollary.** *If M is infinite and* card(alphabet of L) $< 2^{\text{card } M}$, *then "almost all" sets are undefinable.*

Thus, the only way to define all subsets of M is to include a tremendous number of names in the language. For languages which are to describe actual mathematical reasoning this is an unrealistic program. Essentially, any finitely describable collection of modes of expression only allows us to define a countable number of sets. However, it is often technically useful to include in the alphabet, for example, names for all the elements of M.

In the following sections we proceed to study systematically sets of true formulas.

3 Syntactic properties of truth

Let L be a language in $\mathcal{L}_1$, let ϕ be an interpretation of L, and let $T_\phi L$ be the set of ϕ-true formulas. In this section we list some properties of $T_\phi L$ which reflect the logic inherent in languages of $\mathcal{L}_1$ regardless of the specific nature of the interpretation ϕ.

3.1. *The set $T_\phi L$ is complete.* By definition, this means that, for any closed formula P, either P or $\neg P$ lies in $T_\phi L$. This property follows from Corollary 2.11 above.

3.2. *The set $T_\phi L$ does not contain a contradiction*, that is, there is no formula P for which P and $\neg P$ both lie in $T_\phi L$. In fact, $T_\phi L = \{P \mid |P|_\phi = 1\}$, while $|\neg P|_\phi = 1 - |P|_\phi$.

3.3. *The set $T_\phi L$ is closed under the rules of deduction* MP (modus ponens) *and* Gen (*generalization*). By definition, this means that, if P and $P \Rightarrow Q$ lie in $T_\phi L$, then Q also lies in $T_\phi L$; and that, if P lies in $T_\phi L$, then $\forall x P$ lies in $T_\phi L$ for any variable x. The verification is immediate: if $|P|_\phi = 1$ and $|P \Rightarrow Q|_\phi = 1$, then we must have $|Q|_\phi = 1$; if $|P|_\phi(\xi) = 1$ for all ξ, then also $|\forall x P|_\phi(\xi) = 1$. The formula Q is called a *direct consequence of the formulas P and $P \Rightarrow Q$ using the rule of deduction* MP. The formula $\forall x P$ is called a *direct consequence of the formula P using the rule of deduction* Gen.

The intuitive meaning of these rules of deduction is as follows. The rule MP corresponds to the type of argument: "If P is true, and if the truth of P implies the truth of Q, then Q is true." Thus, one might say that the semantics of the expression "if . . . then" in natural languages is divided between the semantics of the *connective* $\Rightarrow$ and the semantics of the *rule of deduction* MP in languages of $\mathcal{L}_1$. Neglecting this point of view often leads to confusion when one attempts to explain the rules for assigning truth values to the formula $P \Rightarrow Q$.

The rule Gen corresponds to the practice in mathematics of writing "identities" or universally true assertions. When we write $(a + b)^2 = a^2 + 2ab + b^2$ or "in a right triangle the square of the hypotenuse is equal to the sum of the squares of the other two sides," the quantifiers $\forall a \, \forall b$ and $\forall$ triangles are omitted. Putting the quantifiers back in does not change the truth values, and has the advantage of freeing the notation for later use.

3.4. *The set $T_\phi L$ contains all tautologies.* To define what a tautology is, we first introduce the notion of a *logical polynomial* over a set of formulas $\mathcal{E}$. This is an element in the least set of formulas which contains $\mathcal{E}$ and is closed with respect to constructing formulas from shorter formulas using logical connectives.

A sequence of formulas $P_1, \ldots, P_n$ and representations of each P_i either in the form Q, where $Q \in \mathfrak{S}$, or in the form $\neg Q$ or $Q_1 * Q_2$, where Q, Q_1, Q_2 lie in $\{P_i, \ldots, P_{i-1}\}$ is called a *representation* of P_n as a logical polynomial over $\mathfrak{S}$. The representation of P_n is not necessarily unique: for example, if $\mathfrak{S} = \{P, Q, P \Rightarrow Q\}$, then $P \Rightarrow Q$ has two representations.

Let $| \ | : \mathfrak{S} \to \{0, 1\}$ be any map. If we are given a representation r of the formula P_n as a logical polynomial over $\mathfrak{S}$, then we can use the formulas in 2.5 to determine $|P_n|_r$ recursively.

A formula P is called a tautology if there exists a set of formulas $\mathfrak{S}$ and a representation r of P as a logical polynomial over $\mathfrak{S}$ such that $|P|_r = 1$ for all maps $| \ | : \mathfrak{S} \to \{0, 1\}$. The property of being a tautology is effectively decidable, since, by syntactically analyzing P we can enumerate all representations of P as a logical polynomial. All tautologies obviously belong to $T_\Phi L$.

Here are our first examples of tautologies:

A0. $P \Rightarrow P$
A1. $P \Rightarrow (Q \Rightarrow P)$
A2. $(P \Rightarrow (Q \Rightarrow R)) \Rightarrow ((P \Rightarrow Q) \Rightarrow (P \Rightarrow R))$
A3. $(\neg Q \Rightarrow \neg P) \Rightarrow ((\neg Q \Rightarrow P) \Rightarrow Q)$
B1. $\neg \square P \Rightarrow P, P \Rightarrow \neg \neg P$
B2. $\neg P \Rightarrow (P \Rightarrow Q)$.

Here P, Q, and R are arbitrary formulas in L; the form in which these tautologies are written makes it clear what representation as a logical polynomial over $\{P, Q, R\}$ is intended.

Thus, tautologies are formulas which are true regardless of the truth or falsity of the component parts (if the notion of component is suitably chosen). B1 is the law of the excluded middle: a double negation is equivalent to the original assertion. B2 is the mechanism by which a contradiction in a set of formulas $\mathfrak{S}$ in L leads to the deducibility of any formula, and thereby destroys the entire system. (See Proposition 4.2 below.)

Example of how a tautology is verified. We give three versions of how to verify that the simple formula A1 is a tautology.

Version (a). By the formulas in 2.5, we have

$$|P \Rightarrow (Q \Rightarrow P)| = 1 - |P| + |P|\,|Q \Rightarrow P|$$
$$= 1 - |P| + |P|(1 - |Q| + |P|\,|Q|) = 1,$$

since $|P|^2 = |P|$.

Version (b). We tabulate $|P \Rightarrow (Q \Rightarrow P)|$ as a function of $|P|$ and $|Q|$:

$\|P\|$	$\|Q\|$	$\|Q \Rightarrow P\|$	$\|P \Rightarrow (Q \Rightarrow P)\|$
0	0	1	1
0	1	0	1
1	0	1	1
1	1	1	1

This is an example of a "truth table."

Version (c). The basic property of the connective $\Rightarrow$ is that $P \Rightarrow Q$ is only false if P is true and Q is false. If $P \Rightarrow (Q \Rightarrow P)$ were false, then P would be true and $Q \Rightarrow P$ would be false; then, in turn, Q would be true and P would be false, a contradiction.

The reader would do well to verify that the more complicated axioms, for example A2, are tautologies, and to decide which of the three versions he prefers.

3.5. *The set* $T_\phi L$ *contains the "logical quantifier axioms,"* that is, the formulas

(a) $\forall x(P \Rightarrow Q) \Rightarrow (P \Rightarrow \forall x Q)$, if all the occurrences of x in P are bound.
(b) $\forall x \neg P \Leftrightarrow \neg \exists x P$.
(c) $\forall x P(x) \Rightarrow P(t)$, if t is free for x in P (*axiom of specialization*). Here we use the notation $P(t)$ for the result of substituting t for each free occurrence of x in P. In all other respects P and Q are arbitrary formulas.

In 3.7 we verify that the formulas in 3.5 are ϕ-true. The intuitive meaning of these formulas is more or less clear. For example, the axiom of specialization means that, if $P(x)$ is true for all x, then $P(t)$ is also true, where t is the name of any object. The condition that t must be free for x is the rule of hygiene for changing notation.

The set

$$\text{Ax } L = \{\text{tautologies of } L\} \cup \{\text{quantifier axioms}\}$$

is called the *set of logical axioms in the language L.*

A set of formulas $\mathcal{E}$ in L will be called *Gödelian* if it is *complete, does not contain a contradiction, is closed with respect to the rules of deduction* MP *and* Gen, *and contains all the logical axioms of L.* The basic conclusion of our discussion is then:

3.6. **Proposition.** *The set of true formulas of L (in any interpretation) is Gödelian.*

In §6 we prove that, conversely, any Gödelian set is a set of true formulas in a suitable interpretation. Thus, the concept of a Gödelian set is the closest approximation to the concept of truth which can be attained "without regard to meaning."

3.7. *Verification that axioms* 3.5 *are true.*

(a) Let R be the formula 3.5(a). We suppose that $|R|(\xi)=0$ for some $\xi \in \overline{M}$ and show that that leads to a contradiction.

In fact, then $|\forall x(P \Rightarrow Q)|(\xi)=1$ and $|P \Rightarrow \forall x\, Q|(\xi)=0$. The second equation implies that $|P|(\xi)=1$ and $|\forall x\, Q|(\xi)=0$. Let ξ' be a variation of ξ along x for which $|Q|(\xi')=0$. Then $|P|(\xi')=|P|(\xi)=1$ by Proposition 2.10, since x does not occur freely in P. Hence, $|P \to Q|(\xi')=0$, which contradicts the relation $|\forall x(P \Rightarrow Q)|(\xi)=1$.

(b) For all $\xi \in \overline{M}$ and for all variations ξ' of ξ along x, we have

$$|\forall x \neg P|(\xi) = \max_{\xi'} |\neg P|(\xi') = 1 - \min_{\xi'} |P|(\xi');$$

$$|\neg \exists x\, P|(\xi) = 1 - \min_{\xi'} |P|(\xi').$$

Hence, the truth values of $\forall x \neg P$ and $\neg\exists x\, P$ coincide, so that $\forall x \neg P \Leftrightarrow \neg \exists x\, P$ is identically true.

(c) Suppose that $|\forall x\, P(x) \Rightarrow P(t)|(\xi)=0$ for some point $\xi \in \overline{M}$. We show that this leads to a contradiction. In fact, then

$$|\forall x\, P(x)|(\xi) = 1, \qquad |P(t)|(\xi) = 0.$$

The first equation implies that $|P(x)|(\xi')=1$ for all variations ξ' or ξ along x. For ξ' we take the variation such that $x^{\xi'} = t^{\xi}$. If we prove that $|P(t)|(\xi) = |P(x)|(\xi')$, then we obtain the desired contradiction.

We prove this by induction on the total number of connectives and quantifiers in P.

(c_1) Let P be an atomic formula $p(t_1, \ldots, t_n)$. Letting $\bar{t}_i$ denote the result of substituting t for each occurrence of x in t_i, we successively obtain:

$$t^{\xi} = x^{\xi'} \quad \text{(by the definition of } \xi'),$$

$$\bar{t}_i^{\xi} = t_i^{\xi'} \quad \text{(by induction on the length of } t_i),$$

$$|P(x)|(\xi') = |p(t_1, \ldots, t_n)|(\xi') = |p(\bar{t}_1, \ldots, \bar{t}_n)|(\xi) = |P(t)|(\xi).$$

(c_2) Let P have the form $\neg Q$ or $Q_1 \mapsto Q_2$, where $\mapsto$ is a connective. Since x does not bind t in P by assumption, the same is true for Q, Q_1, and Q_2, and the necessary induction step is automatic.

(c_3) Finally, let P have the form $\exists y\, Q$ or $\forall y\, Q$. We shall examine the first case; the proof for the second case is analogous.

Subcase 1. $y = x$. Then x is bound in P; therefore, $P(x) = P(t)$, and $|P|(\xi) = |P|(\xi')$ by Proposition 2.10.

Subcase 2. $y \neq x$. The induction assumption has the form: $|Q(t)|(\eta) = |Q(x)|(\eta')$, if η is any point in $\overline{M}$ and η' is a variation of η along x for which $x^{\eta'} = t^{\eta}$. We must show that the following two truth values coincide (where ξ and ξ' are defined as above):

$$|\exists y\, Q(x)|(\xi') = \begin{cases} 1, & \text{if } |Q(x)|(\eta') = 1 \text{ for some variation } \eta' \text{ of } \xi' \text{ along } y, \\ 0, & \text{otherwise.} \end{cases}$$

$$|\exists y\, Q(t)|(\xi) = \begin{cases} 1, & \text{if } |Q(t)|(\eta) = 1 \text{ for some variation } \eta \text{ of } \xi \text{ along } y, \\ 0, & \text{otherwise.} \end{cases}$$

We recall that ξ' is the variation of ξ along x for which $x^{\xi'} = t^{\xi}$.

We first suppose that the second truth value is 1. We choose $\eta \in \overline{M}$ so that $|Q(t)|(\eta) = 1$, and then construct the variation η' of η along x for which $x^{\eta'} = t^{\eta}$. Then, by the induction assumption, $1 = |Q(t)|(\eta) = |Q(x)|(\eta')$. We show that η' is a variation of ξ' along y; this will imply that the first truth value is also 1. In fact, η' was obtained by varying η along x, η was obtained by varying ξ along y, and ξ was obtained by varying ξ' along x. Hence, η' is a variation of ξ' along x and y; we must show the variation along x did not actually take place:

$$x^{\eta'} = x^{\xi'}.$$

But the left-hand side is t^{η} by the definition of η'; the right-hand side is t^{ξ} by the definition of ξ'; and η was obtained by varying ξ along y. Since t is free for x in $P = \exists y\, Q$, it follows that y does not occur in t.

It remains to verify that, if the second truth value is 0, then the first is also 0. The argument is almost the same. If the second truth value is 0, then $|Q(t)|(\eta) = 0$ for all variations η of ξ along y. For each such η we construct η' as in the first part of the proof. As before, we verify that η' is a variation of ξ' along y and, moreover, η' runs through all such variations when η runs through all variations of ξ along y. Hence, the first truth value is also 0.

The proposition is proved. □

Digression: natural logic

1. Logic does not concern itself with the external world, but only with systems for trying to understand it. The logic of one such system—mathematics—is normalized to such an extent that it resembles a rigid stencil, which we can attempt to impose on any other system. But whether or not this stencil fits the system should not be seen as the criterion of suitability or the measure of worth of the system. The physicist's descriptions do not have to form a consistent or coherent whole; his job is to describe nature effectively on certain levels. Natural languages and the spontaneous workings of the mind are even less logical. In general, adherence to logical principles is only a condition for effectiveness in certain narrowly specialized spheres of human endeavor.

Although comparisons between the logic of predicates and the logic of natural languages or their subsystems have no normative force, such comparisons may be interesting and enlightening. Here we give some selected material from linguistics and psychology.

2. B. Russell, K. Döhmann, H. Reichenbach, U. Weinreich, and many others have studied the problem of finding parallels in natural languages for categories which can be formalized in languages of $\mathfrak{L}_1$ and of cataloguing the methods of transmitting these categories. This leads to the grouping of words into so-called *logico-semantic classes*, instead of the traditional division into verbs, nouns, articles, etc. (A. V. Gladkiĭ and I. A. Mel'čuk, *Éléments de linguistique mathématique*, Paris, Dunod, 1972, §6).

For example, the words *sleeps, smart, cry-baby* are parallel to relation symbols (predicates) of rank 1; the words *loves, friendly, sister* correspond to relations of rank 2. For each of them we have atomic formulas, such as "*N* sleeps," "*X* is friendly to *Y*," and so on.

"All, sometimes, something" are quantifier words; while "and, or, but, if . . . then" are, of course, connectives. "The nose, le cadeau" are constants. Nouns are made into constants by using the definite article or its semantic equivalent. In Russian, which does not have definite articles, one must either use the demonstrative articles *etot* (this), *tot* (that), or make it clear from the context that the noun is meant as a constant. The words *nos* (nose), *podarok* (gift) are more like variables which stand for any object satisfying the simple predicate "is a nose," "is a gift." Incidentally, there are other possible interpretations.

The pronoun "he" is, without doubt, a variable. The pronouns "I" and "you" have much more complicated semantics, involving a correlation with who is speaking that does not exist in the speaker-less languages of $\mathfrak{L}_1$. Certain aspects of the first person pronoun are included in the semantics of algorithmic languages. The right type of "memory key" in a program for the IBM 360 will allow the program to change what is contained in any byte in the basic memory region. The memory guard asks "Who is there?", and the program answers, "It is I." Finally, it is even possible in languages of $\mathfrak{L}_1$ to find models for certain types of self-description; see §9–11 and the digression on self-reference.

In Russian, "ili" (or) can be used not only to express the logical $\vee$, but also to express the exclusive "or" and even to express conjunction $\wedge$, as in the sentence "$x^2 > 0$ for $x > 0$ or for $x < 0$" (E. V. Padučeva). In Latin, the functions of exclusive and inclusive "or" are expressed by two different words, *aut* and *vel*. "And" can sometimes express a time sequence: compare the sentences "Jane got married and had a baby" with "Jane had a baby and got married" (S. Kleene). The conjunction $\wedge$ can be expressed in different languages by:

juxtaposition:	Chinese: ma mo—horse and donkey
	Swahili: shika kitabu usome—take a book and read
a preposition:	Russian: Petja s Mašeĭ—Peter and Marsha
a conjunction:	and, i, et

a postpositional particle: Latin: senatus populusque—the senate and the people
two conjunctions: Russian: kak . . . tak.

Döhmann has catalogued the ways of expressing 16 logical polynomials in two variables in several languages of the world.

3. Curious as all this material may be, it should be regarded critically; in such comparisons with logic, the subtleties of usage often elude us. As an example, let us analyze the natural semantics of "if . . . then." We have already mentioned that in languages of $\mathcal{L}_1$ this connective corresponds not only to "$\Rightarrow$," but also to the rule of deduction modus ponens. Moreover, MP more adequately represents the meaning of "if . . . then."

Actually, the rule that any conditional is true if its antecedent is known to be false has almost no parallel in natural logic. Examples of the type "if snow is black, then $2 \times 2 = 5$," which keep cropping up in textbooks, are only capable of confusing the student, since no natural subsystem in our language has expressions with this semantics. A possible exception is certain poetic and expressive formulas with extremely limited usage ("If she be false, *O*, then heaven mocks itself!"). Formal mathematics, in which a single contradiction destroys the entire system, clearly has the features of poetic hyperbole.

Finally, in the logic of predicates there is no place at all for the modal aspect of the use of "if . . . then" in instructions of the type "if this happens, do that." On the other hand, this aspect can easily be expressed by the semantics of the connective "if . . . then . . . else" in algorithmic languages such as Algol. Unless one uses techniques suggested by algorithmic languages, any attempt to find a model for modality in languages based on $\mathcal{L}_1$ is doomed to failure (compare: A. A. Ivin, *The Logic of Norms*, MGU Press, 1973).

4. We have mentioned several times that the choice of the primitive modes of expression in the logic of predicates does not reflect psychological reality. Elementary logical operations, even one-step deductions, may require a highly trained intellect; yet, logically complicated operations can often be performed as a single elementary act of thought even by a damaged brain.

> "Sublieutenant Zasetsky, aged twenty-three, suffered a head injury 2 March 1943 that penetrated the left parieto-occipital area of the cranium. The injury...was further complicated by inflammation that resulted in adhesions of the brain to the meninges and marked changes in the adjacent tissues."

Professor A. R. Luria met Zasetsky at the end of May 1943, and observed his condition for the next 26 years. In this time Zasetsky wrote nearly 3000 pages, describing with agonizing effort his life and illness as he struggled to regain his reason. His notebooks, which provided the material for Luria's book *The Man with a Shattered World* (Basic Books, Inc., New York, 1972, translated by L. Solotaroff), not only show his perseverance and determination, but are also revealing from a psychological point of view.

At first, the destruction of Zasetsky's psyche was overwhelming. The predominant disorder was asematia, the inability to connect symbols with their meaning. Luria describes his first meeting with Zasetsky:

> "'Try reading this page,' I suggested to him.
> 'What's this?...No, I don't know...don't understand...what *is* this?'....
> I suggested he try to do something simple with numbers, like add six and seven.
> 'Seven...six...what's it? No, I can't...just don't know.'"
> The ability to understand the simplest predicates was lost: "'What season is there before winter?' 'Before winter? After winter?...Summer?...Or *something*! No, I can't get it.' 'Before spring?' 'It's spring now...and...and before...I've already forgotten, just can't remember.'"
> Zasetsky lost the ability to interpret the syntactic devices for organizing meaning: "'In the school where Dunya studied a woman worker from the factory came to give a report.' What did this mean to him? Who gave the report—Dunya or the factory worker? And where was Dunya studying? Who came from the factory? Where did she speak?"

This is a fairly difficult example composed by Professor Luria, but here is what Zasetsky himself writes:

> "I also had trouble with expressions like: 'Is an elephant bigger than a fly?' and 'Is a fly bigger than an elephant?' All I could figure out was that a fly is small and an elephant is big, but I didn't understand the words *bigger* and *smaller*. The main problem was I couldn't understand which word they referred to."

What attracts our attention is the complexity of Zasetsky's metalinguistic text describing his linguistic difficulties. The subtlety of the analysis seems incompatible with the crude errors being analyzed. This could be explained by the retrospective nature of the analysis, but the following even more complicated description was written concurrently with the experience of the mental defect being described:

> "Sometimes I'll try to make sense out of those simple questions about the elephant and the fly, decide which is right or wrong. I know that when you rearrange the words, the meaning changes. At first I didn't think it did, it didn't seem to make any difference whether or not you rearranged the words. But after I thought about it a while I noticed that the sense of the four words (*elephant*, *fly*, *smaller*, *larger*) did change when the words were in a different order. But my brain, my memory, can't figure out right away what the word *smaller* (or *larger*) refers to. So I always have to think

> about them for a while.... So sometimes ridiculous expressions like 'a fly is bigger than an elephant' seem right to me, and I have to think about it a while longer."

We can also see how complicated mental abilities were preserved while "simple" ones were lost from examples of Zasetsky's creative imagination, which resemble literary-psychological studies:

> "Say I'm a doctor examining a patient who is seriously ill. I'm terribly worried about him, grieve for him with all my heart. (After all, he's human too, and helpless. I might become ill and also need help. But right now it's him I'm worried about—I'm the sort of person who can't help caring.) But say I'm another kind of doctor—someone who is bored to death with patients and their complaints. I don't know why I took up medicine in the first place, because I don't really want to work and help anyone. I'll do it if there's something in it for me, but what do I care if a patient dies? It's not the first time people have died, and it won't be the last."

All of this shows that there is no basis whatsoever for Rosser's opinion that "once the proof is discovered, and stated in symbolic logic, it can be checked by a moron." The human mind is not at all well suited for analyzing formal texts.

4 Deducibility

4.1. **Definition.** A deduction of a formula P from a set of formulas $\mathfrak{E}$ (in a language L in $\mathfrak{L}_1$) is a finite sequence of formulas $P_1, \ldots, P_n = P$ with the property that for each $i = 1, \ldots, n$ at least one of the following alternative holds:

(a) $P_i \in \mathfrak{E}$;
(b) $\exists j < i$ such that P_i is a direct consequence of P_j using Gen;
(c) $\exists j, k < i$ such that P_i is a direct consequence of P_j and P_k using MP.

We shall write $\mathfrak{E} \vdash P$ to abbreviate "there exists a deduction of P from $\mathfrak{E}$." A deduction of P, together with a precise indication for each $i \leqslant n$ of which of the alternatives (a), (b), (c) and which indices j in case (b) or j, k in case (c) are used to obtain P_i, is called a description of a deduction. A single deduction may have several descriptions.

We usually consider deductions from sets $\mathfrak{E}$ which contain Ax L, the logical axioms of L. The other elements of $\mathfrak{E}$ may be formulas of L which are "guessed" to be true in the standard interpretation; these are called *special axioms* of L. (Examples will be given later in 4.6–4.9.) Such deductions may be considered the formal equivalents of *mathematical proofs* (of a formula $P = P_n$ from the hypotheses $\mathfrak{E}$). This identification is justified for the following reasons:

(a) As shown in 3.3, if $\mathcal{E} \subset T_\phi L$ for some interpretation ϕ, and if $\mathcal{E} \vdash P$, then $P \in T_\phi L$; only true formulas can be deduced from true formulas.

(b) A large amount of experimental work has been done on formalizing mathematical proofs, that is, replacing them by deductions in suitable languages of $\mathfrak{L}_1$, especially $\mathrm{L_1Set}$. This work has shown that for large segments of mathematics, including the foundations of the theory of integers and real numbers, set theory, and so on, proofs can successfully be formalized as deductions within the framework of $\mathfrak{L}_1$. There is much material on this theme in the literature on mathematical logic; see, in particular, Mendelson's book.

(c) Gödel's completeness theorem for the logical modes of expression in $\mathfrak{L}_1$ (see §6) shows that any formula which is not deducible from $\mathcal{E}$ must be false in some model (interpretation) of $\mathcal{E}$.

For further discussion, see "Digression: Proof."

We occasionally consider deductions from another type of sets $\mathcal{E}$. For example, we might remove from $\mathcal{E}$ certain logical axioms, such as the "law of the excluded middle" (B1 in Section 3.4), in order to investigate formally intuitionistic principles. Or we might add to $\mathcal{E}$ a formula which we think is false in order to deduce a contradiction from $\mathcal{E}$; this is the so-called "proof by contradiction."

We now prove some formal aspects of contradiction.

4.2. **Proposition.** *Suppose that* $\mathcal{E}$ *contains all tautologies of type* B.2 *in Subsection* 3.4. *Then the following two properties of* $\mathcal{E}$ *are equivalent*:

(a) *There exists a formula* P *such that* $\mathcal{E} \vdash P$ *and* $\mathcal{E} \vdash \neg P$.
(b) $\mathcal{E} \vdash Q$ *for any formula* Q.

A set $\mathcal{E}$ *with these properties is called inconsistent.*

PROOF. (b)$\Rightarrow$(a) is obvious. Conversely, suppose $\mathcal{E} \vdash P$ and $\mathcal{E} \vdash \neg P$. We first add the formula $\neg P \rightarrow (P \rightarrow Q)$, which is assumed to lie in $\mathcal{E}$, to the descriptions of the two deductions. Then, applying MP twice (to this formula and $\neg P$; then to $P \Rightarrow Q$ and P), we obtain a description of a deduction $\mathcal{E} \vdash Q$. □

4.3. A large part of the theorems of logic consists in proving assertions of the type "$\mathcal{E} \vdash P$" or "it is not true that $\mathcal{E} \vdash P$" for various languages L, sets $\mathcal{E}$, and (classes of) formulas P.

A result of the form $\mathcal{E} \vdash P$ may be proved by presenting a description of a deduction of P from $\mathcal{E}$. However, even in slightly complicated cases, this procedure becomes so long that it is replaced by more or less complete instructions on how to compose such a description. Finally, "$\mathcal{E} \vdash P$" may be proved without presenting even an incomplete description of a deduction of P from $\mathcal{E}$. In this case we "are not proving P, but are proving that a proof of P exists;" see the example in §8 concerning language extensions.

In rare cases a result of the form "it is not true that $\mathcal{E} \vdash P$" can be proved by a purely syntactic argument. But usually such a result is obtained by constructing a model, i.e., an interpretation, in which $\mathcal{E}$ is true and P is false; see the discussion of the continuum problem in Chapters III–IV. If it is true neither that $\mathcal{E} \vdash P$ nor that $\mathcal{E} \vdash \neg P$, we say that P is *independent* of $\mathcal{E}$.

We now give two useful elementary results concerning deductions. It is clear that, compared with usual proofs, deductions are made up of very minor details. The mathematician, as if wearing seven-league boots, covers entire fields of formal deductions in one step.

4.4. **Lemma.** *Suppose that $\mathcal{E}$ contains all tautologies. If $\mathcal{E} \vdash P$ and $\mathcal{E} \vdash Q$, then $\mathcal{E} \vdash P \wedge Q$.*

PROOF. If $P_1, \ldots, P_m$ and $Q_1, \ldots, Q_n$ are deductions of P and Q, respectively, then

$$P_1, \ldots, P_m, Q_1, \ldots, Q_n, P \Rightarrow (Q \Rightarrow (P \wedge Q)), Q \Rightarrow (P \wedge Q), P \wedge Q$$

is a deduction of $P \wedge Q$. The third formula from the end is a tautology; the second formula from the end is a direct consequence of this tautology and $P_m = P$ using MP; and the last formula is a direct consequence of the second to last and $Q_n = Q$ using MP. □

4.5. **Deduction Lemma.** *Suppose that $\mathcal{E} \supset \text{Ax } L$ and P is a closed formula. If $\mathcal{E} \cup \{P\} \vdash Q$, then $\mathcal{E} \vdash P \Rightarrow Q$.*

PROOF. Let $Q_1, \ldots, Q_n = Q$ be a deduction of Q from $\mathcal{E} \cup \{P\}$. We show by induction on n that there exists a deduction of $P \Rightarrow Q$ from $\mathcal{E}$.

(a) $n = 1$. Then either $Q \in \mathcal{E}$, or else $Q = P$. In the first case $P \Rightarrow Q$ is deduced from Q and the tautology $Q \Rightarrow (P \Rightarrow Q)$ using MP. In the second case $P \Rightarrow P$ is a tautology.

(b) $n \geqslant 2$. We assume that the lemma holds for deductions of length $\leqslant n-1$. Then $\mathcal{E} \vdash P \Rightarrow Q_i$ for all $i \leqslant n-1$. Further, we have the following possibilities for $Q_n = Q$: (b_1) $Q \in \mathcal{E}$; (b_2) $Q = P$; (b_3) Q is deduced from Q_i and $Q_j = (Q_i \Rightarrow Q)$ using MP; and (b_4) Q has the form $\forall x\, Q$; for $j \leqslant n-1$. The first two cases are handled in exactly the same way as for $n = 1$.

In case (b_3), $P \Rightarrow Q$ can be deduced from $\mathcal{E}$ in the following way:

(1) deduction of $P \Rightarrow Q$ (induction assumption)
(2) deduction of $P \Rightarrow (Q_i \Rightarrow Q)$ (induction assumption)
(3) $(P \Rightarrow (Q_i \Rightarrow Q)) \Rightarrow ((P \Rightarrow Q_i) \Rightarrow (P \Rightarrow Q))$ (tautology)
(4) $(P \Rightarrow Q_i) \Rightarrow (P \Rightarrow Q)$ (from (2) and (3) using MP)
(5) $P \Rightarrow Q$ (from (1) and (4) using MP).

From now on, arguments of this sort will be presented more briefly, with explicit mention of only the last steps of the induction (here (3), (4), and (5)).

Finally, in case (b_4), we obtain a deduction of $P \Rightarrow \forall x\, Q_j$ from $\mathcal{E}$ if we add the following formulas to the deduction of $P \Rightarrow Q_j$ from $\mathcal{E}$ (which exists by the induction assumption):

$$\forall x (P \Rightarrow Q_j) \qquad \text{(Gen)}$$

$$\forall x (P \Rightarrow Q_j) \Rightarrow (P \Rightarrow \forall x\, Q_j) \qquad \text{(logical quantifier axiom, since } P \text{ is closed)}$$

$$P \Rightarrow \forall x\, Q_j \qquad \text{(MP applied to the two preceding formulas).}$$

The lemma is proved. □

We record for future reference that, in the parts of deductions constructed in Lemmas 4.4 and 4.5, only tautologies of the type A0, A1, and A2 in Subsection 3.4 were used.

We now give some basic examples of special axioms.

Axioms of equality

Let L be a language in $\mathfrak{L}_1$ whose alphabet includes a relation $=$ of rank two. We shall write $t_1 = t_2$ instead of $=(t_1, t_2)$. If P is a formula, x is a variable, and t is a term, we let $P(x, t)$ denote the result of substituting t in P in place of *any or all* of the free occurrences of x in P for which t is free.

4.6. **Proposition.**

(a) *The formulas*

$$t = t; \qquad t_1 = t_2 \Rightarrow t_2 = t_1; \qquad t_1 = t_2 \wedge t_2 = t_3 \Rightarrow t_1 = t_3;$$
$$x = t \rightarrow (P(x, x) \Rightarrow P(x, t))$$

are ϕ-true for any interpretation of L in which $\phi(=)$ is equality.

(b) *All the formulas in* (a) *are deducible from the set*

$$\text{Ax}\, L \cup \{x = x \,|\, x \text{ is a variable}\}$$
$$\cup \{x = y \Rightarrow (P(x, x) \Rightarrow P(x, y)) \,|\, P \text{ is an atomic formula}\}.$$

The formulas in this list, except for Ax L, *are called the axioms of equality.*

(c) *Let ϕ be any interpretation of L in a set M for which the axioms of equality are true. Then $\phi(=)$ is an equivalence relation in M which is compatible with the interpretations of all the relations and operations of L in M. If ϕ' denotes the obvious interpretation of L in the quotient set $M' = M/\phi(=)$, then $\phi'(=)$ is equality, and $T_\phi L = T_{\phi'} L$.*

Proof (sketch)

(a) The ϕ-truth is easily established. We illustrate this by showing that the last formula is ϕ-true. Suppose it were false at a point $\xi \in \overline{M}$. Then $|x = t|(\xi) = 1$, $|P|(\xi) = 1$ and $|P(x, t)|(\xi) = 0$. The first assertion means that $x^\xi = t^\xi$. But then $|P|(\xi) = |P(x, t)|(\xi)$ by Proposition 2.10, contradicting the second and third assertions.

(b) Deduction of $t = t : x = x$ (axiom of equality); $\forall x(x = x)$ (Gen); $\forall x(x = x) \Rightarrow t = t$ (logical axiom of specialization); $t = t$ (MP).

Deduction of $t_1 = t_2 \Leftrightarrow t_2 = t_1$:

(1) $x = y \Rightarrow (x = x \Rightarrow y = x)$ (axiom of equality with $=$ for P)
(2) $Q \Rightarrow ((P \Rightarrow (Q \Rightarrow R)) \Rightarrow (P \Rightarrow R))$, where P is $x = y$, Q is $x = x$, R is $y = x$ (tautology)
(3) $x = x$ (axiom of equality)
(4) $(P \Rightarrow (Q \Rightarrow R)) \Rightarrow (P \Rightarrow R)$ (MP is applied to (2) and (3))
(5) $x = y \Rightarrow y = x$ (MP applied to (1) and (4)).

We then twice apply Gen, the axiom of specialization, and MP, in order to deduce the formula $t_1 = t_2 \Rightarrow t_2 = t_1$ from (5); we replace t_1 by t_2 and t_2 by t_1 to deduce $t_2 = t_1 \Rightarrow t_1 = t_2$; we use Lemma 4.4 to deduce the conjunction of these two formulas; and, finally, the tautology $(t_1 = t_2 \Rightarrow t_2 = t_1) \wedge (t_2 = t_1 \Rightarrow t_1 = t_2) \Rightarrow (t_1 = t_2 \Leftrightarrow t_2 = t_1)$, together with MP, gives the required formula.

The deduction of the third and fourth formulas in (a) will be left to the reader. The existence of a deduction of the fourth formula can be proved by induction on the number of connectives and quantifiers in P. P is represented in the form $\neg Q$, $Q_1 * Q_2$, $\forall x\, Q$, or $\exists x\, Q$; we assume that the formula with Q, Q_1, and Q_2 in place of P has already been deduced, and we complete the deduction for P (see Mendelson, Chapter 2, Proposition 2.25).

(c) If the axioms of equality are ϕ-true, then so are the formulas in (a), since they are deducible. The first three formulas in (a), applied to three different variables x, y, and z, then show that the relation $\phi(=)$ on M is reflexive, symmetric, and transitive. In fact, let X, Y, and Z be any three elements of M, let $\xi \in \overline{M}$ be a point such that $x^\xi = X$, $y^\xi = Y$, and $z^\xi = Z$, and let $\sim$ be the relation $\phi(=)$ on M. The ϕ-truth of the formulas in (a) means that

$$X \sim X; \qquad X \sim Y \Leftrightarrow Y \sim X; \qquad X \sim Y \text{ and } Y \sim Z \Rightarrow X \sim Z.$$

By definition, to say that $\sim$ is compatible with the ϕ-interpretation of all relations and operations on M means the following. Let p be a relation, and let $\phi(p) \subset M^r$ be its interpretation. If $\langle X_1, \ldots, X_r \rangle \in \phi(p)$ and $X_i' \sim X_i$, then $\langle X_1, \ldots, X_i', \ldots, X_r \rangle \in \phi(p)$. Now let f be an operation, and let $\phi(f) : M^r \Rightarrow M$ be its interpretation. If $\phi(f)(X_1, \ldots, X_r) = Y$ and $X_i' \sim X_i$, then $\phi(f)(X_1, \ldots, X_i', \ldots, X_r) = Y' \sim Y$.

We verify this compatibility by using the ϕ-truth of the last formula in 4.6(a) at a suitable point $\xi \in \overline{M}$. Here we take the formulas $p(x_1, \ldots, x_r)$ and $f(x_1, \ldots, x_r) = y$, respectively, for P; we take the variable x_i' for t and the variable x_i for x; and we set $x_i^\xi = X_i$, $x_i'^\xi = X_i'$, and $y^\xi = Y$.

It follows from the compatibility that we can construct an interpretation ϕ' of L in $M' = M/\sim$, such that $\phi'(p) = \phi(p) \bmod \sim$, $\phi'(f) = \phi(f) \bmod \sim$, and $\phi'(=)$ is equality. The last formula in 4.6(a) will then imply that all the ϕ-true formulas remain ϕ'-true, and conversely. $\square$

From now on, when we speak of the special axioms for any language in $\mathcal{L}_1$ having the symbol $=$, we shall without explicit mention always include among them the axioms of equality for $=$. Models in which $=$ is interpreted as equality are called *normal* models.

Special axioms of arithmetic

4.7. **Proposition.** *The following formulas are true in the standard interpretation of* $\mathrm{L_1Ar}$, *and are called the special axioms of* $\mathrm{L_1Ar}$:

(a) *The axioms of equality.*
(b) *The axioms of addition*:

$$x = \bar{0} = x; \qquad x + y = y + x;\ (x + y) + z = x + (y + z);$$
$$x + z = y + z \Rightarrow x = y.$$

(c) *The axioms of multiplication*:

$$x \cdot \bar{0} = \bar{0}; \qquad x \cdot 1 = x; \qquad x \cdot y = y \cdot x; \qquad (x \cdot y) \cdot z = x \cdot (y \cdot z).$$

(d) *The distributive axiom*:

$$x \cdot (y + z) = x \cdot y + x \cdot z.$$

(e) *The axioms of induction*:

$$P(\bar{0}) \wedge \forall x \big(P(x) \Rightarrow P(x + \bar{1})\big) \Rightarrow \forall x\, P(x),$$

where P is any formula in $\mathrm{L_1Ar}$ *having one free variable.*

The proof is trivial and will be left to the reader. We only note that the "proof" that the induction axioms are true itself uses induction.

Remarks

(a) In (b), (c), and (d) above, we have written the usual axioms for a commutative (semi) ring in order to shorten the formal deductions; any informal computation which only uses these axioms can easily be transformed into a formal deduction of the result of the computation in $\mathrm{L_1Ar}$. In Chapter 3 of Mendelson's textbook, he gives an apparently weaker set of axioms, and then shows how to deduce our formulas from them. This takes up 5–6 pages of text, and is basically a tribute to a historical tradition going back to Peano.

(b) The induction axioms are a countable set of formulas in $\mathrm{L_1Ar}$; it is customary to say that 4.7(e) is an *axiom schema*. The corresponding fact in intuitive mathematics is stated as follows; "For any property P of nonnegative integers, if 0 has the property P, and, whenever x has the property P, $x + 1$ also has the property P, then all nonnegative integers have the property P." Here "property of nonnegative integers" means the same as "any subset of the nonnegative integers."

However, in the means of expression of $\mathrm{L_1Ar}$ there is no way to say "any subset." Neither is there any way to say "all properties;" we can only list one-by-one the properties that are definable by formulas in the language. We recall that there are only countably many such properties, while the intuitive interpretation refers to a continuum of properties. Thus, the formal axiom of induction is weaker than the informal one, and is also weaker than the version of this axiom that is obtained by imbedding L_1Ar in L_1Set.

Special axioms of Zermelo–Fraenkel set theory
(see the description of V in the Appendix to Chapter II)

4.8. **Proposition.** *The following formulas are true in the standard interpretation of* L_1Set *in the von Neumann universe* V:

(a) *Axiom of the empty set*: $\forall x \neg(x \in \varnothing)$.
(b) *Axiom of extensionality*: $\forall z(z \in x \Leftrightarrow z \in y) \Leftrightarrow x = y$.
(c) *Axiom of pairing*: $\forall u \, \forall w \, \exists x \, \forall z(z \in x \Leftrightarrow z = u \vee z = w)$.
(d) *Axiom of the union*: $\forall x \, \exists y \, \forall u(\exists z(u \in z \wedge z \in x) \Leftrightarrow u \in y)$.
(e) *Axiom of the power set*: $\forall x \, \exists y \, \forall z(z \subset x \Leftrightarrow z \in y)$, *where* $z \subset x$ *is abbreviated notation for the formula* $\forall u(u \in z \Rightarrow u \in x)$.
(f) *Axiom of regularity*: $\forall x(\neg x = \varnothing \Rightarrow \exists y(y \in x \wedge y \cap x = \varnothing))$, *where* $y \cap x = \varnothing$ *is abbreviated notation for* $\neg \exists z(z \in y \wedge z \in x)$.

Proof and explanations. This is not a complete list of the axioms of Zermelo–Fraenkel; the axiom of infinity, axiom of replacement, and also the axiom of choice, which are more subtle, will be discussed in the next subsection.

(a) The truth of these formulas must, of course, be proved by computing the function $| \ |$ using the rules in 2.4 and 2.5. We do this, for example, for the axiom of extensionality. Let ξ be any point in the interpretation class, and let $X = x^{\xi}$, $Y = y^{\xi}$. We must show that

$$\forall z(z \in x \Leftrightarrow z \in y)|(\xi) = |x = y|(\xi),$$

i.e., that

$$\min_{Z \in V} (|Z \in X| \, |Z \in Y| + (1 - |Z \in X|)(1 - |Z \in Y|)) = |X = Y|,$$

where we have written $|Z \in X|$ instead of $|z \in x|(\xi')$ with $z^{\xi'} = Z$, $x^{\xi'} = X$, and so on. But the left-hand side equals 1 if and only if for every $Z \in V$ either both $Z \in X$ and $Z \in Y$, or else both $Z \notin X$ and $Z \notin Y$, that is, if and only if $X = Y$.

More generally, if we replace V by any subclass $M \subset V$ and restrict the standard interpretation of L_1Set to M, then the same reasoning shows that:

The axiom of extensionality is true in M if and only if for any elements X, $Y \in M$ we have

$$X = Y \Leftrightarrow X \cap M = Y \cap M,$$

i.e., if and only if every element of M is uniquely determined by its elements which lie in M. This result will be used later.

The analogous computations for all the other axioms will be given systematically in a much more difficult context in Chapter III. Hence, at this point we shall only explain how to translate them into argot, as in Chapter I, and why they are fulfilled in V.

(b) The axiom of the empty set does not need special comment. We only remark that, if we interpret L_1Set in a subclass $M \subset V$, then the constant $\varnothing$ may be interpreted as any element $X \in M$ with the property that $X \cap M = \varnothing$, and this axiom will still hold.

(c) The axiom of pairing is true, because, if $U, W \in V_\alpha$, then $\{U, W\} \in \mathcal{P}(V_{\alpha+1})$, so that all pairs lie in V.

(d) The axiom of the union is true, because, if $X \in V$, then the set $Y = \bigcup_{Z \in X} Z$ also lies in V. In fact, if $X \in V_{\alpha+1} = \mathcal{P}(V_\alpha)$, then the elements of X are subsets of V_α, and their union therefore lies in $V_{\alpha+1}$.

(e) The axiom of the power set is true, because if $X \in V$, then $\mathcal{P}(X) \in V$. In fact, if $X \in V_\alpha$, then $X \subset V_\alpha$, and hence $\mathcal{P}(X) \subset \mathcal{P}(V_\alpha) = V_{\alpha+1}$, so that $\mathcal{P}(X) \in V_{\alpha+2}$.

(f) The axiom of regularity is true, because any non-empty set $X \in V$ has an empty intersection with at least one of its elements; in this form the axiom is proved in the Appendix to this chapter.

4.9. The axioms of L_1Set in Subsection 4.8 have one property in common: their simplest model in the standard interpretation is precisely the union $V_{\omega_0} = \bigcup_{n=0}^{\infty} V_n$ of the first ω_0 levels of the von Neumann universe. In other words, this is the set of hereditarily finite sets $X \in V$, i.e., those such that, if $X_n \in X_{n-1} \in \cdots \in X_0 = X$, then all the X_i are finite.

V_{ω_0} is the reliable, familiar world of combinatorics and number theory. Additional principles are needed to force us out of this world. There are two such principles: the axiom of infinity and the axiom schema of replacement.

(a) *Axiom of infinity*:

$$\exists x(\varnothing \in x \wedge \forall y(y \in x \Rightarrow \{y\} \in x)).$$

Here $\{y\} \in x$ is abbreviated notation for $\exists z(z = \{y, y\} \wedge z \in x)$, where the meaning of $z = \{y, y\}$ was explained in 3.7 of Chapter I. This axiom requires that we add to V_{ω_0} some set containing the elements $\varnothing, \{\varnothing\}, \{\{\varnothing\}\}, \ldots$ (a countable sequence). Then, in order to preserve the intuitive version of the axiom of the power set, we must add $\mathcal{P}(X), \mathcal{P}^2(X), \ldots,$ thereby hopelessly leaving the realm of finite sets, countable sets, continua, and so on.

It is a striking fact that none of this is necessary in the formal, as opposed to intuitive, version of set theory, where we can always limit ourselves to hereditarily countable submodels of V. This important fact will be discussed in detail in §7.

(b) *Axiom schema of replacement*. We introduce the following convenient abbreviated notation (in any language of $\mathfrak{L}_1$ having the notion of equality): $\exists! y\ P(y)$ means $\exists y\ P(y) \wedge \forall x\ \forall y (P(x) \wedge P(y) \Rightarrow x = y)$. Thus, this formula is read: "There exists a unique object y with the property P," where we assume that $=$ is interpreted as equality. When other variables besides y occur freely in P, the formula $\exists! y\ P(y)$ is true precisely when P determines y as an "implicit function" of the other variables.

We can now write the replacement axioms. In the formula P below we list all the variables which occur freely in P:

$$\forall z_1 \cdots \forall z_n\ \forall u \big(\forall x (x \in u \Rightarrow \exists! y\ P(x, y, z_1, \ldots, z_n))$$
$$\Rightarrow \exists w\ \forall y \big(y \in w \Leftrightarrow \exists x (x \in u \wedge P(x, y, z_1, \ldots, z_n))\big)\big).$$

The hypothesis says that "P gives y as a function of $x \in u$ (for given values of the parameters $z_1, \ldots, z_n$)"; the conclusion says that "the image of the set u under this function is some set w."

From the standpoint of the formal theory it is worthwhile to note that from this axiom and the axioms of equality are deducible the so-called separation axioms, namely:

$$\forall z_1 \cdots \forall z_n\ \forall x\ \exists y\ \forall u (u \in y \Leftrightarrow u \in x \wedge P(u, z_1, \ldots, z_n)).$$

This says that if we take the class of sets having a property P and intersect it with a set x, we obtain a set.

The replacement axioms should be looked at very carefully. They go beyond the usual, "intuitively obvious" working tools of the topologist and analyst. The axioms assert that, for example, it is impossible to "stretch" an ordinal α too far by means of a function f; for any f we choose, there is always an ordinal β such that all the values $f(\gamma)$, $\gamma \leqslant \alpha$, lie in V_β. In other words, the universe V is incomparably more infinite than any of its levels V_α.

Even if we adopt this axiom, questions remain which are very similar in style, which are beyond the reach of our intuition, and which are not solvable using this and the other axioms. For example, do there exist so-called *inaccessible cardinals* γ? One of the properties of an inaccessible cardinal γ is the following: if f is a function from V_α to V_γ (with $\alpha < \gamma$), then the set of values of f is an element of V_γ. In particular, there is an "upper bound" beyond which ordinals not exceeding γ cannot be "stretched." Do such infinities exist or not?

After thinking about this and related problems, many specialists on the foundations of mathematics have come to the conclusion that such languages of set theory as $\mathrm{L_1Set}$ with a suitable axiom system are the only reality one should work with, and any attempt to make intrinsic sense out of the universe V or similar models is in principle doomed to failure. In particular, the set of formulas in $\mathrm{L_1Set}$ which are true in the standard interpretation is not defined, and we can only talk about formulas which are deducible from the axioms.

But we shall not entirely adopt this point of view for several reasons. The simplest reason is the feeling that a language without an interpretation not only loses its intrinsic justification, but also cannot be used for anything. We cannot even play the "formal game" well unless we master the intuitive concepts which give meaning to the symbols. A language (along with the external world) helps bring order and precision to these intuitive concepts, which, in turn, make us change the language or at least revise our earlier linguistic constructions. But we can never assume that we have achieved complete clarity.

We should understand the need for certain types of self-restraint. However, intellectual asceticism (like all other forms of asceticism) cannot be the lot of many.

(c) *Axiom of choice*:

$$\forall x\big(\neg x = \varnothing \Rightarrow \exists y(\text{"}y \text{ is a function with domain of definition } x\text{"} \\ \wedge \forall u(u \in x \wedge \neg u = \varnothing \Rightarrow \exists w(w \in u \wedge \text{"}\langle u, w\rangle \in y\text{"})))\big).$$

That is, y chooses one element from each nonempty element $u \in x$.

The belief that this axiom is true in V is at least as justified as the belief in the existence of V itself. Over the past fifty years it has become customary for every working mathematician to accept this axiom, and the heated controversies about it at the beginning of the century are now all but forgotten. The interested reader is referred to Chapter II of *Foundations of Set Theory* by Fraenkel and Bar-Hillel (North-Holland, Amsterdam, 1958).

4.10. *General properties of axioms*. Despite the wide variety of concepts reflected in these axioms, each of our sets of axioms for languages in $\mathfrak{L}_1$ (tautologies; Ax L; special axioms of $\mathrm{L_1Ar}$ and $\mathrm{L_1Set}$) have the following informal syntactic characteristics:

(a) An algorithm can be given which tells whether any given expression is an axiom (compare: the syntactic analysis in §1 and the verification of the tautologies in Subsection 3.4).
(b) A finite number of rules can be given for generating the axioms.

It is clear that, *a priori*, property (b) is less restrictive than (a). In fact, an algorithm as in (a) can be transformed into a rule for generating the axioms: "Write out all possible expressions one by one in some order, and take those for which the algorithm gives a positive answer."

It is actually natural to suppose that property (a) should characterize axioms, and property (b) should characterize deducible formulas, no matter how we explicitly describe the axioms and the deducible formulas in a given language. In Part III we make these intuitive ideas into precise definitions and show that (b) is strictly weaker than (a). See also the discussion in Subsection 11.6(c) of this chapter.

Digression: proof

1. A proof only becomes a proof after the social act of "accepting it as a proof." This is as true for mathematics as it is for physics, linguistics, or biology. The evolution of commonly accepted criteria for an argument's being a proof is an almost untouched theme in the history of science. In any case, the ideal for what constitutes a mathematical demonstration of a "nonobvious truth" has remained unchanged since the time of Euclid: we must arrive at such a truth from "obvious" hypotheses, or assertions which have already been proved, by means of a series of explicitly described, "obviously valid" elementary deductions.

Thus, the method of deduction is a method of mathematics *par excellence*. ("Mathematical induction" clearly comes out of the same tradition. Peano's induction principle allows us to write only the first step and the general step of a proof, and is thereby in some sense the first metamathematical principle. This point is observed by the tradition of listing Peano's axiom among the special axioms (see 4.7(e)), but, one way or another, it is one of the archetypes of mathematical thought.)

The longer the deductive argument, the more important it is for all its elementary components to be written in an explicit and normalized fashion. In the last analysis, the amount of initial data in formal mathematics is so small that failure to observe the rules of hygiene in long deductions would lead to the collapse of the system if we did not have external checks on the system. In induction, on the other hand, relatively short deductions are based on a vast amount of initial information. Darwin's theory of evolution is explained to school children, but life is not long enough to judge how persuasive the proofs are. We see a similar situation in comparative linguistics when the features of the so-called protolanguages are reconstructed. In such uses of induction, the "rules of deduction" cannot be so very rigid, despite the critical viewpoint of the neo-grammarians.

2. The above observations concerning the method of deduction are supported by the fact that the notion of a formal deduction in languages of $\mathfrak{L}_1$ is a close approximation to the concept of an ideal mathematical proof. It is therefore enlightening to examine the differences between deductions and the arguments we use in day-to-day practice.

(a) *Reliability of the principles*. Not only the mathematics implicit in the special axioms of L_1Set and L_1Ar, but even the logic of the languages of $\mathfrak{L}_1$ is not accepted by everyone. In particular, Brouwer and others have called into question the law of the excluded middle. From their extremely critical perspective, our "proofs" are at best harmless deductions of nonsense out of falsehood.

The mathematician cannot permit himself to be completely deaf to these criticisms. After thinking about them for a while, he should at least be willing to admit that proofs can have objectively different "degrees of proofness."

(b) *Levels of "proofness.*" Every proof that is written must be approved and accepted by other mathematicians, sometimes by several generations of mathematicians. In the meantime, both the result and the proof itself are liable to be refined and improved. Usually the proof is more or less an outline of a formal deduction in a suitable language. But, as mentioned before, an assertion P is sometimes established by proving that a proof of P exists. This hierarchy of proofs of the existence of proofs can, in principle, be continued indefinitely. We can take down the hierarchy using sophisticated logical and set theoretic principles; however, not everyone might agree with these principles. Papers on constructive mathematics abound with assertions of the type: "there cannot not exist an algorithm which computes x," whereas a classical mathematician would simply say "x exists," or even "x exists and is effectively computable."

(c) *Errors*. The peculiarities of the human mind make it impossible in practice to verify formal deductions, even if we agree that, in principle, such a verification is the ideal form for a proof. Two circumstances act together with perilous effect: formal deductions are much longer than texts in argot, and humans are much slower at reading and comprehending such formal arguments than texts in natural languages.

A proof of a single theorem may take up five, fifteen, or even fifty pages. In the theory of finite groups, the proofs of the two Burnside conjectures occupy nearly five hundred pages apiece. Deligne has estimated that a complete proof of Ramanujan's conjecture assuming only set theory and elementary analysis would take about two thousand pages. The length of the corresponding formal deductions staggers the imagination.

Hence, the absence of errors in a mathematical paper (assuming that none are discovered), as in other natural sciences, is often established indirectly: how well the results correspond to what was generally expected, the use of similar arguments in other papers, examination of small sections of the proof "under the microscope," even the reputation of the author—in short, its reproducibility in the broadest sense of the word. "Incomprehensible" proofs can play a very useful role, since they stimulate the search for more accessible arguments.

The last two decades have seen the appearance of a very powerful method for performing long formal deductions, namely the use of computers. At first glance, it would seem that the status of formal deductions might greatly improve, so that the Leibnizian ideal of being able to verify truth mechanically would become attainable. But the state of affairs is actually much less trivial.

We first give two authoritative opinions on this question by C. L. Siegel and H. P. F. Swinnerton-Dyer. Both opinions relate to the solution by computer of concrete number theoretic problems.

3. The present level of knowledge concerning Fermat's last theorem is as follows. Let p be a prime. It is called regular if it does not divide the numerator of any of the Bernoulli numbers $B_2 = \frac{1}{6}$, $B_4 = \frac{1}{30}, \ldots, B_{p-3}$. Fermat's theorem was proved for regular prime exponents by Kummer. For irregular p there is a series of criteria for Fermat's theorem to hold. These criteria reduce to checking that certain divisibility properties do not hold; if they hold, we must try certain other divisibility properties, and so on. The verification for each p requires extensive computer computations. As of 1955, this was successfully done for all $p < 4002$ (J. L. Selfridge, C. A. Nicol, H. S. Vandiver, *Proc. Nat. Acad. Sci. USA*, 41, 970–973 (1955)).

Let $v(x)$ denote the ratio of the number of irregular primes $\leqslant x$ to the number of regular primes $\leqslant x$. Kummer conjectured that $v(x) \to \frac{1}{2}$ as $x \to \infty$. Siegel (*Nachrichten Ak. Wiss. Göttingen, Math. Phys. Klasse*, 1964, No. 6, 51–57) suggests that $\sqrt{e} - 1$ is a more likely value for the limit, supports this opinion with probabilistic arguments, compares with the data of Selfridge–Nicol–Vandiver, and concludes this discussion with the following unexpected sentence: "In addition, it must be taken into account that the above numerical values for $v(x)$ were obtained using computers, and therefore, strictly speaking, cannot be considered proved"!

4. Siegel's point of view can be explained as a natural reaction to information received secondhand. But the excerpts below are from an article by a professional mathematician and experienced computer programmer (Acta Arithmetica, XVIII, 1971, 371–385). The article is devoted to the following problem:

> "Let L_1, L_2, L_3 be three homogeneous linear forms in u, v, w with real coefficients and determinant Δ; and suppose that the lower bound of $|L_1 L_2 L_3|$ for integer values of u, v, w not all zero is 1." What can be said about the possible value for Δ?
>
> "The corresponding problem for the product of two linear forms is much easier, and was essentially completely solved by Markov. There are countably many possible values of Δ less than 3, each of which has the form
>
> $$\Delta = (9 - 4n^{-2})^{1/2}$$
>
> for some integer n; the first few values of n are 1, 2, 5, 13, 29, and there is an algorithm for constructing all the permissible values of n."

For three forms Davenport (1943) proved that $\Delta = 7$ or $\Delta = 9$ or $\Delta > 9.1$. In Swinnerton–Dyer's paper, all values of $\Delta \leqslant 17$ are computed *under the assumption* that there are only finitely many such values and he gives a list of them: the third value is 148, and the last (the eighteenth) is $\sqrt{2597/9}$. Discussing this result, he makes a very interesting comment:

> "When a theorem has been proved with the help of a computer, it is impossible to give an exposition of the proof which meets the traditional test—that a sufficiently patient reader should be able to work through the

proof and verify that it is correct. Even if one were to print all the programs and all the sets of data used (which in this case would occupy some forty very dull pages) there can be no assurance that a data tape has not been mispunched or misread. Moreover, every modern computer has obscure faults in its software and hardware—which so seldom cause errors that they go undetected for years—and every computer is liable to transient faults. Such errors are rare, but a few of them have probably occurred in the course of the calculations reported here."

The arguments on the positive side are also very curious:

> "However, the calculation consists in effect of looking for a rather small number of needles in a six-dimensional haystack; almost all the calculation is concerned with parts of the haystack which in fact contain no needles, and an error in those parts of the calculation will have no effect on the final results. Despite the possibilities of error, I therefore think it almost certain that the list of permissible $\Delta \leqslant 17$ is complete; and it is inconceivable that an infinity of permissible $\Delta \leqslant 17$ have been overlooked."

His conclusion:

> "Nevertheless, the only way to verify these results (if this were thought worth while) is for the problem to be attacked quite independently, by a different machine. This corresponds exactly to the situation in most experimental sciences."

We note that it is becoming more and more apparent that the processing, and also the storage, of large quantities of information outside the human brain lead to social problems which go far beyond questions of the reliability of mathematical deductions.

5. In conclusion, we quote an impression concerning mechanical proofs, even ones done by hand, which is experienced by many.

After stating a proposition to the effect that "the function $T_{W,\eta_0}\tilde{\theta}$ is correctly defined," a gifted and active young mathematician writes (*Inventiones Math.*, vol. 3, f.3 (1967), 230):

> "The proof of this Proposition is a ghastly but wholly straightforward set of computations. It took me several hours to do every bit and as I was no wiser at the end—except that I knew the definition was correct—I shall omit details here."

The moral: a good proof is one which makes us wiser.

5 Tautologies and Boolean algebras

5.1 **Proposition.** *A finite list, or "basis," of tautologies—logical polynomials in three variables P, Q, R—can be given with the following property.*

Let L be any language in $\mathfrak{L}_1$, and let $\mathfrak{F}$ be the set of all formulas in L which can be obtained from the basis tautologies by substituting all *possible formulas in place of P, Q, R. Then any tautology in L is deducible from $\mathfrak{F}$ using only the rule of deduction* MP.

The choice of the basis tautologies is by no means unique. Our list will consist of the tautologies A0, A1, A2, A3, B1, B2 in Subsection 3.4 and the following tautologies:

C1 $\neg(P\Rightarrow\neg Q)\Rightarrow(P\wedge Q)$, $(P\wedge Q)\Rightarrow\neg(P\Rightarrow\neg Q)$.
C2 $(\neg P\Rightarrow Q)\Rightarrow(P\vee Q)$, $(P\vee Q)\Rightarrow(\neg P\Rightarrow Q)$.
C3 $P\Rightarrow(\neg Q\Rightarrow\neg(P\Rightarrow Q))$.
C4 $(P\Rightarrow Q)\Rightarrow((\neg P\Rightarrow Q)\Rightarrow Q)$.
C5 $(P\Rightarrow Q)\Rightarrow(\neg Q\Rightarrow\neg P)$.
C6 $(P\Rightarrow Q)\Rightarrow((Q\Rightarrow P)\Rightarrow(P\Leftrightarrow Q))$.
C7 $(P\Leftrightarrow Q)\Rightarrow(P\Rightarrow Q)$, $(P\Leftrightarrow Q)\Rightarrow(Q\Rightarrow P)$.

We are not trying to economize on the size of the basis, but rather on the length of the proof of Proposition 5.1; hence, A0–C7 is not the shortest possible list. This does not make any difference for studying the logic of $\mathfrak{L}_1$; but the study of modified logical systems, for example those of the intuitionist type, requires more careful analysis of this list.

PROOF OF PROPOSITION 5.1. Let $\mathcal{E}$ be a finite set of formulas in L, and let P be a logical polynomial (with a fixed representation) over $\mathcal{E}$. For any map $v:\mathcal{E}\to\{0,1\}$, we extend v to P using the same rules that defined the truth function $|\ |$ in Subsection 2.5. We set

$$P^v=\begin{cases} P, & \text{if } v(P)=1\\ \neg P, & \text{if } v(P)=0.\end{cases}$$

5.2. **Fundamental Lemma.** *Let* $\mathcal{E}^v=\{Q^v\,|\,Q\in\mathcal{E}\}$. *Then for any* v *we have*: $\mathfrak{F}\cup\mathcal{E}^v\vdash P^v$ (*using* MP).

This lemma expresses the following idea. It is natural to prove Proposition 5.1 by induction on the length of the tautology. However, the component parts of a tautology themselves might not be tautologies. The operation of taking P to P^v forces any formula to be "v-true" and makes it possible for us to use induction.

5.3. PROOF OF 5.1 ASSUMING THE FUNDAMENTAL LEMMA. Let P be a tautology, so that $P^v=P$ for all v. Set $\mathcal{E}=\{P_1,\ldots,P_r\}$. By the fundamental lemma, $\mathfrak{F}\cup\{P_1^v,\ldots,P_r^v\}\vdash P$ using MP for any v: We show that then $\mathfrak{F}\cup\{P_1^v,\ldots,P_{r-1}^v\}\vdash P$ using MP. Descending induction on r then gives the required assertion (the assumption that P is a logical polynomial in $P_1,\ldots,P_r$ is not used in the induction step).

The Deduction Lemma 4.5 shows that $\mathfrak{F}\cup\{P_1^v,\ldots,P_{r-1}^v\}\vdash(P_r^v\Rightarrow P)$ using MP; to see this we need only examine the proof and notice that the deduction only used MP and the tautologies in $\mathfrak{F}$, since the rule of deduction Gen was not needed.

Since for any v there exists a v' which coincides with v on $P_1, \ldots, P_{r-1}$ but takes a different value on P_r, it follows that: $P_r \Rightarrow P$ and $\neg P_r \Rightarrow P$ are deducible from $\mathfrak{F} \cup \{P_1^v, \ldots, P_{r-1}^v\}$ using MP. On the other hand, the tautology C4: $(P_r \Rightarrow P) \Rightarrow ((\neg P_r \Rightarrow P) \Rightarrow P)$ lies in $\mathfrak{F}$. Applying MP twice, we deduce P. □

5.4. Proof of the Fundamental Lemma. We use induction on the number of connectives in the representation of P as a logical polynomial over $\mathfrak{S}$. If there are no connectives, that is $P \in \mathfrak{S}$, then the assertion is obvious. Otherwise, P has the form $\neg Q$ or $Q_1 * Q_2$, where $*$ is one of the binary connectives.

(a) *The case* $P = \neg Q$. If $v(Q) = 0$, then $Q^v = \neg Q = P = P^v$. That $Q^v = P^v$ is deducible from $\mathfrak{F} \cup \mathfrak{S}^v$ is precisely the induction assumption.

On the other hand, if $v(Q) = 1$, then $Q^v = Q$, $P^v = \neg\neg Q$. Here Q is deducible from $\mathfrak{F} \cup \mathfrak{S}^v$ by the induction assumption, and then the tautology $Q \Rightarrow \neg\neg Q$ in $\mathfrak{F}$ along with MP gives a deduction of P^v.

(b) *The case* $P = Q_1 * Q_2$. For the different connectives and possible values of $v(Q_1)$ and $v(Q_2)$ we first tabulate the formulas for which deductions exist by the induction assumption and the formulas for which we must find deductions. In the columns under $\wedge$ and $\vee$ we give formulas from which $(Q_1 \wedge Q_2)^v$ and $(Q_1 \vee Q_2)^v$, respectively, are deducible using MP and the tautologies in $\mathfrak{F}$ (tautologies C1, C2, and C5). Hence it suffices to find deductions of each of formulas 1–16 from $\mathfrak{F}$ and the pair of formulas on the appropriate row in the second column using MP.

Deduction of formulas 1–16.

		Given: deductions of	Must Find: Deduction of $(Q_1 * Q_2)^v$	
$v(Q_1)$	$v(Q_2)$	Q_1^v and Q_2^v	$\Rightarrow$	$\wedge$
0	0	$\neg Q_1, \neg Q_2$	1. $Q_1 \Rightarrow Q_2$	5. $\neg\neg(Q_1 \Rightarrow \neg Q_2)$
0	1	$\neg Q_1, Q_2$	2. $Q_1 \Rightarrow Q_2$	6. $\neg\neg(Q_1 \Rightarrow \neg Q_2)$
1	0	$Q_1, \neg Q_2$	3. $\neg(Q_1 \Rightarrow Q_2)$	7. $\neg\neg(Q_1 \Rightarrow \neg Q_2)$
1	1	Q_1, Q_2	4. $Q_1 \Rightarrow Q_2$	8. $\neg(Q_1 \Rightarrow \neg Q_2)$

$v(Q_1)$	$v(Q_2)$	Q_1^v and Q_2^v	$\vee$	$\Leftrightarrow$
0	0	$\neg Q_1, \neg Q_2$	9. $\neg(\neg Q_1 \Rightarrow Q_2)$	13. $Q_1 \Leftrightarrow Q_2$
0	1	$\neg Q_1, Q_2$	10. $\neg Q_1 \Rightarrow Q_2$	14. $\neg(Q_1 \Leftrightarrow Q_2)$
1	0	$Q_1, \neg Q_2$	11. $\neg Q_1 \Rightarrow Q_2$	15. $\neg(Q_1 \Leftrightarrow Q_2)$
1	1	Q_1, Q_2	12. $\neg Q_1 \Rightarrow Q_2$	16. $Q_1 \Leftrightarrow Q_2$

Note that if P is deducible then for any Q the formula $Q \Rightarrow P$ is also deducible (tautology A1 and MP) and if $\neg P$ is deducible then for any Q the formula $P \Rightarrow Q$ is deducible (tautology B2 and MP). This immediately yields deductions of 1, 2, 4, 10, and 12. If we remove the double negations in the $\wedge$ column using tautology B1 and MP, we obtain deductions of 5, 6, and 7. And 11 is deducible since by B1 the second column yields a deduction of $\neg\neg Q_1$. In the first and last rows the deductions of 1 and 4 yield deductions of $Q_2 \Rightarrow Q_1$ by symmetry; tautology C6 and MP twice give a deduction of 13 and 16 from $Q_1 \Rightarrow Q_2$ and $Q_2 \Rightarrow Q_1$.

3 is deduced from C3: $Q_1 \Rightarrow (\neg Q_2 \Rightarrow \neg(Q_1 \Rightarrow Q_2))$ and the second column using MP twice.
8 is deduced from C3: $Q_1 \Rightarrow (\neg\neg Q_2 \Rightarrow \neg(Q_1 \Rightarrow \neg Q_2))$ and the second column using MP, applying B1 to Q_2, and again using MP.
9 is deduced from C3: $\neg Q_1 \Rightarrow (\neg Q_2 \Rightarrow \neg(\neg Q_1 \Rightarrow Q_2))$ using MP twice.
15 is deduced from 3 by C7 and C5 and MP twice.
Finally, the deduction of 3 from Q_1 and $\neg Q_2$ yields by symmetry a deduction of $\neg(Q_2 \Rightarrow Q_1)$ from $\neg Q_1$ and Q_2. Hence on the second row the deduction of 14 is analogous to that of 15.

Proposition 5.1 is proved. □

5.5. *Tautologies and probability*. Tautologies are statements which are true independently of the truth or falsity of their "component parts." This assertion still holds even if the components of a tautology are assigned probabilistic truth values $\|P\|$ in the algebra of measurable sets in some probability space.

An example: the tautology $R \vee S \vee \neg R \vee \neg S$—"either it will rain, or it will snow, or it won't rain, or it won't snow"[1]—is a reliable weather forecast despite the great complexity of the meteorological probability space.

For a precise result, it is convenient to use the terminology of Boolean algebras.

5.6. *Boolean algebras*. A Boolean algebra B is a set with an operation of rank one, with two operations $\vee$ and $\wedge$ of rank two, and with two distinguished elements 0 and 1, such that the following axioms hold:

(a) $(A')' = A$ for all $A \in B$;
(b) $\wedge$ and $\vee$ are each associative and commutative;
(c) $\wedge$ and $\vee$ are distributive with respect to one another;
(d) $(a \vee b)' = a' \wedge b'$, $(a \wedge b)' = a' \vee b'$;
(e) $a \vee a = a \wedge a = a$;
(f) $1 \wedge a = a$; $0 \vee a = a$.

EXAMPLES.

(a) B is the set of all subsets of a set M, $'$ is complement, $\wedge$ is intersection, $\vee$ is union, 0 is the empty subset, and 1 is all of M.

[1] A Russian proverb (translator's note).

(b) B is the set of open-and-closed subsets of a topological space M with the same operations.
(c) B is the algebra of measurable subsets (modulo measure zero subsets) of a probability space M with the same operations.

In all of these cases B can be identified with the space of characteristic functions of the corresponding subsets of M (taking the value 1 on the subset and 0 on the complement).

5.7. *Boolean truth functions.* Let B be a Boolean algebra, and let $\mathcal{E}$ be a set of formulas in a language L. Let $\| \ \| : \mathcal{E} \to B$ be any map. We extend this map to the logical polynomials over $\mathcal{E}$ (more precisely, to their representations) by means of the recursive formulas:

$$\begin{aligned}
\|P \Leftrightarrow Q\| &= (\|P\| \wedge \|Q\|) \vee (\|P\|' \wedge \|Q\|'), \\
\|P \Rightarrow Q\| &= \|P\|' \vee \|Q\|, \\
\|P \vee Q\| &= \|P\| \vee \|Q\|, \\
\|P \wedge Q\| &= \|P\| \wedge \|Q\|, \\
\|\neg P\| &= \|P\|'.
\end{aligned}$$

In the case $B = \{0, 1\}$, these formulas coincide with the definitions in 2.5. We note that $\vee$ and $\wedge$ have different meanings in the left- and right-hand sides.

5.8. **Proposition.** *Let the logical polynomial P be a tautology over $\mathcal{E}$. Then for any map $\| \ \| : \mathcal{E} \to B$ to any Boolean algebra B we have $\|P\| = 1$.*

PROOF. An example of a natural map $\| \ \|$ can be obtained as follows: if we are given an interpretation of L in a set M, then the truth functions $|P|(\xi)$ can be considered as the characteristic functions of the definable subsets of the interpretation class $\overline{M}$ (compare §2). Hence, our usual truth functions are essentially Boolean-valued. They are imbedded in the Boolean algebra of all subsets of $\overline{M}$, which decomposes as a direct product of two-point Boolean algebras $\{0, 1\}$. Hence the proposition follows trivially in this case.

In the general case one could use Stone's structure-theorem for Boolean algebras. However, instead of this we shall indicate how to reduce the problem to some simple computations using Proposition 5.1. Because of Proposition 5.1, it suffices to verify that the basis tautologies are $\| \ \|$-true and that $\| \ \|$-truth is preserved when we use MP. For example, if $\|P\| = 1$ and $\|P \Rightarrow Q\| = 1$, then $\|P\|' = 0$ while $\|P\|' \vee \|Q\| = 1$, so that $\|Q\| = 1$ by 5.6(f); this answers the question about MP. The truth values of the basis tautologies are computed in a similar manner using the axioms in 5.6. □

Boolean truth functions will be the basic tool in the presentation of Cohen forcing in Chapter III.

Digression: kennings

1. The process in §5 generates all possible tautologies starting with a finite number of tautologies and using a finite number of rules. It has become very popular in modern linguistics to attempt to find a suitable description of natural languages by means of such generating rules (N. Chomsky and others; see, for example, the book *Éléments de linguistique mathématique* by A. V. Gladkiĭ and I. A. Mel'čuk, Paris, Dunod, 1972).

However, many psychologists consider that this conception has little to do with the actual process of speech. According to one such opinion, real speech has more in common with a game of chance, chasing a fugitive, or a river current near a jagged shoreline. The choice of the next word in a sentence is determined statistically both by a formulating principle (an idea, situation, or psychological state) and by the peculiarities of semantics, grammar, phonetics, and the associative cloud formed by the earlier words.

There is reason to hope that formal grammars are more closely suited to describing special fragments of natural languages which are in some sense more rigidly defined, such as certain language fragments in poetry or law. In these fragments an essential role is played by "prohibitions," which weed out, say, all texts not having a certain rhythmic pattern. Even the most casual attempt at writing poetry reveals the psychological reality of prohibitions in versification. But it is much less obvious that there is a set of generating rules which also has a psychological reality.

2. Yet there has been at least one poetic system in which generating rules occupied an important place. One of the basic elements of skaldic (ancient Icelandic) poetry consisted of special formulas called *kennings*. A kenning is an expression which can replace a single word. For example,

"storm of spears" is a kenning for "battle"

"tree of battle"
"bush of the helmet"
"thrower of swords"
"giver of gold" } are kennings for "warrior" or "man"

"sea of the wagon" is a kenning for "earth"

"fire of war" is a kenning for "gold"

"sky of sand"
"field of seals" } are kennings for "sea," and so on.

A *simple kenning* is a kenning no part of which is a kenning. The examples above are all simple kennings. They play the role of axioms; obviously, only very great poets have the right to create new simple kennings. It falls to the lot of the lesser poets to create new kennings using the rules of deduction. The *rule of deduction* of a new kenning from earlier kennings is as follows: any word in a kenning may be replaced by a (not necessarily simple) kenning for that word. Here is a complicated example of a kenning together with its decomposition into simple kennings (an actual example):

"thrower of the fire of the storm of the witch of the moon of the steed of the ship stables"

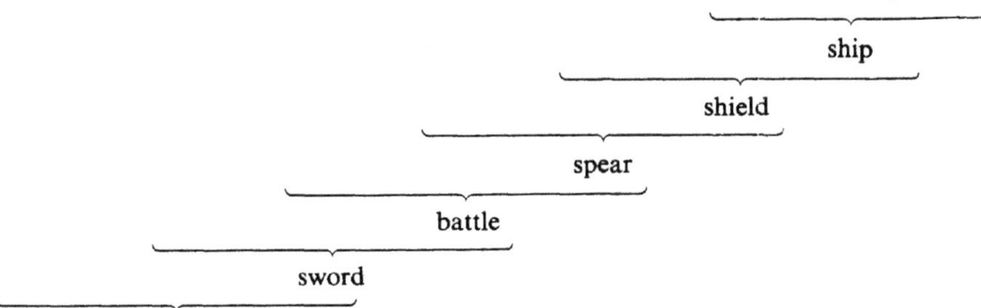

The Soviet poet Leonid Martynov thought of kennings as metaphors (a fundamental error, although an understandable one—kennings and metaphors play completely different structural roles in different poetic systems), and he wrote a poem "Songs of the Skalds" which ends as follows:

...But perhaps the translators have gotten a bit carried away?

No!
In our times, too,
might there not live

some *throwers*
of the fire
of the storm
of the witch
of the moon
of the steed
of the ship stables,

or

squanderers
of the amber
of the cold earth
of the great boar?

Anything is possible!!
And who can be so very sure
That there are no longer songs
which could be called

Surf
of yeast
of the people
of the bones
of the fjord?

Perhaps there really are such songs now,

Who can tell??

After all this, the professional opinion of M. I. Steblin-Kamenskiĭ, whose book *Icelandic Culture* (Leningrad, Nauka, 1967) provided us with the above examples, sounds a little anticlimatic: "As a rule, any kenning for a man or warrior was no richer in content than the pronoun 'he.' "

EXERCISES:

(a) Find the simple kennings from which the last two kennings in Martynov's poem are deduced.
(b) Construct the kennings of maximum length which are deducible from all the simple kennings in the above text. Prove that it is impossible to deduce longer kennings.

6 Gödel's completeness theorem

6.1. Let L be a language in $\mathfrak{L}_1$, let ϕ be an interpretation of L, and let $T_\phi L$ be the set of ϕ-true formulas. In §3 it was shown that the set $T_\phi L$ is *Gödelian*: it is complete, does not contain a contradiction, is closed with respect to deduction, and contains all the logical axioms Ax L. We say that a set of formulas $\mathfrak{E}$ in L is *consistent* if the set of formulas deducible from $\mathfrak{E}$ does not contain a contradiction, i.e., if there is no P such that $\mathfrak{E} \vdash P$ and $\mathfrak{E} \vdash \neg P$; otherwise, we say that $\mathfrak{E}$ is *inconsistent*. The basic purpose of this section is to prove the following converse of the result in §3:

6.2. **Theorem** (Gödel)

(a) *Any Gödelian set T is the set of ϕ-true formulas $T_\phi L$ for a suitable interpretation of L in some set M having cardinality* $\leqslant$ card (*alphabet of* L) + $\aleph_0$. (*Here and below we always mean the cardinality of the alphabet without the variables.*)

(b) *Any set of formulas $\mathfrak{E}$ which contains* Ax L *and is consistent can be imbedded in a Gödelian set.*

The model M which is constructed in the proof consists of expressions in some extension of the alphabet of L, and thus has a somewhat artificial character. In the next section we show that, if we are given some natural interpretation (M, ϕ) of L, then we can find a submodel having cardinality $\leqslant$ card (alphabet of L) + $\aleph_0$.

6.3 **Corollary.** (Deducibility criterion). *Let* $\mathfrak{E} \supset$ Ax L.

(a) *A formula P is deducible from $\mathfrak{E}$ if and only if either $\mathfrak{E}$ is inconsistent, or P is ϕ-true for all models ϕ of the set $\mathfrak{E}$ having cardinality* $\leqslant$ card (*alphabet of* L) + $\aleph_0$.

(b) *A formula P is independent of $\mathfrak{E}$ if and only if both $\mathfrak{E} \cup \{P\}$ and $\mathfrak{E} \cup \{\neg P\}$ are consistent*; *by Theorem* 6.2, *this is true if and only if $\mathfrak{E} \cup \{P\}$ and $\mathfrak{E} \cup \{\neg P\}$ have models.*

In what follows we shall often omit the verification that various formal deductions exist. If the reader wants to fill in such a verification, this can almost always be done more easily using deducibility criterion 6.3 than directly.

Proof of the corollary

(a) If $\mathfrak{S}$ is inconsistent, then any formula can be deduced from $\mathfrak{S}$ (Proposition 4.2). Suppose $\mathfrak{S}$ is consistent and P is ϕ-true for all models of $\mathfrak{S}$. Let $\bar{P} = \forall x_1 \cdots \forall x_n P$ be the "closure" of P. To prove that $\mathfrak{S} \vdash P$ we consider two cases.

(a_1) $\mathfrak{S} \cup \{\neg \bar{P}\}$ is *inconsistent*. Then $\mathfrak{S} \cup \{\neg \bar{P}\} \vdash \bar{P}$, so that, by the Deduction lemma, $\mathfrak{S} \vdash \neg \bar{P} \Rightarrow \bar{P}$. The tautology $(\neg \bar{P} \Rightarrow \bar{P}) \Rightarrow \bar{P}$ and MP give $\mathfrak{S} \vdash \bar{P}$, and then the axiom of specialization and MP give $\mathfrak{S} \vdash P$.

(a_2) $\mathfrak{S} \cup \{\neg \bar{P}\}$ is *consistent*. Then, by Theorem 6.2, the set $\mathfrak{S} \cup \{\neg \bar{P}\}$ has a model. In this model $\mathfrak{S}$ is true and P is false, so that this case is impossible.

(b) Suppose that P is independent of $\mathfrak{S}$, i.e., neither P nor $\neg P$ is deducible. Then, by part (a), there exists a model of $\mathfrak{S}$ in which P is true and a model of $\mathfrak{S}$ in which P is false. The converse is obvious. □

We now proceed to the proof of Gödel's completeness theorem.

6.4. **Definition.** Let $\mathfrak{S}$ be a set of formulas in a language L. The alphabet of L is said to be *sufficient* for $\mathfrak{S}$ if, for each closed formula $\neg \forall x\, P(x)$ in $\mathfrak{S}$ there exists a constant c_P (depending on P) such that the formula

$$R_P\colon\ \neg \forall x\, P(x) \Rightarrow \neg P(c_P)$$

belongs to $\mathfrak{S}$.

The intuitive meaning of R_P is: "If not all x have the property P, then some concrete object c_P can be found which does not have this property." We say that the *alphabet* (rather than $\mathfrak{S}$) is "sufficient" or "insufficient" because, if $\mathfrak{S}$ does not contain enough formulas of the type R_P, we can simply add all the R_P to $\mathfrak{S}$, while if there are not enough constants c_P, we then have to add them to the alphabet of the language.

The plan for proving Theorem 6.2 is as follows. We first prove the Fundamental Lemma:

6.5. **Fundamental Lemma.** *If a set of formulas $\mathfrak{S}$ in a language L is consistent and complete and contains* Ax L, *and if the alphabet of L is sufficient for $\mathfrak{S}$, then $\mathfrak{S}$ has a model with cardinality* $\leqslant$ card (*alphabet of* L) + $\aleph_0$.

The next two lemmas allow us to imbed any consistent $\mathfrak{S}$ in a complete set, or in one for which the alphabet is sufficient.

6.6. **Lemma.** *If* $\mathfrak{S}$ *is consistent and contains* Ax L, *then there exists a consistent and complete set of formulas* $\mathfrak{S}' \supset \mathfrak{S}$.

6.7. **Lemma.** *If* $\mathfrak{S}$ *is consistent and contains* Ax L, *then there exist*:

(a) *a language* L' *whose alphabet is obtained from the alphabet of* L *by adding a set of new constants having cardinality* $\leqslant$ card (*alphabet of* L) + $\aleph_0$.

(b) *a set of formulas* $\mathfrak{S}'$ *in* L' *which is consistent, contains* $\mathfrak{S}$ *and* Ax L', *and has the property that the alphabet of* L' *is sufficient for* $\mathfrak{S}'$.

However, these constructions get in each other's way. If we complete a set $\mathfrak{S}$ for which the alphabet is sufficient, we might obtain a set with an insufficient alphabet; if we add new constants, we increase the overall supply of formulas in the language, and thereby lose the completeness of $\mathfrak{S}$. Hence, we have to alternate the constructions in 6.6 and 6.7 a countable number of times in order to prove our last lemma:

6.8. **Lemma.** *If* $\mathfrak{S} \supset$ Ax L *is consistent, then there exist*:

(a) *a language* $L^{(\infty)}$ *whose alphabet is obtained from the alphabet of* L *by adding a set of new constants having cardinality* $\leqslant$ card (*alphabet of* L) + $\aleph_0$.

(b) *a set of formulas* $\mathfrak{S}^{(\infty)}$ *in* $L^{(\infty)}$ *which is complete and consistent, contains* $\mathfrak{S}$ *and* Ax $L^{(\infty)}$, *and has the property that the alphabet of* $L^{(\infty)}$ *is sufficient for* $\mathfrak{S}^{(\infty)}$.

After Lemma 6.8 is proved, Theorem 6.2 is obtained from the Fundamental Lemma applied to $\mathfrak{S}^{(\infty)}$ if we restrict the resulting model to L and $\mathfrak{S}$.

We now prove the lemmas. The Fundamental Lemma is proved in 6.9, and Lemmas 6.5, 6.6, and 6.7 are proved in Subsections 6.10, 6.11, and 6.12, respectively.

6.9. Proof of the Fundamental Lemma. We begin by explicitly constructing the interpretation ϕ of L which will be our model for $\mathfrak{S}$.

(a) By a *constant term* we mean a term in L which does not contain any symbols for variables. We let $M = \{\bar{t} \mid t \text{ is a constant term}\}$ be a "second copy" of the set of constant terms, and we define the *primary mappings* of the interpretation ϕ of L in M as follows:

$\phi(c) = \bar{c}$ (for any constant c);

$\phi(f)(\bar{t}_1, \ldots, \bar{t}_r) = \overline{f(t_1, \ldots, t_r)}$ (for each operation symbol f of degree r and all constant terms $t_1, \ldots, t_r$);

$\langle \bar{t}_1, \ldots, \bar{t}_r \rangle \in \phi(p)$ if and only if $p(t_1, \ldots, t_r) \in \mathfrak{S}$ (for each relation p of degree r and all constant terms $t_1, \ldots, t_r$).

We now prove the following

(b) **Claim.** Let P be a closed formula. Then $|P|_\phi = 1$ if and only if $P \in \mathfrak{S}$. (This claim implies that ϕ is a model for $\mathfrak{S}$. In fact, if $P \in \mathfrak{S}$ is not closed, then its closure $\forall x_1 \cdots \forall x_n P$ is deducible from $\mathfrak{S}$ using Gen, and hence, since $\mathfrak{S}$ is complete and consistent, $\forall x_1 \cdots \forall x_n P \in \mathfrak{S}$. By the claim, $|\forall x_1 \cdots \forall x_n P|_\phi = 1$, so that $|P|_\phi = 1$.)

Proof of the claim. We use induction on the total number of quantifiers and connectives in P. We shall write $|P|$ instead of $|P|_\phi$.

(b_1) P is an atomic formula $p(t_1, \ldots, t_n)$. The claim follows from the definition of $|P|$ and the list of primary mappings, since the t_i are constant terms (or else P would not be closed).

(b_2) $P = \neg Q$. If $|P| = 1$, then $|Q| = 0$ and $Q \notin \mathfrak{S}$ by the induction assumption applied to Q; since $\mathfrak{S}$ is complete, we have $\neg Q \in \mathfrak{S}$, i.e., $P \in \mathfrak{S}$. On the other hand, if $|P| = 0$, then $|Q| = 1$ and $Q \in \mathfrak{S}$, so that $\neg Q \notin \mathfrak{S}$ since $\mathfrak{S}$ is consistent.

(b_3) $P = (Q_1 \Rightarrow Q_2)$. We first show that if $|P| = 0$ then $P \notin \mathfrak{S}$. In fact, in this case $|Q_1| = 1$ and $|Q_2| = 0$; by the induction assumption, $Q_1 \in \mathfrak{S}$, $Q_2 \notin \mathfrak{S}$; since $\mathfrak{S}$ is complete, $\neg Q_2 \in \mathfrak{S}$; using the tautology $Q_1 \Rightarrow (\neg Q_2 \Rightarrow \neg(Q_1 \Rightarrow Q_2))$ and using MP twice yields $\mathfrak{S} \vdash \neg(Q_1 \Rightarrow Q_2)$. Since $\mathfrak{S}$ is complete and consistent, all closed formulas which are deducible from $\mathfrak{S}$ belong to $\mathfrak{S}$; hence, $\neg(Q_1 \Rightarrow Q_2) = \neg P \in \mathfrak{S}$, so that $P \notin \mathfrak{S}$.

We now show that, if $P \notin \mathfrak{S}$, then $|P| = 0$. In fact, since $\mathfrak{S}$ is complete, we then have $\neg P = \neg(Q_1 \Rightarrow Q_2) \in \mathfrak{S}$. The tautologies $\neg(Q_1 \Rightarrow Q_2) \Rightarrow Q_1$ and $\neg(Q_1 \Rightarrow Q_2) \Rightarrow \neg Q_2$ and MP give $\mathfrak{S} \vdash Q_1$ and $\mathfrak{S} \vdash \neg Q_2$, so that, since $\mathfrak{S}$ is complete and consistent, $Q_1 \in \mathfrak{S}$ and $\neg Q_2 \in \mathfrak{S}$. By the induction assumption, $|Q_1| = 1$ and $|Q_2| = 0$, so that $|P| = |Q_1 \Rightarrow Q_2| = 0$.

(b_4) $P = Q_1 \vee Q_2$ or $Q_1 \wedge Q_2$. Using the tautologies which express $\wedge$ and $\vee$ in terms of $\Rightarrow$ and $\neg$, we can reduce to the previous cases; we omit the details.

(b_5) $P = \forall x Q$. If x does not occur freely in Q, then $|P| = 1$ is equivalent to $|Q| = 1$, i.e., by the induction assumption, to $Q \in \mathfrak{S}$. But $Q \in \mathfrak{S}$ is equivalent to $\forall x\, Q \in \mathfrak{S}$, in one direction using Gen and in the other direction using the axiom of specialization with $t = x$ and then MP.

We now assume that x occurs freely in Q. We first suppose that $|P| = 1$ but $P \notin \mathfrak{S}$, and obtain a contradiction. If $P \notin \mathfrak{S}$, then $\neg P \in \mathfrak{S}$, i.e., $\neg \forall x\, Q(x) \in \mathfrak{S}$. Since the alphabet of L is sufficient for $\mathfrak{S}$, it follows that $\mathfrak{S}$ contains the formula $\neg \forall x\, Q(x) \Rightarrow \neg Q(c_Q)$. Applying MP, we obtain $\mathfrak{S} \vdash \neg Q(c_Q)$; since $\mathfrak{S}$ is consistent, we have $Q(c_Q) \notin \mathfrak{S}$. By the induction assumption, $|Q(c_Q)| = 0$ ($Q(c_Q)$ is closed!). This means that $|Q(x)|(\xi) = 0$ for $\xi \in \overline{M}$ if $x^\xi = \bar{c}_Q$, contradicting the assumption that $|P| = 1$.

We now suppose that $|P| = 0$ but $P \in \mathfrak{S}$, and obtain a contradiction. Since $|P| = 0$, for some $\xi \in \overline{M}$ we have $|Q(x)|(\xi) = 0$. Let t be the constant term for which $x^\xi = \bar{t}$. Clearly t is free for x in Q, so that $0 = |Q(x)|(\xi) =$

$|Q(t)|$. Hence $Q(t) \notin \mathfrak{S}$ by the induction assumption, and $\neg Q(t) \in \mathfrak{S}$ since $\mathfrak{S}$ is complete. On the other hand, if $P \in \mathfrak{S}$, i.e., $\forall x\, Q(x) \in \mathfrak{S}$, then the axiom of specialization $\forall x\, Q(x) \Rightarrow Q(t)$ gives us $\mathfrak{S} \vdash Q(t)$. But, since $\neg Q(t) \in \mathfrak{S}$, this contradicts the consistency of $\mathfrak{S}$.

(b_6) $P = \exists x\, Q$. This reduces to the previous case using the axiom which expresses $\exists$ in terms of $\forall$ and negation; we omit the details. □

6.10. Proof of Lemma 6.6. In order to imbed $\mathfrak{S}$ in a complete and consistent set $\mathfrak{S}'$, we shall have to use Zorn's lemma and the Deduction Lemma for L (see Subsection 4.5 of Chapter II). Zorn's lemma will be applied to the set $\mathcal{C}\mathfrak{S}$ = the set of sets of formulas $\mathfrak{S}'$ in L which contain $\mathfrak{S}$ and are consistent. The set $\mathcal{C}\mathfrak{S}$ is ordered by inclusion.

Verification of the hypothesis of Zorn's lemma. Let $\{\mathfrak{S}'_\alpha\}_{\alpha \in I}$ be a linearly ordered subset of $\mathcal{C}\mathfrak{S}$, i.e., for any α and β we have either $\mathfrak{S}'_\alpha \leqslant \mathfrak{S}'_\beta$ or $\mathfrak{S}'_\beta \leqslant \mathfrak{S}'_\alpha$. Then the union $\cup \mathfrak{S}'_\alpha$ belongs to $\mathcal{C}\mathfrak{S}$. In fact, otherwise $\cup \mathfrak{S}'_\alpha$ would be inconsistent, and there would exist a deduction of a contradiction from a finite number of formulas. Suppose these formulas are contained in $\mathfrak{S}'_{\alpha_1}, \ldots, \mathfrak{S}'_{\alpha_n}$. But one of these sets contains the remaining $n-1$; this set would be inconsistent, contrary to the definition of $\mathcal{C}\mathfrak{S}$.

Proof of Lemma 6.6 from Zorn's lemma. The set $\mathcal{C}\mathfrak{S}$ has a maximal element, i.e., a consistent set $\mathfrak{S}' \supset \mathfrak{S}$ such that if $Q \notin \mathfrak{S}'$ then $\mathfrak{S}' \cup \{Q\}$ is inconsistent. We claim that $\mathfrak{S}'$ is complete. In fact, suppose that there were a closed formula P such that $P \notin \mathfrak{S}'$ and $\neg P \notin \mathfrak{S}'$. Since $\mathfrak{S}'$ is maximal, it follows that $\mathfrak{S}' \cup \{P\} \vdash R$ and $\mathfrak{S}' \cup \{\neg P\} \vdash R$ for any formula R. By the Deduction Lemma, $\mathfrak{S}' \vdash P \Rightarrow R$ and $\mathfrak{S}' \vdash \neg P \Rightarrow R$. Using the tautology $(P \Rightarrow R) \Rightarrow ((\neg P \Rightarrow R) \Rightarrow R)$ and MP, we have $\mathfrak{S}' \vdash R$, contradicting the consistency of $\mathfrak{S}'$. □

6.11. Proof of Lemma 6.7. In constructing a language with a sufficient alphabet for a consistent set of formulas $\mathfrak{S}'$ which contains $\mathfrak{S}$ and Ax L', we proceed in the most natural way.

(a) We add to the alphabet of L a set of new constants whose cardinality is that of the alphabet of $L + \aleph_0$. We obtain a language L'.

(b) We consider the set of formulas $\mathfrak{S} \cup \mathrm{Ax}\, L'$ in the language L', where Ax L' consists of all the logical axioms of L'. We claim that this set of formulas is consistent. In fact, if there were a deduction of a contradiction from $\mathfrak{S} \cup \mathrm{Ax}\, L'$ in L', then the following procedure would transform it into a deduction of a contradiction from $\mathfrak{S}$ in L: take the finite set consisting of all the new constants which occur in the formulas in the deduction and replace these constants by old variables (in L) which do not occur in the formulas in the deduction. It is easily verified that the deduction of a contradiction remains a deduction of a contradiction, and now lies entirely in L.

(c) We consider the set S of formulas $P(x)$ containing one free variable x and such that $\neg\forall x\, P(x) \in \mathfrak{S} \cup \operatorname{Ax} L'$. For each $P(x)$ in S we choose a new constant c_P subject to the following restriction: each c_P can be assigned a natural number, its *rank*, in such a way that if a constant of rank n occurs in $P(x)$ then c_P has rank $> n$. This can be done since $\operatorname{card}(S) \leqslant \operatorname{card}(\text{alphabet of } L') = \operatorname{card}(\text{alphabet of } L) + \aleph_0$. For each $P(x)$ in S define the formula

$$R_P : \neg\forall x\, P(x) \Rightarrow \neg P(c_P)$$

and finally let

$$\mathfrak{S}' = \mathfrak{S} \cup \operatorname{Ax} L' \cup \{R_P | P(x) \in S\}.$$

Call any R_P an R-formula. Note that no R-formula has the form $\neg\forall x\, P(x)$, so that L' is sufficient for $\mathfrak{S}'$. It remains only to verify that $\mathfrak{S}'$ is consistent. If a contradiction were deducible from $\mathfrak{S}'$ then it would be deducible using finitely many R-formulas. At least one R_P among these must be such that c_P does not occur in any of the others: namely, pick c_P of maximal rank. Hence it suffices to verify that if $\mathfrak{S} \cup \operatorname{Ax} L' \cup \mathfrak{R}$ is consistent, where $\mathfrak{R}$ is a set of formulas not containing c_P, then the addition of R_P does not lead to a contradiction.

Suppose $\mathfrak{S} \cup \operatorname{Ax} L' \cup \mathfrak{R} \cup \{R_P\}$ were inconsistent. Then, in particular, we would have a deduction of $\neg R_P$ and, by the Deduction Lemma, $\mathfrak{S} \cup \operatorname{Ax} L' \cup \mathfrak{R} \vdash R_P \Rightarrow \neg R_P$. The tautology $(R_P \Rightarrow \neg R_P) \Rightarrow \neg R_P$ and MP would yield a deduction of $\neg R_P$; that is,

$$\mathfrak{S} \cup Ax\, L' \cup \mathfrak{R} \vdash \neg(\neg\forall x\, P(x) \Rightarrow \neg P(c_P)).$$

Then the tautology $\neg(P \Rightarrow \neg Q) \Rightarrow Q$ and MP would yield a deduction of $P(c_P)$. Transform this deduction by replacing the constant c_P with a variable y which does not occur in the formulas in the deduction. Since c_P does not occur in $\mathfrak{R}$ it is easily verified that the transformation yields a deduction of $P(y)$ from $\mathfrak{S} \cup \operatorname{Ax} L' \cup \mathfrak{R}$. Using Gen, $\mathfrak{S} \cup \operatorname{Ax} L' \cup \mathfrak{R} \vdash \forall y\, P(y)$. But since $\neg\forall x\, P(x) \in \mathfrak{S} \cup \operatorname{Ax} L'$, we have $\mathfrak{S} \cup \operatorname{Ax} L' \vdash \neg\forall y\, P(y)$. Hence $\mathfrak{S} \cup \operatorname{Ax} L' \cup \mathfrak{R}$ is inconsistent, contrary to hypothesis. □

6.12. Proof of Lemma 6.8. Let L be a language in the class $\mathfrak{L}_1$, and let $\mathfrak{S}$ be a set of formulas in L. We imbed $\mathfrak{S}$ in a complete and consistent set $\mathfrak{S}'$, and then apply Lemma 6.7 to $(L, \mathfrak{S}')$. We let L^* and $\mathfrak{S}^*$ denote the resulting language and set of formulas. We further define inductively

$$(L^{(0)}, \mathfrak{S}^{(0)}) = (L, \mathfrak{S}); \qquad (L^{(i+1)}, \mathfrak{S}^{(i+1)}) = (L^{(i)*}, \mathfrak{S}^{(i)*}),$$

and finally

$$L^{(\infty)} = \bigcup_{i=0}^{\infty} L^{(i)}, \qquad \mathfrak{S}^{(\infty)} = \bigcup_{i=0}^{\infty} \mathfrak{S}^{(i)}.$$

The set $\mathfrak{S}^{(\infty)}$ is consistent, since any deduction of a contradiction would be obtained "at some finite level," and all the $\mathfrak{S}^{(i)}$ are consistent. It is complete, since every closed formula in $L^{(\infty)}$ is written in the alphabet of $L^{(i)}$ for some i, and $\mathcal{E}^{(i+1)}$ contains the completion of $\mathcal{E}^{(i)}$ in $L^{(i)}$. Finally, the alphabet of $L^{(\infty)}$ is sufficient for $\mathcal{E}^{(\infty)}$ by the same argument.

This completes the proof of the lemmas. □

6.13. Deduction of Theorem 6.2 from the lemmas. Let T be a Gödelian set of formulas in L. Applying Lemma 6.8 to T, we imbed (L, T) in $(L^{(\infty)}, T^{(\infty)})$, where the pair $(L^{(\infty)}, T^{(\infty)})$ satisfies Lemma 6.5. Let $\phi^{(\infty)}$ be an interpretation of $L^{(\infty)}$ such as must exist by Lemma 6.5. The cardinality of $M^{(\infty)}$ does not exceed card (alphabet of L) $+ \aleph_0$. The restriction ϕ of $\phi^{(\infty)}$ to L satisfies the condition $T \subset T_\phi L$. We prove that $T = T_\phi L$. In fact, let $P \in T_\phi L$. If P is closed, then $P \in T$, since either P or $\neg P$ lies in T by completeness, and $\neg P \notin T$ because P is ϕ-true. If P is not closed, and $x_1, \ldots, x_n$ are the variables which occur freely in P, then $\forall x_n\, P$ is closed and belongs to T. By the axiom of specialization, P is deducible from $T \cup \{\mathrm{A}x_1 \cdots \forall x_n\, P\}$, so that $P \in T$, since T is closed under deduction. This proves the first assertion of the theorem.

The second assertion follows from the analogous argument applied to $\mathcal{E}$ instead of T. We find a model ϕ for $\mathcal{E}$; then $\mathcal{E} \subset T_\phi L$ and $T_\phi L$ is Gödelian. □

6.14. In conclusion, we note that, if the alphabet of L contains a symbol $=$ for which the axioms of equality are included in $\mathcal{E}$ (or T), then there exists a normal interpretation which satisfies Theorem 6.2 and takes $=$ into equality. To prove this, we take the above model M and divide out by the equivalence relation $\phi(=)$, as in Subsection 4.6.

7 Countable models and Skolem's paradox

> "I know what you're thinking about," said Tweedledum: "but it isn't so, nohow."
> "Contrariwise," continued Tweedledee, "if it was so, it might be; and if it were so, it would be: but as it isn't, it ain't. That's logic."
>
> Lewis Carroll, Through the Looking Glass

7.1. In this section we discuss the technique of "cutting down" models, in particular, models for L_1Set. Let L be a language in $\mathfrak{L}_1$, let $M \subset N$ be two sets (or classes in V), and let ϕ and ψ be interpretations of L in M and N, respectively, which are compatible in the obvious sense, so that ψ is an extension of ϕ. We have a natural imbedding of interpretation classes $\overline{M} \subset \overline{N}$.

7.2. **Definition.** A formula P in L is called (M, N)-absolute if for all $\xi \in \overline{M}$ we have

$$|P|_M(\xi) = |P|_N(\xi).$$

(We write $|\ |_M$ instead of $|\ |_\phi$, and so on.)

The property of being absolute is usually used as follows: if P is absolute, and is also N-true, then it is automatically M-true. A formula P often fails to be absolute for the following reason: a formula $P = \exists x\, Q(x)$ can be N-true, so that N has an object with the property Q, but not M-true, because no such object lies in M. The proof of the following assertion shows how to handle this situation.

7.3. **Proposition.** *Let $\mathfrak{S}$ be a set of formulas in L, let ψ be an interpretation of L in N, and let $M_0 \subset N$ be a subset. Then there exists a set M, $M_0 \subset M \subset N$, having cardinality $\leqslant \operatorname{card} M_0 + \operatorname{card} \mathfrak{S} + \aleph_0$, such that all the formulas in $\mathfrak{S}$ are (M, N)-absolute.*

7.4. **Corollary** (Löwenheim–Skolem). *If the alphabet of L is countable and N is a model for $\mathfrak{S}$, then N has a countable submodel for $\mathfrak{S}$.*

The corollary follows from Proposition 7.3 if we construct a countable submodel with respect to which *all* the formulas of L are absolute, and, in particular, in which all formulas which were true before remain true.

Proof of 7.3. Suppose the set $M_i \subset N$, $i \geqslant 0$, has already been defined. Set

$$M_{i+1} = M_i \cup \left\{ x^{\xi'} | \xi' = \xi'(x, P, \xi) \right\},$$

where x runs through the variables in L, P runs through the subformulas of the formulas in $\mathfrak{S}$, and ξ runs through the points of $\overline{M_i}$, and where, for each fixed triple (x, P, ξ), $\xi'(x, P, \xi)$ is any one variation of ξ along x for which $|P|_N(\xi') = 1$ if such a variation exists; otherwise the triple does not make any contribution to M_{i+1}.

Further set $M = \bigcup_{i=0}^{\infty} M_i$. M clearly has the desired cardinality. We now show that all subformulas of the formulas in $\mathfrak{S}$ are (M, N)-absolute. We use induction on the number of quantifiers and connectives in the formula. The result is obvious for atomic formulas; the inductive step when a new formula is constructed using a connective is also clear. The quantifier $\forall$ reduces to $\exists$ in the usual way.

Thus, suppose P is absolute. We show that $\exists x\, P$ is also absolute. It suffices to consider the case when x occurs freely in P. For $\xi \in \overline{M}$ we have:

$$|\exists x\, P|_N(\xi) = \begin{cases} 1, & \text{if there exists a variation } \xi' \in \overline{N} \text{ of } \xi \text{ along } x \\ & \text{with } |P|_N(\xi') = 1, \\ 0, & \text{otherwise.} \end{cases}$$

$$|\exists x\, P|_M(\xi) = \begin{cases} 1, & \text{if there exists a variation } \xi'' \in \overline{M} \text{ of } \xi \text{ along } x \\ & \text{with } |P|_M(\xi'') = 1, \\ 0, & \text{otherwise.} \end{cases}$$

But the conditions on the right are equivalent. In fact, there exists a variation η of the point ξ along variables which do not occur freely in P, such that $\eta \in \overline{M_i}$ for some i. Then in the case $|\exists x\, P|_N(\xi) = |\exists x\, P|_N(\eta) = 1$ there is a $\xi' \in \overline{N}$ with $|P|_N(\xi') = 1 \Rightarrow$ there is an $\eta' \in \overline{M}_{i+1}$ with $|P|_N(\eta') = 1$, where η' is a variation of η along x, by the construction of M_{i+1}. This completes the proof. □

7.5. We now apply Corollary 7.4 to the standard interpretation of $\mathrm{L_1Set}$ in the von Neumann universe V and the set $\mathfrak{S}$ of Zermelo–Fraenkel axioms. We obtain a countable model N for this axiom system, but this model has one defect: if $X \in N$, some elements in X might not themselves belong to N, i.e., $\in$ is not necessarily transitive. The following result of Mostowski shows how to replace N by a transitive countable model.

Let $N \subset V$ be a subclass, and let $\varepsilon \subset N \times N$ be a binary relation. We shall write $X \varepsilon Y$ instead of $\langle X, Y \rangle \in \varepsilon$. For any $X \in N$ we set

$$[X] = \{ Y \mid Y \varepsilon X \}.$$

Suppose that $[X] \varepsilon V$ for all $X \in N$, i.e., each $[X]$ is a set rather than a class. We consider the interpretation ϕ of $\mathrm{L_1Set}$ in the class N for which $\phi(\in)$ is ε and $\phi(=)$ is equality.

7.6. **Proposition** (Mostowski). *Suppose that the axiom of extensionality and the axiom of the empty set are ϕ-true, and that N does not contain any infinite chain $\cdots X_n \varepsilon X_{n-1} \varepsilon \cdots \varepsilon X_1 \varepsilon X_0$. Then there exists a unique transitive class $M \subset V$ and a unique isomorphism $f : (N, \varepsilon) \tilde{\to} (M, \in)$.*

If we apply this proposition to the countable model $(N, \in)$ for the Zermelo-Fraenkel axioms in subsection 7.5, we obtain a transitive countable model $(M, \in)$, that is, a "small universe." (The condition that all ε-chains are finite even holds in V, as well as in N; $[X]$ is the subset $X \cap N \subset X$, and hence is an element of V.)

7.7. PROOF OF PROPOSITION 7.6. Using transfinite induction, for every ordinal α we construct sets $N_\alpha \subset N$, $M_\alpha \subset V$ and compatible isomorphisms $f_\alpha : (N_\alpha, \varepsilon|_{N_\alpha}) \tilde{\to} (M_\alpha, \in|_{M_\alpha})$, and we show that $\bigcup N_\alpha = N$.

(a) Since the axiom of extensionality is ϕ-true and $\phi(=)$ is equality, we easily obtain $X_1 = X_2 \Leftrightarrow [X_1] = [X_2]$ for all $X_1, X_2 \in N$. Let $\varnothing_N \in N$ be the interpretation of the constant $\varnothing$ of the language $\mathrm{L_1Set}$. Since the axiom of the empty set is ϕ-true, we may conclude that $\varnothing_N$ is the unique element of N for which $[\varnothing_N] = \varnothing \in V$. We set

$$N_0 = \{\varnothing_N\}, \qquad M_0 = \{\varnothing\}, \qquad f_0(\varnothing_N) = \varnothing.$$

(b) *Recursive construction.* Let α be an ordinal. Suppose that N_α, M_α and f_α have already been constructed. We set:

$$N_{\alpha+1} = \{X \in N \mid [X] \subset N_\alpha \wedge X \notin N_\alpha\} \cup N_\alpha;$$

$$f_{\alpha+1}(X) = \{f_\alpha(Y) \mid Y \in [X]\}, \quad \text{for } X \in N_{\alpha+1} \setminus N_\alpha; \qquad f_{\alpha+1}|_{N_\alpha} = f_\alpha;$$

$$M_{\alpha+1} = \text{image of } f_{\alpha+1} = \text{range of } f_{\alpha+1}.$$

If β is a limiting ordinal, we set $N_\beta = \bigcup_{\alpha<\beta} N_\alpha, M_\beta = \bigcup_{\alpha<\beta} M_\alpha$, and $f_\beta = \bigcup_{\alpha<\beta} f_\alpha$. Finally, we set $M = \bigcup M_\alpha$ and $f = \bigcup f_\alpha$, where the union is taken over all the ordinals.

(c) *Inductive proof.* We verify that for each α

(c_1) N_α *is a set, i.e.,* $N_\alpha \in V$.
(c_2) M_α *is a transitive subset of* V.
(c_3) f_α *is an isomorphism of* N_α *with* M_α *taking* ε *to* $\in$.
(c_4) $N = \bigcup_\alpha N_\alpha$.

Assertions (c_1)–(c_3) are obvious for $\alpha = 0$. If they hold for all $\alpha < \beta$ and if β is a limiting ordinal, then they also hold for β. It remains to check the step from α to $\alpha + 1$.

(c_1) [ ] is obviously a function from $N_{\alpha+1} \setminus N_\alpha$ to $\mathcal{P}(N_\alpha)$; since the axiom of extensionality is true, there exists an inverse function. Its image $N_{\alpha+1} \setminus N_\alpha$ is a set, since N_α, and therefore $\mathcal{P}(N_\alpha)$, are sets by the induction assumption.

(c_2) Any element in $M_{\alpha+1} \setminus M_\alpha$ has the form $\{f_\alpha(Y) \mid Y \in [X]\}$, where $X \in N_{\alpha+1} \setminus N_\alpha$. But then $[X] \subset N_\alpha$. Hence, an element $f_\alpha(Y)$ of this element of $M_{\alpha+1} \setminus M_\alpha$ belongs to the image of f_α, i.e., to the set $M_\alpha \subset M_{\alpha+1}$. This proves the transitivity of $M_{\alpha+1}$.

(c_3) We first verify that $f_{\alpha+1}$ is a bijection. The surjectivity is obvious; using the induction assumption, we see that it suffices to verify injectivity on $N_{\alpha+1} \setminus N_\alpha$. But if $X_1, X_2 \in N_{\alpha+1} \setminus N_\alpha$ and $f_{\alpha+1}(X_1) = f_{\alpha+1}(X_2)$, then

$$\{f_\alpha(Y) \mid Y \in [X_1]\} = \{f_\alpha(Y) \mid Y \in [X_2]\}.$$

Since f_α is injective, we obtain $[X_1] = [X_2]$, so that $X_1 = X_2$.

We then find:

$$Y \varepsilon X \Leftrightarrow Y \in [X] \Leftrightarrow f_\alpha(Y) \in f_{\alpha+1}(X),$$

so that for $X \in N_{\alpha+1} \setminus N_\alpha$ the relation $Y \varepsilon X$ goes to $f_{\alpha+1}(Y) \in f_{\alpha+1}(X)$. This is clearly sufficient to complete the induction.

(c_4) Finally, we verify that $N = \bigcup N_\alpha$. Let $N' = N \setminus \bigcup N_\alpha$; we suppose that N' is nonempty and show that this leads to a contradiction. If there existed an $X \in N'$ such that $[X] \cap N' = \varnothing$, then we would have $[X] \cap N \subset \bigcup N_\alpha$; then $[X] \subset N_{\alpha_0}$ for some α_0, so that $X \in N_{\alpha_0+1}$, contradicting the assumption that $X \in N \setminus \bigcup N_\alpha$. On the other hand, if we had $[X_0] \cap N' \neq \varnothing$ for all $X_0 \in N'$, then, successively choosing $X_{n+1} \in [X_n] \cap N'$, we would obtain an infinite chain $X_{n+1} \varepsilon X_n \varepsilon X_{n-1} \varepsilon \ldots \varepsilon X_0$, contradicting the hypothesis of the theorem.

(d) Suppose we have two transitive subclasses M and M', and an isomorphism $g : (M, \in) \tilde{\to} (M', \in)$. We set $M_\alpha = V_\alpha \cap M$ and $M'_\alpha = V_\alpha \cap M'$. An obvious induction on α then shows that g is the identity map. The proposition is proved. □

7.8. *Skolem's paradox*. Let M be a transitive countable model for the Zermelo–Fraenkel axioms. Then the following formulas are M-true:

the axiom of infinity;
the power set axiom;
Cantor's theorem that there is no mapping of x onto $\mathcal{P}(x)$ for any set x (this theorem is deducible from the Zermelo–Fraenkel axioms).

Since $\mathcal{P}(X)$ is uncountable when X is countably infinite, the content of the assertion that the power set axiom is true in the countable model M must be very different from the content of the assertion that this axiom is V-true. In fact, in $\mathrm{L_1Set}$ let "$y = \mathcal{P}(x)$" be abbreviated notation for the formula $\forall z(\text{"}z \subset x\text{"} \Leftrightarrow z \in y)$. Let $\xi \in \overline{M}$, $x^\xi = X \in M$, and $y^\xi = Y \in M$. Then we easily see that

$$|\text{"}y = \mathcal{P}(x)\text{"}|_M(\xi) = 1 \Leftrightarrow Y = \{Z \mid Z \subset X \wedge Z \in M\},$$

i.e., $\mathcal{P}(X)_M = \mathcal{P}(X) \cap M$ plays the role of $\mathcal{P}(X)$ in M. Here $\mathcal{P}(X)_M$ is at most countably infinite, since M is countable; so, from the usual point of view, there exists a mapping of a countably infinite set X onto $\mathcal{P}(X)_M$. This does not contradict Cantor's theorem, because the M-truth of Cantor's theorem merely means that there are no (graphs of) such mappings in the model M. Such graphs may exist outside of M, but, if we add such a graph to M (along with everything that must be added for the axioms to remain true), we thereby increase M, and at the same time $\mathcal{P}(X)_M$, and the mapping stops being onto.

All such ways statements of set theory change their meaning in countable models are customarily referred to as Skolem's paradox.

Cohen was the first who was able to use the properties of countable models to prove the nondeducibility of the continuum hypothesis. In his models sets of "M-intermediate" cardinality lie between ω_0 and $\mathcal{P}(\omega_0)_M$, although from an external point of view both ω_0 and $\mathcal{P}(\omega_0)_M$, along with all the other sets, are simply countable. Cohen introduced fundamentally new ideas of relativizing the very notion of truth, and it is only with the benefit of hindsight that we can so easily understnad the situation in his models. For details, see Chapter III.

Skolem himself, and other specialists on the foundations of mathematics, were willing to work with countably infinite sets, but not with larger infinities. They considered Skolem's paradox to be a manifestation of the relative character of set theoretic concepts. In particular, they considered that there exist "different continua" $\mathcal{P}(\omega_0)_M$, none of which coincides with the "real" $\mathcal{P}(\omega_0)$.

From the point of view of the topologist or analyst, for whom the continuum is a working reality, the existence of countable models means that formal language has limitations as a means of imitating intuitive reasoning. We encountered similar limitations when discussing the formal axioms of induction in §4.

For the psychologist or philosopher, perhaps the most interesting aspect of the situation is that any mathematician can understand the viewpoint of another mathematician (without having to agree with it). This means that what mathematician A says, though demonstrably incapable of conveying unambiguous information about the continuum, nevertheless is capable of bringing the brain of mathematician B to the point where it forms an idea of the continuum which adequately represents the idea in A's brain. Then B is still free to reject this idea.

"I know what you're thinking about," said Tweedledum: "but it isn't so, nohow."

8 Language extensions

8.1. In this section we study the formal version of "introducing new notation." Here we only consider names of new functions and constants which are "demonstrably definable" in the language. Adding such names to the alphabet shortens formulas and formal deductions, but does not increase the set of deducible formulas—this will be the fundamental theorem of this section.

Of course, in practice, abbreviated notation and well-chosen new names can immediately make accessible to our intuition entire areas of mathematical facts that were previously inaccessible. One of the best known examples are the groups introduced by Galois to study equations. In 1924, commenting on the attempt to curb the inflation in Germany by introducing a new unit of currency, the Rentenmark, Hilbert remarked skeptically: "A problem cannot be solved by renaming the independent variable." But, as his biographer Constance Reid noted, Hilbert was wrong: the economic situation gradually stabilized.

We start with the following data.

8.2. Let L' be a language in $\mathfrak{L}_1$ with equality and with an infinite set of variables, and let $P'(x)$ be a formula in L' in which x occurs freely. We recall that the abbreviated notation $\exists! x\ P'(x)$ (read: "there exists a unique x with the property P'") stands for the formula

$$\exists x\ P'(x) \wedge \forall x\ \forall y (P'(x) \wedge P'(y) \Rightarrow x = y).$$

Let $\mathfrak{E}'$ be a set of formulas in L' which contains Ax L', the axioms of equality, and perhaps some special axioms. Suppose that the formula $\exists! x\ P'(x, y_1, \ldots, y_n)$ is deducible from $\mathfrak{E}'$, where P' has no free variables other than $x, y_1, \ldots, y_n$. Intuitively, this means that P' defines x as an implicit function of $y_1, \ldots, y_n$, and in the informal text we can introduce a new notation for this function, say, $x = f(y_1, \ldots, y_n)$, and then always use that notation. Now we give the formal version of this procedure.

8.3. **Proposition.** *Under the conditions in* 8.2, *let L denote the language in* $\mathfrak{L}_1$ *whose alphabet is obtained from the alphabet of L′ by adding a new operation symbol f of degree n if* $n \geqslant 1$, *or a constant f if* $n = 0$. *Let* $\mathfrak{E}$ *be the smallest set of formulas in L containing* Ax *L, the axioms of equality,* $\mathfrak{E}'$, *and the formula* $P'(f(y_1, \ldots, y_n), y_1, \ldots, y_n)$.

Then there exists an explicitly describable map from the set of formulas of the (richer) language L to the set of formulas of the (poorer) language L′ which correlates with each Q a translation Q′ and which has the following properties:

(a) *If f does not occur in Q, then the translation of Q coincides with Q.*

(b) *If Q is deducible from* $\mathfrak{E}$ *in L, then Q′ is deducible from* $\mathfrak{E}'$ *in L′. In particular, the set of formulas in L′ which are deducible from* $\mathfrak{E}'$ *in L′ coincides with the set of formulas in L which do not contain f and are deducible from* $\mathfrak{E}$ *in L.*

Proof.

Translation of formulas. Suppose $n \geqslant 1$. (The case $n = 0$ is analogous, and is simpler, so we shall omit it.) The first effect of adding f is to increase the set of terms: L includes terms of the form $f(t_1, \ldots, t_n)$, where f can occur in $t_1, \ldots, t_n$, and so on. In order to decrease the number of references to f, we must say "$f(t_1, \ldots, t_n)$" in a roundabout way: "that x for which $P(x, t_1, \ldots, t_n)$." This is the basic idea behind the translation of formulas. We now give a precise inductive definition.

(a) A term $f(t_1, \ldots, t_n)$ is called a simple f-term if f does not occur in $t_1, \ldots, t_n$.

(b) Let Q be an atomic formula in L. If f does not occur in Q, we let Q be its own translation. If f occurs in Q, then there exists a simple f-term $f(t_1, \ldots, t_n)$ which occurs in Q. We take the very first occurrence of a simple f-term in Q, then take a variable symbol x which does not occur in Q, substitute it in place of this occurrence, thereby obtaining a formula Q^*, and finally construct the formula

$$Q'_{(1)}: \quad \exists x (P(x, t_1, \ldots, t_n) \wedge Q^*(x)).$$

We apply this procedure to $Q'_{(1)}$ to obtain $Q'_{(2)}$, and so on. After a finite number of steps we obtain a formula $Q'_{(i)} = Q'$ in which f does not occur. This Q' is the translation of Q.

(c) If Q is not an atomic formula, it has the form $\neg Q_1$ or $Q_1 * Q_2$ (where $*$ is a connective), or else $\forall y\, Q_1$ or $\exists y\, Q_1$. In all cases Q is translated automatically using the translations of Q, Q_1, Q_2, i.e., by "adding prime" to the component parts.

Translation of deductions. The problem is the following: Let $Q_1, \ldots, Q_n = Q$ be a deduction of Q from $\mathfrak{S}$, and let Q' be the translation of Q. We must construct a deduction of Q' from $\mathfrak{S}'$. The most obvious idea is to write the sequence of translations $Q'_1, \ldots, Q'_n$. Why isn't this a deduction of Q' from $\mathfrak{S}'$, since MP and Gen are translated in a trivial way, and tautologies are translated as tautologies? Because, for example, the logical axiom $\forall x\, R(x) \Rightarrow R(f)$ might appear in this sequence, and this formula stops being an axiom after it is translated, if f occurs in R. Hence, we must fill in the sequence $Q'_1, \ldots, Q'_n$ by adding deductions from $\mathfrak{S}'$ of certain of its terms. This is a rather cumbersome combinatoric procedure, which one can read in §74 of Kleene's book *Introduction to Metamathematics* (Van Nostrand, New York–Toronto, 1952). (The moral of the story is that new notation really does economize on time and space.)

Instead of using this procedure, we shall give an ineffective proof that $\mathfrak{S}' \vdash Q'$ using the deducibility criterion in 6.3. We state this criterion once more:

(a) *If Q' is true in any model of $\mathfrak{S}'$, then* $\mathfrak{S}' \vdash Q'$. Since $\mathfrak{S}'$ contains the axioms of equality, we can slightly strengthen this as follows:

(b) *If Q' is true in any normal model of $\mathfrak{S}'$ then Q' is true in any model of $\mathfrak{S}'$.*

Recall that $=$ is interpreted as equality in a normal model. On the other hand, in §4 we showed that in any model $=$ is interpreted as an equivalence relation which is compatible with the interpretation of all the constants, functions, and relations. Factoring out by this equivalence relation leads to a normal model, in which the truth values of all the formulas remain as before.

(c) *The normal models of $\mathfrak{S}'$ (in the language L') coincide with the normal models of $\mathfrak{S}$ (in the language L).*

More precisely, we can give the following natural one-to-one correspondence between them which preserves the truth function. We shall limit ourselves to the case $n \geqslant 1$. Let ϕ be a normal interpretation of L' in M for which $|Q'|_\phi = 1$ for all $Q' \in \mathfrak{S}'$. In particular, since $\mathfrak{S}' \vdash \exists! x\, P'$, we have

$$|\exists! x\, P'(x, y_1, \ldots, y_n)|_\phi = 1.$$

Computing the truth value on the left at a point $\xi \in \overline{M}$ and using the normality of the model, we then find that to every n-tuple $\langle y_1^\xi, \ldots, y_n^\xi \rangle \in$

M^n there corresponds a unique $x^{\xi'} \in M$ such that $|P'(x^{\xi'}, y_1^{\xi}, \ldots, y_n^{\xi})|_\phi = 1$ (this is not the standard notation, but the meaning is clear). We now interpret the symbol f (which is the new symbol in the language L) as the function $M^n \to M$ which takes $\langle y_1^{\xi}, \ldots, y_n^{\xi} \rangle$ to $x^{\xi'}$. We obviously obtain a normal model for $\mathfrak{S}$ in L.

Conversely, any normal model for $\mathfrak{S}$ can be restricted to L' to obtain a normal model for $\mathfrak{S}'$.

(d) *If Q is deducible from $\mathfrak{S}$ in L, then Q' is true in any normal model for $\mathfrak{S}'$.*

PROOF. Q is true in any model ϕ for $\mathfrak{S}$. To prove that Q' is true, we begin with atomic formulas Q which contain f. In the notation in the first part of the proof (translation of formulas), we construct Q^* and then $Q'_{(1)} = \exists x(\underline{P}(x, t_1, \ldots, t_n) \wedge Q^*(x))$. To verify that $|Q'_{(1)}|_\phi = 1$, for each point $\xi \in M$ we must find a variation ξ' of ξ along x for which

$$|P|_\phi(\xi') = 1 \quad \text{and} \quad |Q^*(x)|_\phi(\xi') = 1.$$

We determine $x^{\xi'}$ from the condition $|P(x^{\xi'}, t_1^{\xi}, \ldots, t_n^{\xi})|_\phi = 1$. The description in (c) of the interpretation of f shows that we now have $|Q^*|_\phi(\xi') = |Q|_\phi(\xi) = 1$.

Thus, truth is preserved in going from Q to $Q'_{(1)}$. Repeating this procedure, we find that Q' is true for atomic formulas Q. Finally, the truth of Q' in the general case is proved by induction on the number of connectives and quantifiers. Combining the results (a)–(d), we then obtain $\mathfrak{S}' \vdash\!\!\!\to Q'$, which completes the proof of Proposition 8.3. □

8.4. EXAMPLES

(a) In L$_1$Set the following formula is deducible from the axioms of extensionality and pairing (and also the axioms of equality and the logical axioms):

$$\exists! x \, \forall z (z \in x \Leftrightarrow z = u \vee z = v).$$

Using Proposition 8.3, we see that we may add to L$_1$Set a new degree 2 function symbol $\{\ \}$, "unordered pair," without changing the set of formulas in L$_1$Set which are deducible from the Zermelo–Fraenkel axioms. Therefore, without hesitation we may use not only the abbreviated notation "$x = \{u, w\}$" as before, but also terms which are put together using the symbol $\{\ \}$. In particular, (here the use of $\{\ \}$ is not normalized, but is in agreement with tradition):

(b) We can introduce notation for the finite ordinals

$$\varnothing, \{\varnothing\}, \{\varnothing, \{\varnothing\}\}, \cdots$$

as terms in their own right in our language extension, and then imbed formal arithmetic in formal set theory.

(c) After deducing the formula

$$\exists! x(\text{“}x \text{ is an ordinal”} \wedge \text{“}x \text{ is not finite”} \wedge \text{“}\forall \text{ ordinal } y < x, y \text{ is finite”})$$

from the Zermelo–Fraenkel axioms, we can introduce a new constant ω_0, and then continue to introduce names of more and more ordinals which are demonstrably uniquely characterized by formulas in $\mathrm{L_1Set}$ (or in language extensions which are formed in the same way).

We shall make use of this new freedom of action in Chapter III.

9 Undefinability of truth: the language *SELF*

9.1. When modeled in formal languages, arguments of the "Liar Paradox" type lead to important theorems on the limitations of the modes of expression and proof in these languages. The best known of these theorems are Tarski's theorem on the undefinability of the set of true formulas and Gödel's theorem on the impossibility of effectively axiomatizing arithmetic.

The next three sections are devoted to Tarski's theorem. Our presentation is based on an excellent article by Smullyan (Languages in which self-reference is possible, J. Symb. Logic. vol. 22, no. 1 (1957), 55–67).

In this section we describe the extremely elementary language SELF (which does not belong to $\mathfrak{L}_1$), which was designed to illustrate self-reference and which graphically demonstrates the idea of such a construction. In §10 we introduce the language SAr, which is just as expressive as $\mathrm{L_1Ar}$, but does not belong to $\mathfrak{L}_1$. Its syntax is close to that of SELF, which greatly simplifies proofs. Finally, in §11 we use a method of Smullyan to prove Tarski's theorem for SAr.

9.2. *The language SELF* (*S*mullyan's *E*asy *L*anguage *F*or self-reference)

The *alphabet* of SELF. E, $*$ (symmetric quotes), r (relation of degree 1), $\neg$ (negation).

The *syntax* of SELF. The distinguished expressions are: labels, displays, formulas, and names. The *label* of any expression P is $*P*$ ("P in quotes"). The *display* of any expression P is $P*P*$ ("something with a label"). *Formulas* are expressions of the form $rE\ldots E*P*$ or $\neg rE\ldots E*P*$, where E appears $k \geqslant 0$ times after r. We use the abbreviated notation rE^k*P* and $\neg rE^k*P*$ for formulas. Finally, we introduce the binary relation "is the name of" on the set of all distinguished expressions. This relation is defined recursively:

(a) The label of P is a name of P.
(b) If P is a name of Q, then EP is a name of the display of Q, i.e., a name of the expression $Q*Q*$.

9.3. *Remarks*

(a) If P is a name of Q, then the display of Q has at least two different names: EP and $*Q*Q**$. Thus, an expression can have several names. But, conversely, an expression is uniquely determined if we know its name; names all have the form $E^k * P *$, $k \geqslant 0$. We shall write $N(Q)$ in place of "one of the names of Q."

(b) Every formula has the form $rN(Q)$ or $\neg rN(Q)$. In 9.4 we interpret such a formula as the statement, "The expression Q has (or does not have) the property R," and it is natural that the formula, in saying something about Q, "calls Q by name."

(c) The expression $E * E *$ is one of two possible names for itself. In exactly the same way, the formula $rE * rE *$ "says something about itself" (see 9.5). The language SELF was constructed precisely in order to produce these effects of self-reference with the fewest possible modes of expression.

9.4. *The standard interpretations*. In order to give one of the standard interpretations of the language SELF, we choose any set (property) R of expressions of the language and introduce the truth function $|\ |_R$ on the formulas by stipulating

$$1 - |\neg rN(Q)|_R = |rN(Q)|_R = \begin{cases} 1, & \text{if } Q \in R \\ 0, & \text{otherwise.} \end{cases}$$

We say that a formula is R-true (R-false) if the value of $|\ |_R$ on the formula equals 1 (resp. 0).

9.5. **Undefinability Theorem.** *For any property R*

$$R \cap \{\textit{formulas}\} \neq \begin{cases} R\textit{-true formulas} \\ R\textit{-false formulas.} \end{cases}$$

Proof.

(a) The formula $Q = \neg rE * \neg rE *$ is R-true$\Leftrightarrow rE * \neg rE *$ is R-false$\Leftrightarrow Q \notin R$, since $E * \neg rE *$ is a name of the display of $\neg rE$, i.e., a name of Q. Thus, Q cannot both lie in R and be true, which proves the first part of the theorem. The connection with the Liar paradox becomes clear if we note that Q says about itself: "I do not have the property R."

(b) Analogously, the formula $rE * rE *$ says about itself: "I have the property R," and so cannot both lie in R and be R-false. □

10 Smullyan's language of arithmetic

10.1. In this section we describe the language of arithmetic SAr and its standard interpretation. The main difference between SAr and L_1Ar is that in SAr we are allowed to form "class terms"—names of certain sets of natural numbers. More precisely, if $P(x)$ is a formula in SAr with one free variable x, then the expression $x(P(x))$ in SAr names the set $\{x \in N | P(x)$ is true$\}$, and the expression $x(P(x))\bar{k}$, where the term $\bar{k}$ is a name for an integer $k \geqslant 1$, is a name for the statement "k satisfies P." The greater richness of the modes of expression in SAr, as opposed to L_1Ar, does not increase the class of subsets in $\bigcup_{r \geqslant 1} N^r$ which are definable by formulas. But it brings the syntax of SAr so close to that of SELF that we can imitate the proof of Theorem 9.5.

In addition, the alphabet of SAr is somewhat altered and shortened in comparison with the alphabet of L_1Ar, but this is only done in order to simplify the description of the syntax. These changes do not make the logic of SAr any poorer.

10.2. *The alphabet of* SAr: x (a variable); $'$ (used to form a countable set of variables $x, x', x'', \ldots$); $\cdot$ (multiplication, a degree 2 operation); $\uparrow$ (raising to a power, a degree 2 operation, as in Algol); $=$ (equality); $\downarrow$ (a connective, the conjunction of negations); (,) (parentheses); and $\bar{1}$ (the constant one).

10.3. *The syntax and interpretation of* SAr. Because we are allowed to form the class terms $x(P(x))$ and the formulas $x(P(x))\bar{k}$, the syntax is more complicated than in languages of $\mathfrak{L}_1$. We use induction on the integer $i \geqslant 0$ to define two sequences of sets of expressions: Tm_{2i} (terms of rank $\leqslant 2i$) and Fl_{2i+1} (formulas of rank $\leqslant 2i+1$). (Using double induction—on the rank of the term or formula, and, within the set Tm_{2i} or Fl_{2i+1}, on the length of the term or formula—one can prove a unique reading lemma; this lemma is the basis for defining free and bound occurrences of variables and truth functions. However, since there is nothing new here beyond what was done in §1, we leave the details to the reader.)

Along with our description of the syntax, we give a parallel description of the standard interpretation of SAr in N. In order to interpret expressions with free variables, we must fix a point $\xi \in N^N = N \times N \times \ldots$, which we shall identify with the corresponding infinite vector with natural number coordinates. Here the value of the kth variable $(x'^{\cdots'})^\xi$ ($k-1$ primes) is in the kth place in the vector.

(a_0) *Tm_0 is the set of numerical terms i.e., the least set of expressions which contains the variables $x, x', x'', \ldots$ and the names of the natural numbers $\bar{1}, \overline{11}, \overline{111}, \ldots$ and is closed with respect to forming the expressions $(t_1)\cdot(t_2)$ and $(t_1)\uparrow(t_2)$, where $t_i \in Tm_0$.*

Instead of $x'^{\cdots'}$ ($k-1$ primes) we shall write x_k, and instead of $\bar{1}\ldots\bar{1}$ ($k \geqslant 1$ ones) we shall write $\bar{k}$. The term $\bar{k}$ is interpreted as k (not depending on ξ); x_k^ξ is interpreted as the kth coordinate of ξ; and, if $t_1^\xi, t_2^\xi \in N$ have already been determined, then $[(t_1)\cdot(t_2)]^\xi = t_1^\xi t_2^\xi$ and $[(t_1)\uparrow(t_2)]^\xi = (t_1^\xi)^{t_2^\xi}$. The occurrences of the expressions $x_k = x'^{\cdots'}$ in any term in Tm_0 are obviously independent of one another. All such occurrences are considered free.

(b_0) *Fl_1 is the least set of expressions which contains all expressions of the form $t_1 = t_2$ (where $t_i \in Tm_0$) and is closed with respect to forming the* *expressions* $(P_1)\downarrow(P_2)$, *where* $P_i \in Fl_1$. In other words, Fl_1 is the logical closure of the set of atomic formulas $\{t_1 = t_2 | t_i \in Tm_0\}$.

Choosing a point ξ determines a truth value for any formula $P \in Fl_1$ by induction on the number of times $\downarrow$ occurs:

$$|t_1 = t_2|(\xi) = \begin{cases} 1, & \text{if } t_1^\xi = t_2^\xi, \\ 0, & \text{otherwise;} \end{cases}$$

$$|(P_1)\downarrow(P_2)|(\xi) = \begin{cases} 1, & \text{if } |P_1|(\xi) = |P_2|(\xi) = 0, \\ 0, & \text{otherwise.} \end{cases}$$

All occurrences of variables in elements of Fl_1 are independent of one another, and are considered free.

Now let $i \geqslant 1$, and suppose that the sets Tm_{2k-2}, Fl_{2k-1} are already defined for $k \leqslant i$ along with the interpretations and the division into free and bound occurrences of variables. We define the next sets Tm_{2i} and Fl_{2i+1} as follows.

(a_i) Tm_{2i} consists of the class terms of rank $\leqslant 2i$:

$$Tm_{2i-2} \cup \{x_k(P) | k \geqslant 1, P \in Fl_{2i-1}\}$$

(Tm_0 need not be included when $i = 1$). These elements have the following interpretation:

$$(x_k(P))^\xi = \left\{ \begin{array}{l} x_k^{\xi'} | \xi' \text{ runs through the variations of } \xi \text{ along } x_k \\ \text{for which } |P|(\xi') = 1 \end{array} \right\}.$$

All occurrences of the variable x_k in $x_k(P)$ are considered bound, and the occurrences of other variables remain the same (free or bound) as in P.

(b_i) Fl_{2i+1} *is the logical closure of the set of expressions*

$$Fl_{2i-1} \cup \{x_k(P) = x_k(Q) | k \geqslant 1;\ P, Q \in Fl_{2i-1}\} \cup \{T\bar{k} | k \geqslant 1, T \in Tm_{2i}\},$$

The *truth function* is defined as follows: if we set $x_k(P) = T_1$ and $x_k(Q) = T_2$, then

$$|x_k(P) = x_k(Q)|(\xi) = \begin{cases} 1, & \text{if } T_1^\xi = T_2^\xi \text{ as subsets of } N, \\ 0, & \text{otherwise;} \end{cases}$$

$$|T\bar{k}|(\xi) = \begin{cases} 1, & \text{if } k \in T^\xi, \\ 0, & \text{otherwise.} \end{cases}$$

The function $|\ |$ is extended to the logical closure in the same way as in b_0). All occurrences of variables in $x_k(P) = x_k(Q)$ and in $T\bar{k}$ are the same (free or bound) as in the corresponding class term. Composition using the connective $\downarrow$ does not change the nature of the occurrence. As in subsection 2.10, one can prove that $|P|(\xi)$ only depends on the ξ-values of the variables which have free occurrences in the formula $P \in \bigcup_{i=0}^{\infty} Fl_{2i+1}$.

This finishes the description of the syntax and semantics of SAr.

In conclusion, we show that the classes of sets in $\bigcup_{r \geq 1} N^r$ which are definable by formulas in L_1Ar and in SAr coincide. This result is not used in the proof of Tarski's theorem in the next section. However, the result itself and the method of proof are instructive, and we shall return to these ideas in Part III of the book.

Let L_1Ar have a countable set of variables. If we denote them by $x_1, x_2, \ldots, x_n, \ldots$ and identify x_i with $x'^{\cdots\prime}$ ($i-1$ primes), we can also identify the interpretation classes for L_1Ar and SAr in the obvious way. Our claim that the classes of definable sets coincide is then an immediate consequence of the following stronger fact:

10.4. **Proposition.** *Two translation mappings*

$$\{\text{formulas of } L_1Ar\} \rightleftarrows \{\text{formulas of SAr}\}$$

can be explicitly defined with the following properties:

(a) *At every point ξ the truth values of any formula and its translation coincide.*

(b) *The sets of free variables of any formula and its translation coincide.*

We note that the mappings we define will not be inverse to each other!

PROOF.

(a) *The translation from* L_1Ar *to* SAr. The translation of a formula P will be denoted "P". We first translate atomic formulas, and then use induction on the length. The alphabet of SAr does not have addition, but it has both multiplication and raising to a power, so that in place of $z = x + y$ we can write $2^z = 2^x \cdot 2^y$.

(a_1) *Atomic formulas.* They have the form $t_1 = t_2$. By "carrying out the operations," we replace every nonzero term in L_1Ar by a "normalized term," i.e., a polynomial of the form $\Sigma x_i^{i_1} \cdots x_n^{i_n}$, where the monomials are written in the form $(\ldots(x_1 \cdot x_1) \cdot \ldots \cdot x_1) \cdot x_2) \ldots)$, then arranged in lexicographical order, and finally separated by parentheses: $(\ldots((m_1 + m_2) + m_3) + \ldots)$. It is clear how to correlate such a term t to the term "$\bar{2}\uparrow t$" in SAr. For example, "$\bar{2}\uparrow((x_1)\cdot(x_1) + x_2)$" is $(\bar{2}\uparrow(x_1)\cdot(x_1))\cdot(\bar{2}\uparrow(x_2))$. By definition, the translation "$\bar{2}\uparrow\bar{0}$" is $\bar{1}$. Then we define the translation of the formula $t_1 = t_2$ to be "$2\uparrow t_1$" = "$2\uparrow t_2$." It is clear that such a formula and its translation have the same variables and are true at the same points ξ.

(a_2) If "Q," "Q_1," and "Q_2" have already been defined, then "$\neg Q$" is defined as "Q"$\downarrow$"Q". We similarly construct "$Q_1 * Q_2$" for the other connectives (see "Digression: Syntax" in Chapter I).

(a_3) If "Q" has already been defined, then "$\forall x_k\, Q$" is defined as

$$x_k(\text{“}Q\text{”}) = x_k(x_k = x_k).$$

Both the formula and its translation are true at a point ξ if and only if Q (and "Q") are true at all variations ξ' of ξ along x_k. They also have the same free variables, since, by induction, we may assume that this is the case for Q and "Q."

(a_4) By definition, "$\exists x_k\, Q$" coincides with "$\neg \forall x_k \neg Q$."

(b) *The translation from* SAr *to* L_1Ar. As before, we let "P" denote the translation of a formula P, although this time P will be a formula in SAr and "P" will be a formula in L_1Ar.

There is a subtle point here, namely, how to translate $x_1 = x_2 \uparrow x_3$. It will be shown in Part II of the book that such a translation exists, and can even be taken in the form $\exists x_4 \cdots \exists x_n\, p(x_1, x_2, x_3, x_4, \ldots, x_n)$, where p is an atomic formula in L_1Ar. Here we shall take this fact on faith, and choose a translation "$x_1 = x_2 \uparrow x_3$" once and for all.

(b_1) *Translation of formulas in* Fl_0. The following rules give an inductive definition:

"$t_1 = t_2$" has exactly the same form if $t_1, t_2 \in \{\text{variables}\} \cup \{\bar{1}, \bar{1}\bar{1}, \ldots\}$ (of course, in the sense that $x'^{\cdots\prime}$ is replaced by x_k and $\bar{1} \cdots \bar{1}$ is replaced by $(\cdots(\bar{1} + \bar{1}) + \bar{1}) + \cdots))$. "$x_k = t_1 \cdot t_2$" has the form $\exists x_i\, \exists x_j$("$x_i = t_1$"$\wedge$"$x_j = t_2$"$\wedge x_k = x_i \cdot x_j$) and "$x_k = t_1 \uparrow t_2$" has the form $\exists x_i\, \exists x_j$("$x_i = t_1$"$\wedge$"$x_j = t_2$"$\wedge$"$x_k = x_i \uparrow x_j$"), where x_i and x_j are the first two variables not occurring in t_1 or t_2. We similarly translate formulas with the left and right-hand sides permuted, and also with $\bar{1} \cdots \bar{1}$ instead of x_k. We further stipulate that "$t_1 = t_2$" has the form $\exists x_i$("$x_i = t_1$"$\wedge$"$x_i = t_2$"), where x_i is the first variable not occurring in t_1 or t_2, and where we only assume that neither t_1 nor t_2 is a variable or $\bar{1} \cdots \bar{1}$. It is clear that the truth function and the set of free variables are preserved under these translations.

(b_2) Suppose that the formulas in Fl_{2i-1} have already been translated. Let

$$\text{"}x_k(P_1) = x_k(P_2)\text{" be } \forall x_k(\text{"}P_1\text{"} \Leftrightarrow \text{"}P_2\text{"}), \quad \text{and}$$
$$\text{"}x_k(P)\bar{n}\text{" be "}P\text{"}(\bar{n}),$$

where on the right $\bar{n} = (\cdots(\bar{1} + \bar{1}) + \bar{1}) + \cdots)$ is substituted in place of all free occurrences of x_k in "P." This completes the proof. □

11 Undefinability of truth: Tarski's theorem

11.1. The language SAr is interpreted in N, and not in the set of its own formulas the way SELF is. In order to be able to determine the set of definable formulas, we number formulas by (certain) integers as follows.

We number the symbols of the alphabet (of which there are nine) from 1 to 9 in any order, *as long as* $\bar{1}$ *corresponds to* 9. We then set (here $a_i \in \{\text{alphabet of SAr}\}$ and $v(a_i)$ is the number of a_i):

$$\text{number}\,(a_1 \cdots a_k) = n(a_1 \cdots a_k) = \sum_{i=1}^{k} v(a_i) 10^{k-i} + 1.$$

In other words, we obtain the number of an expression by replacing all of its symbols by the corresponding decimal digits ($\bar{1}$ is replaced by 9), then reading the resulting number in the decimal system and adding 1. It is clear that an expression can be reconstructed in a unique way if we know its number.

The name in SAr of the number of an expression P, i.e., $\bar{1} \cdots \bar{1}$($n(P)$ times), is called the *label* of P. As in SELF, we shall denote the label of P by $*\, P\, *$ (but now this is abbreviated notation). We call the expression $P * P *$ the *display* of P.

11.2. **Definition.** Let $P(x)$ be a formula in SAr with one free variable x.

(a) An expression Q satisfies P if the number of Q lies in the set $\{k | P(\bar{k}) \text{ is true}\}$.

(b) An expression Q is displayed in P if the display of Q satisfies P.

11.3. **Lemma.** *Let $P(x)$ be as in 11.2. Let $P_E(x)$ denote the formula $P((x)\cdot((\overline{10})\uparrow(x)))$ (i.e., the term "$x10^x$" is substituted in place of all free occurrences of x). Then the set of expressions satisfying P_E coincides with the set of expressions displayed in P.*

PROOF. If Q has number k, then the display of Q has number $k \cdot 10^k$ (which is why $\bar{1}$ has number nine!):

$$n(Q * Q *) = \underbrace{n(Q \;\; 1 \cdots 1 \;\;)}_{n(Q) \text{ times}}$$

$$= (n(Q)-1)10^{n(Q)} + \underbrace{9 \cdots 9}_{n(Q) \text{ times}} + 1 = n(Q)10^{n(Q)}.$$

Hence, $n(Q)$ satisfies P_E if and only if $n(Q * Q *)$ satisfies P. □

11.4. **Theorem.** *For any formula $P(x)$ as in 11.2, we have:*

$$\textit{the set of formulas satisfying } P \neq \begin{cases} \textit{the set of true formulas} \\ \textit{the set of false formulas}. \end{cases}$$

PROOF. We consider the Tarski–Smullyan formula $S : xP_E * xP_E *$. According to the definitions, we have (recall that xP_E is a class term and $* \, xP_E \, *$ is the name of a number): S is true $\Leftrightarrow xP_E$ satisfies $P_E \Leftrightarrow xP_E$ is displayed in P (by Lemma 11.3) $\Leftrightarrow$ the display of xP_E satisfies $P \Leftrightarrow S$ satisfies P. Hence, S is either not false and satisfies P, or else is false and does not satisfy P. Therefore, the set of formulas satisfying P cannot coincide with the set of false formulas. As in §9, the formula S says, "I satisfy P."

Similarly, the formula

$$x((P)\downarrow(P))_E * x((P)\downarrow(P))_E *$$

says, "I do not satisfy P," and thus either satisfies P or is true, but not both. The theorem is proved. □

11.5. Of course, Lemma 11.3 is pure magic. The decimal system really has nothing to do with all this, and $\bar{1}$ did not really have to be number nine, but this way everything is much prettier.

More generally, let ?Ar be any language of arithmetic with a finite alphabet containing the alphabet of SAr. Let the rules for forming distinguished expressions and the standard interpretation of formulas in ?Ar be an arbitrary extension of the rules in SAr. We only require that the terms and formulas in SAr keep their earlier meaning, and that, for any formula $P(x)$ in ?Ar with a free variable x, the expression $x(P(x))\bar{k}$ must be a formula in ?Ar and be interpreted by the same recipe as in SAr. (For example, we might add to SAr the + sign, the connectives, and the quantifiers, and then allow formulas to be constructed by the rules of $\mathcal{L}_1$ as well, thereby imbedding L_1Ar in ?Ar.)

Then the Undefinability theorem 11.4 *holds for* ?Ar.

We must choose the numbering as follows: if m is the number of elements in the alphabet of ?Ar and ν is a numbering of the symbols for which $\nu(\bar{1}) = m$, then

$$n(a_1 \cdots a_k) = \sum_{i=1}^{k} \nu(a_i)(m+1)^{k-i} + 1.$$

Then, using the same conventions as before, we have

$$\begin{aligned} n(Q * Q *) &= n(Q\ \underbrace{\bar{1} \cdots \bar{1}}_{n(Q)\text{ times}}\ ) \\ &= (n(Q) - 1)(m+1)^{n(Q)} + m \sum_{j=0}^{n(Q)-1} (m+1)^j + 1 \\ &= n(Q)(m+1)^{n(Q)}. \end{aligned}$$

Defining $P_E(x)$ as $P((x)\cdot((\overline{m+1})\uparrow(x)))$, without any further alterations we obtain Lemma 11.3 and Tarski's theorem for ?Ar.

11.6. *Remarks*

(a) If Tarski's theorem were not true, and there were a formula $P(x)$ such that $\{Q \mid Q$ is a formula and $P(\overline{n(Q)})$ is true$\}$ coincided with the set of all true formulas of arithmetic, then this would mean that all number theoretic questions would reduce to a series of problems all of the same type. Instead of asking, "Is assertion number n true?" we could ask, "Is

$P(\bar{n})$ true?" Although such an all-encompassing problem could still be rather complicated (in a certain sense even "infinitely complicated," see Part III), Tarski's theorem says that arithmetic has much more diversity than could be contained in any such single problem.

(b) We still have reason to suspect that perhaps everything worked out this way because we could "cleverly" number the formulas. This is not the case; the results in Part III will imply that Tarski's theorem remains true for any numbering in which a formula and its number can be effectively reconstructed from one another.

(c) It is natural to ask whether the set of numbers of *provable*, or *deducible*, formulas is definable (for some set of axioms and rules of deduction, for example in SAr). The answer is *yes* this set is *definable*. We shall give some intuitive considerations in this direction, which anticipate the systematic theory in Part III.

However we define the notion of provability, it is natural to expect it to have the following property: *there exists an algorithm (for example, a computer program) which for any text of the given language determines whether this text is a proof and, if so, of what formula.*

We now write a program which constructs the texts in the language in lexicographical order, verifies whether each one is a proof, and, when it is, computes the number of the formula it proves. Roughly speaking, the graph of the function (number of a proof)$\mapsto$ (number of the formula proved) is definable in L_1Ar because machine logic and arithmetic are imbedded in L_1Ar. Hence, the set of numbers of provable formulas is definable in L_1Ar, in SAr, or in any language ?Ar as in 11.5.

Combining this discussion with Tarski's theorem, we obtain the following form of Gödel's theorem:

11.7. **Gödel's Incompleteness Theorem for Arithmetic.** *In any language of arithmetic of type* ?Ar, *and for any definition of deducibility in which the set of (numbers of) deducible formulas is definable,*

$$\{\textit{true formulas}\} \neq \{\textit{deducible formulas}\}.$$

In Part III we discuss more general formulations of this theorem and other versions of the proof, and we give a detailed verification of the principle in 11.6(c) for deductions in L_1Ar.

Digression: self-reference

In natural languages it is only recently that linguists have taken note of the so-called "performative" statements. The characteristic feature of such a statement is *self-reference*, which can be defined as the ability to "refer to a reality that it creates itself, because it is stated under circumstances which make it into an act" (E. Benveniste, La Philosophie analytique et le langage, *Les Et. Philos.*, No. 1 (1963) 9). Examples of performative statements include: "I solemnly swear," the saying of which constitutes the act of swearing; "I proclaim a general mobilization," and "I appoint you director," when these two statements come from an authority that has the power to carry out the respective acts. If we look carefully at the semantics of performative statements, we find an imperative nuance, even though it is expressed by the declarative mood of the verb.

In this connection, it is interesting to compare the role of self-reference in formal and algorithmic languages (see also subsection 1.2 of Chapter I). In formal languages (and, in general, in descriptive languages), self-reference leads to logical circles, to paradoxes, or, if we try to avoid logical circles, to demonstrations of certain inadequacies of the language. On the other hand, in algorithmic languages (and, in general, in control languages and systems), self-reference is the most important device for turning a finite program into a process that is potentially arbitrarily long ("loops"); it takes part in the control instructions (feedback), and is among the fundamental possibilities of the system.

A similar dichotomy can also be found in psychological behavior—compare with the distinction between introspection and self-improvement.

Finally, self-reference can play a role in the genetic causality of aging processes (of biological and social systems). A self-regenerating cycle, when repeated many times, leads to erosion at the place of generation.

12 Quantum logic

12.1. The last section of this chapter is devoted to certain physical facts and to the mathematical constructions which have been developed to describe them. In particular, we discuss von Neumann's theorem that it is impossible to introduce hidden variables into the quantum mechanical picture of the world. This material, while not completely traditional for a course in logic, is relevant here for two reasons.

In the first place, von Neumann's theorem is a vivid example of a metaphysical assertion. It is concerned with properties of the language, rather than with the subatomic world described by the language, and thus is analogous to, for example, Tarski's theorem in metamathematics. This is why it occupies an isolated position in physics, and why we are interested in it here.

In the second place, analyzing quantum mechanical phenomena reveals a profound divergence between the internal logical structures of the macroworld and the microworld. Although explanations of these differences by means of natural language and natural logic are agonizingly difficult and, in the last analysis, always leave one feeling unsatisfied, these attempts to explain continue. The development of the foundations of physics in the twentieth century has taught us a serious lesson. Creating and understanding these foundations turned out to have very little to do with the epistemological abstractions which were of such importance to the twentieth century critics of the foundations of mathematics: finiteness, consistency, constructibility, and, in general, the Cartesian notion of intuitive clarity. Instead, completely unforeseen principles moved into the spotlight: complementarity, and a nonclassical, probabilistic truth function. The electron is infinite, capricious, and free, and does not at all share our love for algorithms.

The following exposition is based on the article by S. Kochen and E. P. Specker in *J. Math. Mech.*, vol. 17, no. 1 (1967), 59–87. Subsections 12.9–12.16 contain pure algebra and formally do not depend on the preceding semi-physical considerations.

12.2. *The atom of orthohelium.* We now describe certain characteristics of the behavior of the physical system "an atom of orthohelium in the state $n = 2$, $l = 0$, $s = 1$." Such a helium atom is in an excited state: its two electrons are on the second energy level, and their spin is pointed in the same direction. Nevertheless, the state is meta-stable, because, in order to fall to the first energy level, the electrons must turn their spins in opposite directions (parahelium); this creates a certain stability.

Spin is a physical quantity which is expressed in the same units as the "angular momentum." The total spin of our system (in atomic units: $h = 2\pi$) is represented by a unit vector in physical three-dimensional space. As a first approximation we may think of it as changing with time but having instantaneous values which can be measured. (The inadequacy of this picture will soon be demonstrated.)

An experiment for the purpose of measuring the instantaneous value of the spin of our system could consist of turning on a magnetic field having a specified geometry and registering the shift in energy levels (spectral lines) of the atom. Each outcome of such an experiment can be precisely interpreted as a measurement of the projection of the spin on some axis, which is uniquely determined by the geometry of the field. We shall identify these directions with points of the unit sphere S^2.

Quantum mechanics makes the following positive assertions concerning measurements of the spin of orthohelium. The following quantities are measurable:

(a) the projection $s(\alpha, t)$ of the spin in the direction $\alpha \in S^2$ at the moment of time t;

(b) the lengths $|s|(\alpha_i, t)$, $i = 1, 2, 3$, of three projections of the spin in three pairwise orthogonal directions $\{\alpha_1, \alpha_2, \alpha_3\} \subset S^2$ (a "frame") at the time t. The predictions concerning the results of these measurements are as follows:

(c) $s(\alpha, t)$ is a random variable which can only take the values $-1, 0, 1$. (The probabilities of these values can be predicted from the results of the previous measurements, but this is not essential for us here.)

(d) $\sum_{i=1}^{3} |s|(\alpha_i, t) = 2$ for any frame $\{\alpha_1, \alpha_2, \alpha_3\}$ and any t.

12.3. *Attempt at a classical interpretation.* This could consist in adopting the following hypotheses A and B:

A. There is a certain space Ω of "hidden variables" or "internal states" of the system and a function $s(\alpha, t; \omega)$, $\omega \in \Omega$, such that, if the system is in the state ω at time t, then $s(\alpha, t; \omega)$ is the "true value of the projection of the spin on the α-axis" at this moment.

B. The probabilistic aspect of the predictions in 12.2(c) results from our not knowing the exact values of $\omega = \omega(t)$, so that for some measure $d\mu(\omega)$ we have

$$\text{mathematical expectation of } s(\alpha, t) = \int_{\Omega} s(\alpha, t; \omega)\, d\mu(\omega),$$

and similarly for $|s|$.

Generalizing, we might suppose that Ω does not only depend on the system itself but also on the arrangement for measuring the spin; μ may depend on the time, and so on. However, all of these possibilities actually contradict the predictions in 12.2(c) for the following startling reason.

12.4. **Proposition** (Kochen, Specker). *There does not exist a mapping $S^2 \to \{0, 1\}$ such that for every frame $\{\alpha_1, \alpha_2, \alpha_3\}$ this mapping takes the value zero on precisely one of the directions α_i. Moreover, it is possible to construct a finite system $\Gamma \subset S^2$ of 117 points with the following property. For any mapping $k : \Gamma \to \{0, 1\}$ either there is a frame $\{\alpha_1, \alpha_2, \alpha_3\} \subset \Gamma$ on which k does take the value 0 exactly once, or else there is a pair of perpendicular directions $\{\alpha_1, \alpha_2\} \subset \Gamma$ on which k equals 0.*

Here we note that adopting both the assertions in 12.2 and the hypotheses in 12.3 would allow us to construct such a mapping of the sphere. In fact, it would be sufficient to consider

$$S^2 \to \{0, 1\}: \quad \alpha \mapsto |s|(\alpha, t; \omega)$$

for fixed t and ω. By 12(c), $|s|$ only takes the values 0 and 1, and, by 12(d), it takes the value 1 twice and 0 once on any frame $\{\alpha_1, \alpha_2, \alpha_3\}$.

We prove Proposition 12.4 in subsections 12.12–12.15, and now proceed to a more systematic study of "quantum logic." We shall adhere to our customary and useful dualism between "language and interpretation," although these categories are much less formalized and are harder to distinguish from each other in physics.

12.5. *The language of nonrelativistic quantum mechanics.* We have a somewhat unusual situation in that quantum mechanics does not really have its own language. More precisely, to describe a physical system S such as a "free electron," "atom of helium in a magnetic field," etc., quantum mechanics uses a certain fragment of the language of functional analysis, "oriented on describing S." Assuming that the reader is familiar with functional analysis, we shall limit ourselves to a glossary of the most frequently used terms. We also give some synonyms used by physicists to indicate the "physical sense," i.e., the interpretation, which will be considered separately in our text.

(a) *A separable complex Hilbert space* $\mathcal{H}_S$. Here we are also interested in its one-dimensional subspaces and its vectors of length one. A synonym for the former is the (pure) states, and for the latter is the (normalized) ψ-functions, or, more precisely, the instantaneous values of the ψ-functions.

(b) *Unitary representations of* $\mathbf{R}$ *in* $\mathcal{H}_S$: $t \mapsto U_t = e^{-iH_S t}$. For synonyms we have $t \mapsto U_t$ is the dynamic group; t is the time; and the infinitesimal generator H_S (which is a self-adjoint operator) is the dynamic operator, or Hamiltonian, of S.

(c) *Schrödinger equation*: $\partial\psi_t/\partial t = -iH_S\psi_t$. It is satisfied by the ψ-functions $\psi_t = e^{-H_S t}$, which evolve with time.

(d) *Self-adjoint operators in* $\mathcal{H}_S$. Synonym: the observables of the system. The operator H_S is an energy observable. The discrete spectrum of H_S gives us the energy levels of S. We shall be especially interested in the orthogonal projection observables. Here the pure states $\mathbf{C}_\psi \subset \mathcal{H}_S$ are in one-to-one correspondence with the projections P_ψ onto the corresponding subspace.

Another important class of projections is constructed using the spectral decomposition theorem. Let $A = \int_{-\infty}^{\infty} \lambda \, dP_A(\lambda)$. Then the projection $P_A(U)$ is defined for any Borel subset $U \subset \mathbf{R}$. In the simplest cases its image is spanned by the vectors in $\mathcal{H}_S$ which are eigenvectors for A with eigenvalues in U.

Projection observables are also called "questions" (Mackey) or "Eigenschaften" (von Neumann).

(e) *Commuting operators*. Synonym: compatible (or simultaneously measurable) observables. For unbounded operators A and B, whose formal commutator may have an empty domain of definition, we define commutativity to mean that $P_A(U_1)$ and $P_B(U_2)$ commute for all Borel sets $U_1, U_2 \subset \mathbf{R}$.

(f) *Unitary representations in* $\mathcal{H}_S$ *of various groups, such as* $SO(3)$, $SU(2)$, S_n, *etc*. Synonym: symmetries of the system S (if the representations commute with the Hamiltonian H_S), or approximate symmetries (if $H_S = H_0 + H_1$, where the representations commute with H_0, and H_1 is a "small perturbation").

12.7. Example. Let S be "an electron in the electric field of a proton" (where we disregard the motion of the proton, the spin, and the relativistic effects). Here:

$\mathcal{H}_S = L^2(E^3)$ consists of the square integrable complex functions in the Euclidean "physical coordinate space of the electron."

H_S is the self-adjoint extension of the operator

$$-\frac{h}{4\pi m}\Delta - \frac{1}{h}\frac{e^2}{r},$$

where h is Planck's constant, m is the mass of the electron, e is its charge, and r is its distance from the origin (where the proton is).

The energy levels (the discrete spectrum of H_S) are: $E_n = -(2\pi^2 me^4/h^2)/(1/n^2)$, $n = 1, 2, 3, \ldots$. The eigenfunctions ψ corresponding to the points of this spectrum are the states of an electron in a hydrogen atom. The energy level $n = 1$ corresponds to the unexcited state, and the other values of n correspond to excited states. The positive semiaxis is the continuous spectrum of H_S; in states with positive electron energy, "the hydrogen atom is ionized."

The most important observables of the electron are: the operators of multiplication by the three coordinate functions x_j (the coordinate observables), and the self-adjoint extension of the operators $p_j = (h/2\pi i)(\partial/\partial x_j)$ (the momentum projection observables). The operators x_j and p_j do not commute, so that the x_j-coordinate and the projection of the momentum on the x_j-axis are not simultaneously measurable.

The system S is spherically symmetric. The natural representation of SO(3) in $L^2(E^3)$ commutes with H_S. The restriction of this representation to the subspace of $\mathcal{H}_S$ corresponding to the discrete spectrum of H_S in a natural way splits into a direct sum of representations corresponding to a given energy level E_n. This E_n-subspace, in turn, splits into a direct sum of representations of SO(3) on spherical polynomials of degree $j = 0, 1, 2, \ldots, n-1$ with multiplicity one. If the ψ-function of the electron belongs to the level E_n and the subspace corresponding to the representation of SO(3) on spherical polynomials of degree j, we say that n and j are the principal and orbital quantum numbers, respectively, of the electron's state in the hydrogen atom.

The above text is typical of what might be found in a physics textbook. The "language" is mixed with the "metalanguage" which gives the standard interpretation of the language. We now describe them separately and more systematically.

12.8. *The interpretation.* A very important aspect of the interpretation which we shall not discuss here is the list of informal recipes for choosing $\mathcal{H}_S$, H_S, and the observables corresponding to a given system S. These "units of expression" are often chosen in two stages: a classical description is chosen, and then the "rules of quantization" are applied to it. This procedure might be "approximate" in the sense that certain circumstances are not taken into account (such as the spin in 12.7).

Suppose that $\mathcal{H}_S$ and H_S have already been chosen. The most characteristic peculiarity of the interpretation of quantum language is that it is "two-layered." Part of the mathematical statements are interpreted as assertions about a "freely evolving system," and part are interpreted as assertions about the results of observations on this system.

(a) *Freely evolving system.* It is generally believed that the system's ψ-function $\psi_t \in \mathcal{H}_S$ gives (within the framework of a given approximation) maximally complete information about the state of the system at time t. As long as no one looks in on the system, ψ_t evolves as $e^{-iH_S t}\psi_0$, starting from the initial state ψ_0. (How do we know ψ_0? See subsection 12.8(c) below.)

(b) *Observation.* Suppose we want to measure the instantaneous value of some physical quantity for our system S at the moment t. This quantity corresponds to an observable A. (How do we know the form of A? See the beginning of 12.8.) For simplicity we suppose that A has a discrete spectrum with all multiplicities one. The predictions of what will be observed are as follows.

If $A\psi_t = a\psi_t$, then a will be the value of the observable A at the time t for the system S in the state with ψ-function ψ_t.

In the general case, let $\psi_A^{(i)}$, $i = 1, 2, \ldots,$ be an orthonormal basis for $\mathcal{H}_S$ consisting of eigenvectors for A. We expand ψ_t with respect to this basis: $\psi_t = \sum_{i=1}^{\infty} \alpha^{(i)}(t)\psi_A^{(i)}$. Let $A\psi_A^{(i)} = a_i\psi_A^{(i)}$. Then the result of measuring A will be a random variable taking the value a_i with probability $|\alpha^{(i)}(t)|^2$. (It is easy to see that the mathematical expectation of this random variable is $(A\psi_t, \psi_t)$. This formula holds for all A. More generally, the probability of A falling in a Borel subset $U \subset \mathbf{R}$ is equal to $(P_A(U)\psi_t, \psi_t)$, where $P_A(U)$ was defined in 12.5(d).)

(c) *System evolving after observation.* With the same assumptions as before, the ψ-function of the system after the observation is determined by the result of the observation. If we registered the value a_i for A at the time t_0, then, starting from $\psi_A^{(i)}$ at t_0, S evolves until the next observation completely independently of how it evolved before.

Thus, the result of the observation lets us know the form of the ψ-function *after* the observation, but it tells us nothing about the ψ-function *before* the observation. Hence, physicists often say that registering the value $\psi_A^{(i)}$ *prepares* the system in the state $\psi_A^{(i)}$ at the time t_0. Another synonym: at the moment of observation the ψ-function of the system *reduces* to $\psi_A^{(i)}$.

If we were able simultaneously to register the values of two observables, then we would prepare the system with a ψ-function which is an eigenfunction for both observables. Since noncommuting observables always have different eigenvectors, in general the values of such variables are not simultaneously measurable.

12.9. *Quantum logic.* We now investigate the algebraic framework of quantum logic. We start with the following analogous situation.

Suppose we are given a formal language in $\mathfrak{L}_1$ having one variable and an interpretation of this language in a set M where this variable takes values. Then we can distinguish the Boolean algebra B of definable sets in

M (see §3). The conjunction of formulas corresponds to the Boolean intersection of the sets that define them, and so on. By definition, $N \in B$ if we can ask in the language, "Does the value of the variable belong to N?". The algebra B is the most important invariant of the pair {language, interpretation}.

We now consider the language of quantum mechanics, oriented on describing a system S. We shall exclude the time aspect by *fixing* a moment of time to which all statements about the state of the system refer. Then the "state of the system" will be the only variable in the language. It takes values in the set of lines in the Hilbert space $\mathcal{H}_S$. The only questions to which we can give a yes or no answer are those of the form: "Does the state of the system belong to a given closed subspace of $\mathcal{H}_S$?". It is the closed subspaces of $\mathcal{H}_S$ which form the analogy of the Boolean algebra B. The conjunction of questions corresponds to the intersection of subspaces and the disjunction corresponds to their sum, but both operations can only be performed when the corresponding projection observables commute. Only in this case are the Boolean identities fulfilled.

We axiomatize the situation as follows:

12.10. **Definition.** A partial Boolean algebra is a set B together with the following structures on B:

(a) A reflexive and symmetric binary relation $*$ called "compatible measurability." Instead of $(a, b) \in *$ we write $a * b$.
(b) Partial binary operations $\vee$ and $\wedge$ and a unary operation $'$.
(c) Two elements 0 and $1 \in B$.

These structures must satisfy the following axioms:

(d) The relation $*$ is closed with respect to the operations $\wedge$, $\vee$, and $'$: if a_1, a_2, and a_3 are pairwise compatibly measurable, then $(a_1 \wedge a_2) * a_3$, $(a_1 \vee a_2) * a_3$, and $a_1' * a_3$; in addition, $a * 0$ and $a * 1$ for all $a \in B$.
(e) If a_1, a_2, and a_3 are pairwise compatibly measurable, then together with 0 and 1 they generate a Boolean algebra relative to the operations $\vee$, $\wedge$, and $'$.

12.11. Example. Let $\mathcal{H}$ be a Hilbert space (possibly real and finite dimensional). The partial Boolean algebra $B(\mathcal{H})$ is defined as the set of closed subspaces of $\mathcal{H}$ with the following structures:

(a) $a * b$ if and only if there exist three pairwise orthogonal closed subspaces $c, d, e \in \mathcal{H}$ such that $a = c \oplus d$ and $b = e \oplus d$. The motivation for this definition is that this condition is equivalent to commutativity of the projections onto a and b.
(b) $a \wedge b =$ the intersection of a and b.
(c) $a \vee b =$ the sum of a and b.

(d) a' = the orthogonal complement of a.
(e) $0 = \{0\}$ and $1 = \mathcal{H}$.

One form for the theorem that there are no hidden variables is as follows.

12.12. **Theorem.** *If* dim $\mathcal{H} \geqslant 3$, *then* $B(\mathcal{H})$ *cannot be imbedded in a Boolean algebra in such a way that the operations are preserved.*

This result can be strengthened formally in various ways: see §5 of Kochen and Specker, and also N. Fierler, M. Schlessinger, *Duke Math. J.*, vol. 32, No. 2 (1965), 251–262. We shall not dwell on this here.

Proof. We choose a real Euclidean space $E^3 \subset \mathcal{H}$ and show that even $B(E^3)$ cannot be imbedded in a Boolean algebra. Otherwise there would exist a homomorphism of the partial Boolean algebra $B(E^3)$ onto the two-element Boolean algebra $\{0, 1\}$, since, for any pair of elements in any Boolean algebra, there exists a homomorphism onto $\{0, 1\}$ which separates them.

Let h be such a homomorphism. If $a_1, a_2, a_3 \in E^3$ are pairwise orthogonal lines, then $h(a_i \wedge a_j) = h(a_i) \wedge h(a_j) = 0$ for $i \neq j$. Hence, in any pair of orthogonal lines, at least one of the pair must go to 0 under h. Furthermore, $h(a_1 \vee a_2 \vee a_3) = h(a_1) \vee h(a_2) \vee h(a_3) = h(E^3) = 1$. Hence, in any frame exactly one of the lines goes to 1.

If we map the points of the unit sphere S^2 onto the lines joining them to the origin and then apply h, we obtain a mapping of S^2 with the property in Proposition 12.4 (where we only have to switch the roles of 0 and 1). We prove that no such map exists even on a certain subset consisting of 117 points on S^2. The latter stronger result is combinatorially elegant and physically meaningful: a physicist might raise objections to asking to be able to measure the projection of the spin of orthohelium simultaneously in *all* directions, independently of the question of whether or not hidden variables are possible. In fact, we only need finitely many directions to show the futility of such an attempted measurement.

Consider a finite graph. By a *realization* of the graph on S^2 we mean any imbedding of the set of its vertices in S^2 for which the distance between the endpoints of any edge equals 90°.

12.13. **Lemma.** *Let* α *and* β *be points on* S^2 *such that the sine of the angle between them* $\in [0, \frac{1}{3}]$. *Then there exists a realization of the following graph* Γ_1 *in which* a_0 *goes to* α *and* a_9 *goes to* β.

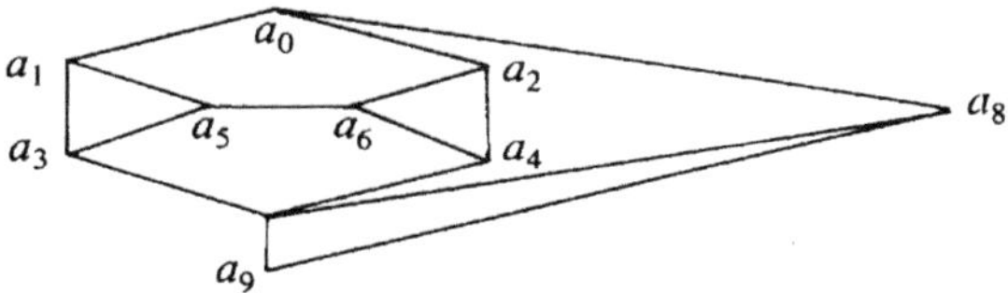

PROOF. Let $\bar{x}, \bar{y}, \bar{z}$ be an ort on S^2. We take a_5 to $\bar{x}$ and a_6 to $\bar{z}$. For certain $\xi, \eta \in \mathbf{R}$ (to be chosen later), we set

$$a_1 \mapsto \frac{\bar{y} + \xi\bar{z}}{\sqrt{1+\xi^2}}, \qquad a_2 \mapsto \frac{\bar{x} + \eta\bar{y}}{\sqrt{1+\eta^2}}.$$

Then the images of a_3 and a_4 are determined up to a sign by the property of being orthogonal to (a_1, a_5) and (a_2, a_6), and we choose

$$a_3 \mapsto \frac{\xi\bar{y} - \bar{z}}{\sqrt{1+\xi^2}}, \qquad a_4 \mapsto \frac{\eta\bar{x} - \bar{y}}{\sqrt{1+\eta^2}}.$$

We similarly set

$$a_0 \mapsto \frac{\xi\eta\bar{x} - \xi\bar{y} + \bar{z}}{\sqrt{1+\xi^2+\xi^2\eta^2}}, \qquad a_7 \mapsto \frac{\bar{x} + \eta\bar{y} + \xi\eta\bar{z}}{\sqrt{1+\eta^2+\xi^2\eta^2}},$$

and, finally, a_8 and a_9 are determined up to a sign. The sine of the angle between a_0 and a_9 is easy to compute: it equals

$$\xi\eta/\sqrt{(1+\xi^2+\xi^2\eta^2)(1+\eta^2+\xi^2\eta^2)}\ .$$

As ξ and η vary, this expression takes on all values in $[0, \frac{1}{3}]$. □

12.14. **Lemma.** *Consider the graph Γ_2 which is obtained from Figure 1 by identifying the vertices $a = p_0$, $b = q_0$, and $c = r_0$ (the apparent intersections of the edges inside the circle are not vertices). This graph is realized on S^2.*

PROOF. For $0 \leqslant k \leqslant 4$ set

$$p_k \mapsto \cos\frac{\pi k}{10}\cdot\bar{x} + \sin\frac{\pi k}{10}\cdot\bar{y},$$

$$q_k \mapsto \cos\frac{\pi k}{10}\cdot\bar{y} + \sin\frac{\pi k}{10}\cdot\bar{z},$$

$$r_k \mapsto \sin\frac{\pi k}{10}\cdot\bar{x} + \cos\frac{\pi k}{10}\cdot\bar{z}.$$

Since $\sin(\pi/10) < \frac{1}{3}$, we can first extend this map to a realization of the subgraph between the points p_0, p_1 and r_0 by using the preceding lemma. Rotating the resulting realization around r_0 so as to take (p_0, p_1) to (p_1, p_2), (p_2, p_3), . . . , we obtain a realization of the "lower arc" and r_0. By similarly rotating around the images of p_0 and q_0, we obtain a realization of the other two arcs as well. □

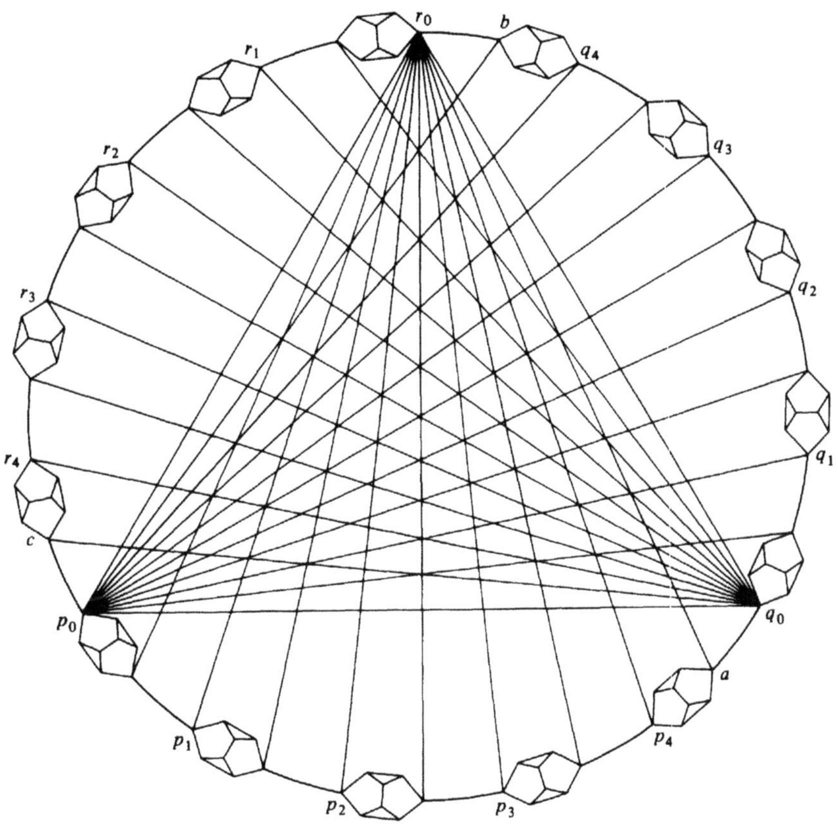

Figure 1.

12.15. End of the proof of Proposition 12.4 and Theorem 12.12. Consider an arbitrary map k of the vertices of the graph Γ_2 to $\{0, 1\}$. Suppose that exactly one vertex in each triangle goes to 1 and at least one of the two vertices on each edge goes to 0. In the triangle $\{p_0, r_0, q_0\}$ suppose that p_0 goes to 1. We consider the copy of the graph Γ_1 between the vertices p_0, r_0, and p_1, which we identify with a_0, a_8, and a_9, respectively.

We must have $k(p_1) = k(a_9) = 1$. In fact, if we had $k(a_9) = 0$, then we would also have $k(a_7) = 1$, and then $k(a_1) = k(a_2) = k(a_3) = k(a_4) = 0$, and $k(a_5) = k(a_6) = 1$, which is a contradiction.

We now return to Γ_2. Since $k(p_0) = k(p_1) = 1$ we similarly find that $k(p_2) = 1$, and then $k(p_3) = k(p_4) = k(q_0) = 1$. But $k(q_0) = 1$ contradicts the fact that $k(p_0) = 1$. This completes the proof. □

12.16. *Quantum tautologies*. This theme has been largely neglected. We give a counterexample due to Kochen and Specker and formulate some recent results of Gelfand and Ponomarev.

(a) *Counterexample.* This consists of the following: it is possible to give a logical polynomial in 117 variables which represents a classical tautology but which is defined and takes the value 0 in the partial Boolean algebra $B(E^3)$ for some values of the variables. This is simply another aspect of the impossibility of imbedding $B(E^3)$ in a Boolean algebra.

In fact, let $P(p, q, r)$ be a logical polynomial in three variables which takes the truth value 1 when exactly one of $|p|$, $|q|$, and $|r|$ is 1. We may assume that only the connectives $\vee$, $\wedge$ and $\neg$ occur in P. Similarly, let $Q(p, q) = \neg p \vee \neg q$. Then Q takes the value 1 when at least one of $|p|$, $|q|$ is 0. We index the vertices of Γ_2 from 1 to 117 and set

$$R(p_1, \ldots, p_{117}) = \neg\Big(\bigwedge_{\{i,j,k\}} P(p_i, p_j, p_k) \bigwedge_{\{r,s\}} Q(p_r, p_s)\Big).$$

The first $\bigwedge$ is taken over all triples $\{i, j, k\}$ corresponding to triangles in Γ_2, and the second $\bigwedge$ is taken over all pairs $\{r, s\}$ corresponding to edges. The argument in 12.15 shows that for any mapping $\{p_1, \ldots, p_{117}\} \to \{0, 1\}$ at least one of the Boolean factors takes the value 0. Hence R is a classical tautology.

But if we substitute for p_i the line from the origin to the image of the ith vertex in a fixed realization of Γ_2, then we obtain for the value of R the element $0 \in B(E^3)$. In fact, if p_r and p_s are orthogonal, then $p_r' \vee p_s' = E^3$. Similarly, if p_i, p_j, and p_k are orthogonal, then $P(p_i, p_j, p_k) = 1 \in B(E^3)$. The latter assertion is verified as follows: if we set

$$a + b = (a \wedge b') \vee (a' \wedge b),$$

then we may take

$$P(p, q, r) = p + q + r + p \wedge q \wedge r$$

(for any arrangement of parentheses on the right), so that

$$P(p_i, p_j, p_k) = p_i \oplus p_j \oplus p_k = E^3.$$

(b) *Results of Gelfand and Ponomarev.* We start with the following observation. The operations $\wedge$, $\vee$ and $'$ are actually defined everywhere on the set $B(\mathcal{H})$ of closed subspaces of the Hilbert space $\mathcal{H}$, although they do not satisfy the Boolean axioms, and, if we ignore the compatible measurability relation $*$, it seems as if they no longer have physical meaning.

Nevertheless, it is also natural to investigate these structures, which were first introduced into the logic of quantum mechanics by G. Birkhoff and J. von Neumann (*Annals of Math.* vol. 37 (1936), 823–843). Here is how these structures are axiomatized:

Definition. A *modular structure* L is a set with binary operations $\wedge$ and $\vee$ which satisfy the following conditions:

(a) $\wedge$ and $\vee$ are associative and commutative;

(b) $a \wedge a = a \vee a = a$ for all $a \in L$;
(c) If $a \wedge b = b$, then $(a \vee c) \wedge b = b \vee (c \wedge b)$ (the "modular identity").

Birkhoff and von Neumann also require an "orthogonal complement" operation to exist with the usual axioms, but we shall omit this here.

We note that the modular identity is only fulfilled universally in $B(\mathcal{H})$ if $\mathcal{H}$ is finite dimensional. It is also fulfilled for triples a, b, c whose elements have finite dimension or codimension in $\mathcal{H}$.

I. M. Gelfand and V. A. Ponomarev (*Uspehi mat. nauk*, vol. XXIX (1974), No. 6 (180), 3–58) have studied the linear representations of free modular structures with r generators in $B(\mathcal{H})$ for finite dimensional spaces over arbitrary fields. Such a representation is called indecomposable if it does not split into a direct sum of representations in $B(\mathcal{H}_1) \oplus B(\mathcal{H}_2)$.

Definition. A modular question is an element of a free modular structure which takes the value 0 or 1 for any indecomposable finite dimensional representation.

One of the main results of Gelfand and Ponomarev is the construction of a very nontrivial countable series of modular questions. We shall only formulate these results here.

Let L^n be a free modular structure with n generators $\{a_1, \ldots, a_n\}$. We set $I = \{1, \ldots, n\}$. A sequence $\alpha = (i_1, \ldots, i_l)$ of length $l \geqslant 1$ of elements of I is called *admissible* if it does not have any identical neighboring entries. A sequence $\beta = (k_1, \ldots, k_{l-1})$ of length $l - 1$ of elements of I is called subordinate to α if it is admissible and if $\forall_j \leqslant l - 1$, $k_j \notin \{i_j, i_{j+1}\}$. For admissible α we inductively define

$$a_\alpha = a_{i_1 \ldots i_l} = a_{i_1} \wedge (\vee_\beta a_\beta),$$

where β runs through all sequences subordinate to α. Further, for $t \in \{1, \ldots, n\}$ we define

$$A_t(l) = \bigvee_\alpha a_\alpha,$$

where α runs through all admissible sequences of length l with last entry t. Finally, we set

$$H_t(l) = \bigvee_{j \neq t} A_j(l).$$

The substructure in L^n *generated by the elements* $H_1(l), \ldots, H_n(l)$ *consists entirely of modular questions* for all $l \geqslant 1$.

This is a difficult result. It is relatively easy to prove that this substructure is a Boolean algebra consisting of 2^n elements. If we substitute the elements in this Boolean algebra for the variables in the usual Boolean tautologies, we obtain "quantum tautologies," but to see this we must consider structures with complements.

It is not yet clear whether this algebra leads to nontrivial physics. Perhaps one should combine it with the techniques in the representation theory of symmetry groups.

12.17. *The orthohelium atom revisited.* In conclusion, we return to the orthohelium atom S and show how the material in 12.2 looks from a more general vantage point.

(a) *Choice of* $\mathcal{H}_S$. As explained in 12.7, an electron without spin corresponds to the space $L^2(E^3)$. If we want to take the spin into account, we must introduce a "two-component" ψ-function, i.e., use the space $L^2(E^3) \otimes \mathbf{C}^2$. The system of two electrons in helium is described by ψ-functions in the tensor square of this space. However, by Pauli's principle, the ψ-function of this system must behave antisymmetrically when the electrons corresponding to the two parts of the tensor square are permuted. Hence, we finally obtain $\mathcal{H}_S = \Lambda^2(L^2(E^3) \otimes \mathbf{C}^2)$.

(b) *Choice of* H_S. This is a difficult problem, because each electron moves in the variable electromagnetic field created by the nucleus and the other electron. The principal term in the Hamiltonian corresponds to the spherically symmetric constant potential obtained by averaging over time. The remainder is treated as a small perturbation. We give the approximate form of the ψ-function of orthohelium, more precisely, of the element in $\Lambda^2(L^2(E^3))$ corresponding to the projection of $\mathcal{H}_S$ onto the subspace of the unit projection of the spin:

$$\psi \approx e^{-k(r_1+r_2)}\big[(C_1 + C_2(r_1 + r_2) + C_4 r_{12} + C_5 r_{12}(r_1 + r_2)\sinh C_0(r_1 - r_2)$$
$$+ (r_1 - r_2)(C_3 + C_0 r_{12})\cosh C_0(r_1 - r_2))\big]$$

where $r_i = (\Sigma_{j=1}^3 x_{ij}^2)^{1/2}$, $i = 1, 2$; $r_{12} = (\Sigma_{j=1}^3 (x_{1j} - x_{2j})^2)^{1/2}$, and the constants $k, C_1, \ldots, C_6$ are found experimentally. (E. U. Condon and G. H. Shortley, *The Theory of Atomic Spectra*, Cambridge University Press, London, 1935.)

(c) *Approximate symmetries*. The group SU(2) acts on the space $\mathcal{H}_S$: on $L^2(E^3)$ through the quotient group SO(3), and on $\mathbf{C}^2$ by the standard representation. This is the group of approximate symmetries of the system. The ψ-function of orthohelium is "not too far" from the subspace corresponding to a suitable representation of SU(2), so we may speak of the principal (n), orbital (j), and other quantum numbers of the state, as in the case of a hydrogen atom.

(d) *Spin*. The total angular momentum operator $\mathcal{J}$ commutes with the Hamiltonian H_S. In the state $n = 2$ and $j = 1$, its eigenvalue is 2 (in atomic units). The eigensubspace $N \subset \mathcal{H}_S$ corresponding to this eigenvalue is three-dimensional. Further, the squared spin projection operators $\mathcal{J}_x^2$, $\mathcal{J}_y^2$, $\mathcal{J}_z^2$ commute in pairs (this is a peculiarity of spin 1). Letting P denote the projection of $\mathcal{H}_S$ onto N, we are then able to imbed the partial Boolean algebra $B(E^3)$ in $B(\mathcal{H}_S)$ by letting a line $\alpha \subset E^3$ correspond to the image in $\mathcal{H}_S$ of the operator $P\mathcal{J}_\alpha^2$. This takes the place of the somewhat naive picture in 12.2.

Appendix: The von Neumann universe

1. The premises of "naive" Cantorian set theory reduce to the following: a set may consist of any distinguishable elements (of the physical or intellectual world); a set is uniquely determined by its elements, and any property determines a set, namely, the set of objects which have this property.

However, the formal language of set theory $\mathrm{L_1Set}$ was introduced in order to describe a more restricted class of sets (a *universe*). Part of these restrictions come from considerations of convenience, and part come from the desire to avoid the so-called paradoxes. This gives an "upper bound" for our classes. We give a "lower bound" by asking that the class of sets be closed with respect to all mathematical constructions needed for certain (ideally, "all") parts of intuitive mathematics.

2. Following Zermelo, von Neumann, and others, we consider two basic restrictions on sets.

(a) All elements of sets must themselves be sets. In particular, since any chain $X_0 \in X_1 \in X_2 \in \cdots$ in the von Neumann universe V must terminate (see below), it follows that the last element in such a chain must be the empty set. Thus, all the sets in V are constructed "from nothing."

(b) The assumption that every collection of sets, even sets as in (a), is again a set in V, immediately leads to contradictions (Burali–Forti, Russell, and others). In particular, the collection of all sets in the universe is not itself an element of V. Hence, we must give a sufficiently complete description of which operations do not take us outside of V. The two basic formal languages of set theory—that of Gödel–Bernays and that of Zermelo–Fraenkel—differ in the choice of objects over which the variable symbols are to range under the standard interpretation of the language in V. In the Zermelo–Fraenkel language (our $\mathrm{L_1Set}$), they range over the sets in V. In the Gödel–Bernays language, they name classes (collections of sets in V) which "are not necessarily sets," and the property of "being a set" is specially defined as the property of "being an element of another class." The Gödel–Bernays language is studied in Chapter 4 of Mendelson's book.

In this section we describe the von Neumann universe using the customary terminology of intuitive mathematics. The relationship of this construction to formalism will be discussed in subsection 18.

3. *The first levels.* The von Neumann universe is constructed inductively, starting from the empty set, by successively applying the "set of all

subsets" or "power set" operation $\mathcal{P}$. In this way:

$$V_0 = \varnothing,$$
$$V_1 = \mathcal{P}(\varnothing) = \{\varnothing\},$$
$$V_2 = \mathcal{P}(V_1) = \{\varnothing, \{\varnothing\}\},$$
$$\vdots$$
$$V_{n+1} = \mathcal{P}(V_n),$$
$$\vdots \; .$$

It is easy to see that $V_n \subset V_{n+1}$ (later this will be proved in complete generality). The level V_n consists of

$$2^{2^{\cdot^{\cdot^{2}}}} \quad (n-1 \text{ twos})$$

finite sets, whose elements are also finite sets, and so on.

We cannot go beyond finite sets unless we regard all the V_n as "already constructed" and apply $\mathcal{P}$ to the union of the V_n. We set

$$V_{\omega_0} = \bigcup_{n=0}^{\infty} V_n,$$
$$V_{\omega_0+1} = \mathcal{P}(V_{\omega_0})$$
$$\vdots$$

The indices which we now use for the levels are the names of the first infinite ordinals. This remarkable idea of transfinite iteration of such constructions is due to Cantor, who first applied it to study trigonometric series, and then investigated it systematically, finding in it the key to the infinite.

In the next two subsections our sets will temporarily be Cantorian sets. We shall return to V after developing some properties of ordinals.

4. *Ordinals*. Let X be any set on which we are given a binary relation $<$. We consider the following properties of this relation:

(a) *$Y \nless Y$, for all $Y \in X$; if $Y_1 < Y_2$ and $Y_2 < Y_3$, then $Y_1 < Y_3$.*
(b) *For any $Y, Z \in X$, either $Y < Z$ or $Z < Y$, or else $Y = Z$.*
(c) *Every nonempty subset of X has a least element (in the sense of $<$).*

The relation $<$ is a *partial ordering* of X if it satisfies (a), a *linear ordering* of X if it satisfies (a) and (b), and a *well-ordering* of X if it satisfies all three conditions (a), (b), and (c).

Let $(X, <)$ be a well-ordering. The *initial segment* $\hat{Y}$ determined by an element $Y \in X$ is the well-ordered set $(Z, <)$, where $Z = \{Y' \mid Y' < Y\}$. As is customary when speaking about a well-ordered set, we shall omit the explicit indication of the ordering if it is clear from the context.

5. **Lemma.** *Let X and Y be two well-ordered sets. Then exactly one of the following alternatives holds:*

(a) *X and Y are isomorphic.*
(b) *X is isomorphic to an initial segment in Y.*
(c) *Y is isomorphic to an initial segment in X.*

In each case the isomorphism is uniquely determined.

PROOF. We divide the argument into several steps.

(a) Let X be well-ordered, and let $f: X \to X$ be a monotonic map, i.e., $Z_1 < Z_2 \Rightarrow f(Z_1) < f(Z_2)$. Then for all $Z \in X$ we have $f(Z) \geqslant Z$. In fact, among the elements not having this property there would have to be a least element Z_0. But $f(Z_0) < Z_0$ and the monotonicity of f imply that $f(f(Z_0)) < f(Z_0)$, so that we would have an even smaller element in the set of elements not having the desired property.

(b) Therefore X is not isomorphic to any of its initial segments $\hat{X}_1$: if $f: X \stackrel{\sim}{\Rightarrow} \hat{X}_1$, then $f(X_1) < X_1$.

(c) Now let X and Y be well-ordered. We set $f = \{\langle X_1, Y_1 \rangle | X_1 \in X, Y_1 \in Y$, and there exists an isomorphism of $\hat{X}_1$ with $\hat{Y}_1\}$. First of all, f is the graph of a one-to-one mapping of $\mathrm{pr}_1 f$ onto $\mathrm{pr}_2 f$. In fact, if $X_1 \neq X_2$, say $X_1 < X_2$, then by (b) $\hat{X}_1$ is not isomorphic to $\hat{X}_2$; by symmetry, the same holds for f^{-1}. It is also clear from this that f and f^{-1} are monotonic. Further, if $X_1 \in \mathrm{pr}_1 f$ and $X_2 < X_1$, then $X_2 \in \mathrm{pr}_1 f$, and similarly for $\mathrm{pr}_2 f$. Finally, we show that either $\mathrm{pr}_1 f = X$, or else $\mathrm{pr}_2 f = Y$. Otherwise, there would exist a minimal element X_1 in $X \setminus \mathrm{pr}_1 f$ and a minimal element Y_1 in $Y \setminus \mathrm{pr}_2 f$. But, by the preceding paragraph, f induces an isomorphism of $\hat{X}_1$ with $\hat{Y}_1$. By the definition of f, we then have $\langle X_1, Y_1 \rangle \in f$, a contradiction.

(d) All of this means that either f is an isomorphism (more precisely, the graph of an isomorphism) of the set X onto Y or an initial segment in Y, or else f^{-1} is an isomorphism of Y onto X or an initial segment of X. It is clear from the definition of f that the graph of any other isomorphism must be contained in the graph of f, so we have uniqueness. The lemma is proved. □

As a preliminary definition, we can now consider the class of all well-ordered sets isomorphic to some fixed totally ordered set X, and call that class an ordinal. Two ordinals α and β satisfy the relation $\alpha = \beta$, $\alpha < \beta$, or $\alpha > \beta$ depending on which of the alternatives in Lemma 5 holds for representatives $X \in \alpha$ and $Y \in \beta$ (this obviously does not depend on the choice of representatives).

The next step is, naturally, to consider "all" ordinals as a class and show that $<$ induces a well-ordering on this class, thereby giving a universal well-ordering. However, an unnecessary difficulty arises here: the class of well-ordered sets isomorphic to a fixed X is extremely large, and so the class of ordinals must be a "class of classes," which needlessly complicates matters. An elegant technical discovery, due to von Neumann, removes this difficulty: instead of a vast number of possible orderings imposed on X from outside, we consider a single relation given by internal properties. Recall that a set X is transitive if $Z \in X$ whenever $Z \in Y \in X$ for some Y.

6. **Definition.** An ordinal is a transitive set X of sets which is well-ordered by the relation $\in$ between its elements.

7. **Theorem.**

(a) *The class of ordinals* On *is well-ordered by the relation* $\alpha \in \beta$ (*which we shall also write* $\alpha < \beta$).

(b) *Any well-ordered set is isomorphic to a unique ordinal* α, *and also to a unique initial segment of ordinals* (*those less than* $\alpha \cup \{\alpha\}$).

Proof.

(a) We must verify conditions (a), (b), and (c) of subsection 4. The first of them follows immediately from the definition.

To prove the second condition, we consider two ordinals α and β. By Lemma 5, there exists an isomorphism f of one of them, say α, onto either β or an initial segment of β. We show that then $\alpha = \beta$ or $\alpha \in \beta$. To do this, we prove that $f(\gamma) = \gamma$ for all $\gamma \in \alpha$. In fact, if γ_1 is the minimal element with $f(\gamma_1) \neq \gamma_1$, then $f(\gamma_2) = f(\gamma_2)$ for all $\gamma_2 \in \gamma_1$. Since f is an isomorphic imbedding of α with respect to the ordering $\in$, and since γ_1 and $f(\gamma_1)$ are sets, we have $f(\gamma_1) = \{f(\gamma_2) | \gamma_2 \in \gamma_1\} = \{\gamma_2 | \gamma_2 \in \gamma_1\} = \gamma_1$, which contradicts the choice of γ_1. The same argument shows that $f(\alpha) = \alpha$, from which the condition follows.

Finally, let C be a nonempty class of ordinals, and let $\alpha \in C$. If α is not the least element in C, then the least element in the intersection $\alpha \cap C$ will be the least element in C.

(b) Let X be a well-ordered set. Let S denote the set of ordinals which are isomorphic to some initial segment in X. S is nonempty, since, for example, the ordinal $\{\varnothing\}$ is isomorphic to the segment consisting of the least element of X. It is easy to see that the set $\beta = \bigcup_{\alpha \in S} \alpha$ is an ordinal. We claim β is isomorphic to X. In fact, if this were not the case, then β would be isomorphic to an initial segment in X, say $\hat{X}_1$. But then the ordinals $\beta \cup \{\beta\}$, which is larger than β, would be isomorphic to the initial segment $\hat{X}_1 \cup \{X_1\}$, contradicting the definition of β. □

We now give the elementary properties of ordinals.

8. (a) The finite ordinals are the "natural numbers" (and zero) in the first levels of the universe V. Thus, we shall write:

$$0 = \varnothing, \quad 1 = \{\varnothing\}, \quad 2 = \{\varnothing, \{\varnothing\}\}, \quad 3 = \{\varnothing, \{\varnothing\}, \{\varnothing, \{\varnothing\}\}\}, \ldots .$$

(b) The ordinal which immediately follows a given α is $\alpha \cup \{\alpha\}$. It is also denoted $\alpha + 1$, which agrees with the notation in (a) in the case of finite α.

(c) An ordinal β is called a *limit* ordinal if $\beta \neq \varnothing$ and $\beta \neq \alpha + 1$ for any α. The first limit ordinal ω_0 is isomorphic as a totally ordered set to $\{0, 1, 2, 3, \ldots, n, \ldots\}$. If α is a limit ordinal, then $\alpha = \bigcup_{\beta<\alpha} \beta$. The converse is also true.

Ordinals are mainly used for three purposes: proofs using (transfinite) induction, constructions using (transfinite) recursion, and measuring cardinalities. Here are the basic principles.

9. *Transfinite induction*. Let C be a class of ordinals for which

(a) $\varnothing \in C$.
(b) If $\alpha \in C$, then $\alpha + 1 \in C$.
(c) If a set of ordinals $\{\alpha_i\}$ is contained as a subset in C, then $\cup \alpha_i \in C$.

Then C contains all ordinals.

In fact, otherwise there would exist a least ordinal not in C, but this could not be the empty set by (a), a limit ordinal by (c), or any other ordinal by (b). In concrete applications, the verification of (a) and (c) are often trivial and are omitted.

10. *Transfinite recursion*. Let G be a function of sets (it will actually be sufficient to assume that G is defined on all sets in the universe) whose values are sets. Then there exists a unique function F on the ordinals such that

$$F(\alpha) = G\,(\text{the set of values of } F \text{ on the elements of } \alpha).$$

In fact, this equality uniquely determines $F(0) = G(\varnothing)$, and then $F(1) = G(\{F(0)\})$, $F(2) = G(\{F(0), F(1)\})$, and so on. Thus, if we consider the class C of ordinals α for which we can define F with the required property on the initial segment of ordinals $< \alpha$, then C satisfies the conditions 9(a)–(c), and therefore contains all the ordinals. Uniqueness follows similarly (if $F \neq F'$, consider the least α with $F(\alpha) \neq F'(\alpha)$).

11. *Measuring cardinalities*. Different ordinals can have the same cardinality. For example, all the ordinals ω_0, $\omega_0 + 1$, $\omega_0 + 2, \ldots$ (and many more after them!) are countable. However, jumps in cardinality occur arbitrarily far out.

An ordinal which does not have the same cardinality as any lower ordinal is called a cardinal. All finite ordinals and ω_0 are cardinals. Clearly, any infinite cardinal is a limit ordinal. Further, any set has the same cardinality as some cardinal, and, in fact, a unique one (see §1 of Chapter III). The infinite cardinals form a totally ordered class, which is naturally indexed by ordinals. Thus:

$$\begin{aligned}\omega_0 &= \text{the first countable ordinal;}\\ \omega_1 &= \text{the first ordinal of cardinality} > \omega_0\\ &= \text{the set of all finite and countable ordinals;}\\ \omega_2 &= \text{the first ordinal of cardinality} > \omega_1\\ &= \text{the set of all ordinals of cardinality} \leqslant \omega_1,\end{aligned}$$

and so on.

We can now give our fundamental definition.

12. **Definition.** The (von Neumann) universe V is the class of sets $\bigcup_{\alpha \in \mathrm{On}} V_\alpha$, where the set V_α is defined by the following transfinite recursion:

$$\begin{aligned}V_0 &= \varnothing\\ V_{\alpha+1} &= \mathcal{P}(V_\alpha)\\ V_\alpha &= \bigcup_{\beta<\alpha} V_\beta, \quad \text{if } \alpha \text{ is a limit ordinal.}\end{aligned}$$

We give some elementary properties of the universe V.

13. *Each of the sets V_α is transitive: if $Y \in X \in V_\alpha$ then $Y \in V_\alpha$. (In other words, $V_\alpha \subset V_{\alpha+1}$.)*

Suppose that this were not true. Then there would exist a least ordinal α with $V_\alpha \not\subset V_{\alpha+1}$, where $\alpha \geqslant 2$. If α is not a limiting ordinal, $\alpha = \beta + 1$, $Y \in X \in V_\alpha$, and $Y \notin V_\alpha$, then we obtain a contradiction as follows: $X \in V_{\beta+1} = \mathcal{P}(V_\beta) \Rightarrow X \subset V_\beta \Rightarrow Y \in V_\beta \Rightarrow Y \in V_{\beta+1} = V_\alpha$, since for β it is still true that $V_\beta \subset V_{\beta+1}$ by our choice of α. If α is a limit ordinal, the argument is analogous (find $\gamma < \alpha$ with $Y \in X \in V_\gamma$ and $Y \notin V_\alpha$). □

We define the *rank* of any set $X \in V$ as follows: rank $X = \alpha$ if α is the least ordinal such that $X \in V_{\alpha+1}$. If $Y \in X$ then, *rank* $X \geqslant$ *rank* $Y + 1$.

14. *All ordinals belong to V, and rank $\alpha = \alpha$.*

We first show that $\alpha \in V_{\alpha+1}$ for all ordinals α. This is true for $\alpha = 0$. Suppose that α is the least ordinal with $\alpha \notin V_{\alpha+1}$. If $\alpha = \beta + 1$, then $\beta \in V_{\beta+1}$, so that β and $\{\beta\} \in V_{\beta+2} = \mathcal{P}(V_{\beta+1})$, and hence $\alpha = \beta + 1 = \beta \cup \{\beta\} \in V_{\beta+2} = V_{\alpha+1}$, a contradiction. On the other hand, if α is a limit ordinal, then $\alpha = \bigcup_{\beta<\alpha} \beta$ and $\beta \in V_{\beta+1} \subset V_\alpha$ by the choice of α, so that $\alpha = \bigcup_{\beta<\alpha} \beta \subset \bigcup_{\beta<\alpha} V_\beta = V_\alpha$, and $\alpha \in \mathcal{P}(V_\alpha) = V_{\alpha+1}$, a contradiction. Therefore, *rank* $\alpha \leqslant \alpha$. We similarly prove strict equality. □

15. The universe V is closed with respect to the standard set operations: difference, union, intersection, forming $\mathcal{P}(X)$ and $\bigcup_{Y \in X} Y$, and "collecting" sets indexed by any set: $\{X_Y \mid Y \in Z\}$. In particular, if $X, Y \in V_\alpha$, then the pair $\{X, Y\} \in V_{\alpha+1}$. We write $\{X\}$ in place of $\{X, X\}$.

16. Direct products, relations, and functions can also be defined as elements of V using a device of Kuratowski. The intuitive notion of an ordered pair of sets $X, Y \in V$ is realized by means of the set

$$\langle X, Y\rangle = \{\{X\}, \{X, Y\}\} \in V.$$

As elements of V, ordered pairs are characterized by the following properties: an ordered pair is set of two elements X' and Y', one of which is a subset of the other (say $X' \subset Y'$); if $X' \subset Y'$, then $X' = \{X\}$ is a one-element set, and X is called the *first term of the pair*; Y' is a set of at most two elements, and its element Y which is different from X (if it exists) or X itself (otherwise) is called the *second term of the pair*. Thus, $\langle X, Y\rangle = \langle X'', Y''\rangle$ if and only if $X = X''$ and $Y = Y''$, which justifies the name "ordered pair."

We emphasize that this definition is introduced so that the direct product construction does not leave the universe V, and so that a set corresponding to a direct product can be described in terms of the relation $\in$, i.e., in the language L_1Set.

An ordered n-tuple of sets is defined as

$$\langle X_1, \ldots, X_n\rangle = \langle \cdots \langle\langle X_1, X_2\rangle, X_3\rangle \cdots \rangle.$$

We define the direct product of two sets as

$$X \times Y = \{\langle U, W\rangle | U \in X, W \in Y\}.$$

Similarly,

$$X_1 \times \cdots \times X_n = (\cdots ((X_1 \times X_2) \times X_3) \times \cdots).$$

We note that, in general, $(X \times Y) \times Z \neq X \times (Y \times Z)$; we only have a canonical one-to-one correspondence between these two sets. But it is usually harmless to take the liberty of identifying the two sets and writing $X \times Y \times Z$.

A *binary* relation (or correspondence) r is a set (or class) all of whose elements are ordered pairs. If $r \in V$ is a relation, then its *domain of definition* $\mathrm{dom}(r)$ is the class of all first terms in the elements of r, and the *range of values* $\mathrm{rng}(r)$ is the class of all second terms.

A function is a binary relation in which each element is uniquely determined by its first term. Thus, functions which are maps of sets in V are identified with their graphs. If f is a function, we often write $W = f(U)$ instead of $\langle U, W\rangle \in f$. In addition, we set

$$f^{-1}(X) = \{Y | f(Y) \in X\},$$
$$f|_X = \{\langle U, W\rangle \in f | U \in X\}.$$

A *family* $\{X_Y | Y \in Z\}$ as an element of V is defined to be a function consisting of pairs $\{\langle Y, X_Y\rangle | Y \in Z\}$, and so on.

We again emphasize that the most important feature of these definitions is that we do not introduce any new objects besides elements of V, or any new relations other than those expressible in terms of $\in$. It should also be noted that, in accordance with the usual ("extensional") notion, a *property* of the elements of a set $X \in V$ is a subset $Y \subset X$ (consisting of all elements with this property). Thus, $Y \in V$, so that properties, properties of properties, properties of sets of properties, . . . (with transfinite iteration) are elements of V.

The "universe" V has earned its name.

17. Finally, we show that a chain $X_1 \in X_2 \in \cdots$ of elements of V must terminate (of course, with the empty set).

We prove that, if X is nonempty, then there exists a $Y \in X$ with $Y \cap X = \varnothing$ (the desired result is obtained if we apply this to the set X of terms in the chain). In fact, let Y be the element of least rank in X (which exists because the ranks, since they are ordinals, are well-ordered). If we had $X \cap Y \neq \varnothing$, then any element $Z \in X \cap Y$ would have lower rank than Y, a contradiction. □

18. *Connection with the axioms of* L_1Set. The point of view adopted in this book is as follows.

The intuitive notion of a set, to which we appealed when constructing the universe V, is the primary material. The language L_1Set was devised in order to write formal texts based on this material which are equivalent to our intuitive arguments concerning V. The axioms of L_1Set (including the logical axioms) are obtained as a result of analyzing intuitive proofs. Our criterion for the completeness of this list is that we can write a formal deduction which translates any intuitive proof. The fact that we are able to do this must be proved by a rather large compendium of formal texts, which can be found in other books on logic. In particular, in L_1Set we can write the formula "$\forall x\ \exists$ordinal α $(x \in V_\alpha)$" and deduce it from the axioms. This formula is the formal expression of our restriction to sets in V.

The question of the formal consistency of the Zermelo–Fraenkel axioms must remain a matter of faith, unless and until a formal inconsistency is demonstrated. So far all the proofs which have been based on these axioms have never led to a contradiction; rather, they have opened up before us the rich world of classical and modern mathematics. This world has a certain reality and life of its own, which little depends on the formalisms called upon to describe it.

The discovery of a contradiction in any of various formalisms, even if it should occur, would merely serve to clarify, refine, and perhaps reconstruct certain of our ideas, but would not lead to their downfall, as has happened several times in the past.

CHAPTER III

The continuum problem and forcing

1 The problem: results, ideas

1.1. Cantor introduced two fundamental ideas in the theory of infinite sets: he discovered (or invented?) the scale of cardinalities of infinite sets, and gave a proof that this scale is unbounded. We recall that two sets M and N are said to *have the same cardinality* (card M = card N) if there exists a one-to-one correspondence between them. We write card $M \leqslant$ card N if M has the same cardinality as a subset of N. We say that M and N are *comparable* if either card $M \leqslant$ card N or card $N \leqslant$ card M. We write card $M >$ card N if card $M \geqslant$ card N but M and N do not have the same cardinality.

1.2. **Theorem** (Cantor, Schröder, Bernstein, Zermelo)

(a) *Any two sets are comparable. If both* card $M \leqslant$ card N *and* card $N \leqslant$ card M, *then* card M = card N. *In other words, the cardinalities are linearly ordered.*

(b) *Let* $\mathcal{P}(M)$ *be the set of all subsets of* M. *Then* card $\mathcal{P}(M) >$ card M. *In particular, there does not exist a largest cardinality.*

(c) *In any class of cardinalities there is a least cardinality. In other words, the cardinalities are well-ordered.*

Proof.

(a) Suppose M has the same cardinality as the subset $M' \subset N$ and N has the same cardinality as the subset $N_1 \subset M \cong M'$. We identify M with M'. We then have three sets $N_1 \subset M \subset N$ and a one-to-one correspondence $f: N \to N_1$. We must construct a one-to-one correspondence $g: N \to M$.

Here is an explicit definition of such a map:

$$g(x) = \begin{cases} f(x), & \text{if } x \in f^n(N) \setminus f^n(M) \text{ for some } n \geq 0, \\ x, & \text{otherwise.} \end{cases}$$

Here $f^n(y) = f(f(\cdots f(y) \cdots))$ (n times); $f^n(N) = \{f^n(y) | y \in N\}$, and $f^0(y) = y$. We leave the verification that g has the required properties to the reader.

To prove that any two sets are comparable, it is sufficient to show that any set can be well-ordered, since Lemma 5 of the Appendix to Chapter II implies that well-ordered sets are comparable to each other. Let M be any set. For every nonempty subset $N \subset M$ choose an element $c(N) \in N$. We call a well-ordering $<$ of a subset $M' \subset M$ *admissible* (with respect to c) if $c(M \setminus \hat{X}) = X$ for all $X \in M'$, where $\hat{X} = \{Y | Y \in M', Y < X\}$.

We claim that, if $M' \neq M''$ are two subsets of M having admissible well-orderings, then one set is an initial segment of the other, and the orderings are compatible. In fact, as in subsection 7 (a) of the Appendix to Chapter II, we prove that the canonical isomorphism f of, say, M' with an initial segment of M'' is the identity imbedding: if $f(X) \neq X$ and X is the least element with this property, then

$$f(\hat{X}) = \hat{X}, X = c(M \setminus \hat{X}) \Rightarrow X = c(M \setminus f(\hat{X})) = f(X),$$

which is a contradiction.

It is now easy to see that the union M' of all subsets of M which have a well-ordering admissible with respect to c itself has an admissible ordering; moreover, M' coincides with M, since otherwise we could imbed M' in $M' \cup \{c(M \setminus M')\}$.

In particular, it follows that any set has the same cardinality as some ordinal, and hence the same cardinality as a unique cardinal. This justifies the use of the term "cardinality" and the use of cardinals as our standard scale of cardinalities (see subsection 11 of the Appendix to Chapter II).

(b) Since $\mathcal{P}(M)$ contains all the one-element subsets of M, we have card $\mathcal{P}(M) \geq$ card M. In addition, any map $f: M \to \mathcal{P}(M)$ cannot be one-to-one (or even onto). In fact, we set

$$N = \{z | z \notin f(z)\} \in \mathcal{P}(M),$$

and show that N is not contained in the image of f. If there existed an $n \in M$ such that $N = f(n)$, we would immediately obtain a contradiction by considering the relationship of n to N:

$$n \in N \Rightarrow n \in f(n) \Rightarrow n \notin N \quad \text{by the definition of } N;$$

$$n \notin N \Rightarrow n \notin f(n) \Rightarrow n \in N \quad \text{by the definition of } N.$$

This is Cantor's famous "diagonal process."

(c) The well-ordering of the cardinals is established at the same time as their comparability in the first stage of the theory of ordinals (see the Appendix to Chapter II). □

1.3. *Remark*. This proof of the lemma that any set can be well-ordered is essentially due to Zermelo. It was probably what prompted the most severe objections to the axiom of choice. The intuitive idea behind the proof reduces to a recipe for choosing one element after another from the set M until all of M is exhausted. In this form it is immediately apparent that the prescription is "physically" unthinkable, and to many of Zermelo's contemporaries the whole proof seemed to be nothing but a trick. For example, the idea of "first" choosing an element $c(N)$ in each subset $N \subset M$ met with the following objection of Lebesgue. If the elements we choose are not characterized by any special properties, how do we know that we are always thinking about the same elements throughout the proof? But today, except for specialists in the foundations of mathematics, hardly any working mathematicians share these doubts.

We now formulate the basic problem that will concern us during the next two chapters. We shall write card $\mathcal{P}(M) = 2^{\mathrm{card}\ M}$, in analogy to the finite case. The continuum is 2^{ω_0}.

1.4. *The continuum problem*. What place does the continuum occupy on the scale of cardinalities?

By Theorem 1.2(b), we have $2^{\omega_0} > \omega_0$. Hence, in any case, $2^{\omega_0} \geqslant \omega_1$. On the other hand, if $2^{\omega_0} > \omega_1$, $2^{\omega_0} > \omega_2, \dots, 2^{\omega_0} > \omega_n, \dots$ for any n, then we would have $2^{\omega_0} > \omega_{\omega_0}$, since the continuum cannot be a union of countably many subsets of lower cardinality (König).

1.5. *The Continuum Hypothesis* (CH). $2^{\omega_0} = \omega_1$.

The Generalized Continuum Hypothesis asserts that $2^{\mathrm{card}\ M}$ comes immediately after card M for any infinite M. Here is almost everything we know about this question:

1.6. **Theorem**

(a) *The negation of the Continuum Hypothesis cannot be deduced from the other axioms of set theory, if those axioms are consistent* (*Gödel*).

(b) *The Continuum Hypothesis cannot be deduced from the other axioms of set theory, if those axioms are consistent* (*Cohen*).

The same holds true for the Generalized Continuum Hypothesis.

If we grant that the axioms of set theory and the logical means of expression and deduction in $\mathrm{L_1Set}$, which are implicit in the statement of Theorem 1.6, actually exhaust the apparatus for constructing proofs in modern mathematics, then we can say that the continuum problem is the only known example of an absolutely undecidable problem. Although Gödel's incompleteness theorem provides concrete examples of undecidable propositions in any formal system having reasonable properties, these examples can be decided in an "obvious" way in some higher system. The situation with the continuum problem seems much more difficult. If we agree that it is a meaningful question, then it can only be decided by introducing a new principle of proof. Various possibilities for doing this have been discussed, but none of the suggested new axioms for set theory seem sufficiently convincing or, more important, sufficiently useful in "real" mathematics. In the hundred years since the introduction of transfinite induction, not a single new method of constructing sets has come into common use. Incidentally, the basic idea in Gödel's proof of Theorem 1.6(a) actually consists in verifying that all the old methods allow us to construct at most ω_1 subsets of ω_0 (or, equivalently, at most ω_1 real numbers).

1.7. *Gödel's idea.* Gödel considers the basic set-theoretic operations—forming pairs, products, complements, sums, and so on—and constructs the class of all sets which are obtained by transfinite iteration of these operations, starting from $\varnothing$. Such sets are called *constructible* sets. It is *a priori* completely unclear whether or not all subsets of $\{0, 1, 2, \ldots\}$ are constructible, or, more generally, whether or not all sets in the universe V are constructible. (It turns out that this problem is formally undecidable to the same extent as the continuum problem.) But we find that, within the class of constructible sets, the number of subsets of $\{0, 1, 2, \ldots\}$ is equal to ω_1—most likely because we have omitted a vast number of nonconstructible sets. Meanwhile, all the axioms of set theory, restricted to this class, are true (in a reasonable meaning of "true"), as are all deductions from these axioms. Hence the negation of the CH is not deducible, since it is false in this model. The next chapter will be devoted to Gödel's theorem.

1.8. *Cohen's idea.* We shall present this idea in the version due to Scott and Solovay. First we give its application to a certain simplified problem, concerned with a language weaker than $\mathrm{L_1Set}$; then in §§4–8 we present the application to $\mathrm{L_1Set}$. For another version of Cohen's idea, see §9.

We shall discuss the CH in the form: *there does not exist a subset of the real numbers* $\mathbf{R}$ *whose cardinality is strictly between that of* $\{0, 1, 2, \ldots\}$ *and that of* $\mathbf{R}$. In fact, if we had $2^{\omega_0} > \omega_1$, then any subset of $\mathbf{R}$ of cardinality ω_1 would have such an intermediate cardinality.

In order to show that this assertion is not deducible, which is equivalent to Cohen's theorem, it suffices to construct a model of the real numbers in which all the axioms and all propositions deducible from them are fulfilled and in which a set of intermediate cardinality exists. This model will be the set $\bar{R}$ of random variables on a very big probability space Ω. For a suitable choice of Ω, $\bar{R}$ will be so big that within the model there exists a set of intermediate cardinality, containing $\bar{N}$ (the integers of the model) and contained in $\bar{R}$ (the continuum of the model).

Of course, it cannot be quite this simple; there must be some obstacle to carrying out this program. The obstacle is that almost all the properties of $\mathbf{R}$, including most of the axioms, turn out to be false for $\bar{R}$, so that $\bar{R}$ cannot be a model for $\mathbf{R}$ in the usual sense of the word. Cohen's basic idea was to develop a method for overcoming this difficulty. He replaced the property of an assertion being true by another property, which we shall temporarily call "truth" in quotes, and which has the necessary formal properties. Namely, all the axioms of **R** are "true" in $\bar{R}$, all deductions from "true" assertions using the rules of logic again lead to "true" assertions, and the CH is not "true," and hence is not deducible from the axioms. We now show in greater detail how this is done.

1.9. Let I be a set of cardinality $> \omega_1$. We set

$$\Omega = [0, 1]^I, \quad \text{with Lebesgue measure},$$

$$\bar{R} = \text{the set of random variables on } \Omega$$

$$= \text{the set of measurable real-valued functions on } \Omega.$$

1.10. **Theorem**

(a) *All the axioms of the real numbers and all deductions from them are "true" for* $\bar{R}$.

(b) *The* CH *is not "true" for* $\bar{R}$.

Here we say that an assertion P about random variables $\bar{x}, \bar{y}, \cdots \in \bar{R}$ is "true" if the following condition is fulfilled:

> for each point $\omega \in \Omega$ we consider the values $\bar{x}(\omega), \bar{y}(\omega), \cdots$ of the random variables $\bar{x}, \bar{y}, \cdots$ and form the assertion P_ω about these ordinary real numbers; then for almost all $\omega \in \Omega$ (i.e., all but a set of measure 0) P_ω is true in the usual sense of the word.

Briefly, "truth" means experimental truth with probability one.

Example. Let P be the assertion that "**R** has no zero divisors," i.e., "if $x, y \in \mathbf{R}$ are such that $xy = 0$, then either $x = 0$ or $y = 0$." Then the assertion "$\bar{R}$ has no zero divisors" is, of course, not true. However, it is "true" because: if $\bar{x}, \bar{y} \in \bar{R}$ are such that $\overline{xy} = 0$, then for almost all $\omega \in \Omega$ either $\bar{x}(\omega) = 0$ or $\bar{y}(\omega) = 0$.

1.11. In order to give a precise meaning to the definition of "truth" and learn how to verify effectively the "truth" of rather complicated assertions, we must introduce a formal language, in this case the language of real numbers. This formal language is a mathematical object, and the precise formulation of Theorem 1.10 will concern this object, and not **R** or $\bar{R}$ at all.

The connection between this language and **R** is given by a system of informal recipes which tell how to translate the usual intuitive texts about **R** into this language, and by a system of theorems which tell us that the translation is always possible and that the recipes are faithful to the informal texts. The role of $\bar{R}$ is reduced to that of auxiliary construction which is used to define and compute a special "truth" function on the formulas of the language. Thus we see the role of logic in the program.

1.12. A detailed proof of Theorem 1.10 would be rather lengthy and nontrivial for several reasons. In the first place, a certain amount of space must be devoted to describing the formal language and the axioms of **R** in this language. We must then verify that all the axioms are "true" and that the CH is not "true"—this amounts to one or two dozen verifications, each of which involves an inductive argument with infinite sums and products in the Boolean algebra of measurable sets in Ω. However, the most serious difficulties arise because the meaning of every assertion changes considerably in going from **R** to $\bar{R}$, and not always in a convenient direction. We shall illustrate this qualitative aspect by attempting to explain why the CH is not "true," and why this is nontrivial.

As we have said, we want to construct a subset $\bar{M}$ of $\bar{R}$ having cardinality intermediate between the cardinality of $\bar{N}$ and the cardinality of $\bar{R}$. We do this as follows: for any $i \in I$, let the random variable $\bar{x}_i : [0, 1]^I \to [0, I]$ be the ith projection. Choose a subset $\mathcal{J} \subset I$ such that $\omega_0 < \text{card } \mathcal{J} < \text{card } I$ (this is possible if I is large), and set

$$\bar{M} = \{x_j | j \in \mathcal{J}\} \subset \bar{R}.$$

Then card $\bar{N} <$ card $\bar{M} <$ card $\bar{R}$ is true in the usual meaning of the word. However, we must show that the corresponding assertion is "true" in our Pickwickian sense. But then the role of the integers is assumed by the "locally integral" random variables (whose values are integral with probability one), and these random variables can have cardinality much greater than ω_0. Thus, the required lower estimate for card $\bar{M}$ becomes much more serious. Similarly, if we formalize our naive description of $\bar{M}$ and then interpret it in $\bar{R}$, then $\bar{M}$ takes on a new meaning, and leads to a much larger set than the "real" $\bar{M}$. Thus, it is also unclear that the upper inequality for card $\bar{M}$ still holds. It seems almost miraculous that everything eventually falls into place.

The plan for the rest of the chapter is as follows. In §2 and §3 we give a (shortened) exposition for the second-order language of real numbers of this abbreviated version of the theorem that the CH is not deducible. If the reader is only interested in the complete proof for L_1Set, he may skip to §4, where we introduce the Boolean-valued "universe of random sets," which takes the place of V. In §§5–7 we verify that the Zermelo–Fraenkel axioms are "true," and in §8 we verify that the CH is "false." Finally, in §9 we discuss Cohen's original method, which is more syntactic and involves somewhat different intuitive ideas.

2 A language of real analysis

2.1. In this section we describe a formal language based on the theory of real numbers. In particular, this means that the variables x, y, z will be considered as names of real numbers. However, if we try to use a first-order language to formulate the assertions we are interested in, such as the Continuum Hypothesis CH, or even the completeness axiom (which differentiates the real numbers from the rational numbers), we find that we are not able to do this. In fact, in these assertions we need to refer to arbitrary subsets (or relations of degree one) of the real numbers, whereas first-order languages do not have symbols for variable relations (compare with subsection 3.17 of Chapter I).

This leads us to consider the second-order language L_2Real, which is the most economical language in which the axioms and the CH can be expressed. We shall give a brief description of this language, for the most part only noting those features which show the connections with the real numbers and those which are peculiar to second-order languages.

2.2. *The language* L_2Real. The alphabet consists of the variable symbols $x, y, z, \ldots$; the symbols for degree 1 functions $f, g, h, \ldots$; the constants 0 and 1; the degree 2 operations $+$ and $\cdot$; the degree 2 relations $=$ and $\leqslant$; and the same connectives, quantifiers, and parentheses as in languages of $\mathfrak{L}_1$. The *terms* are $x, y, z, \ldots$ and 0 and 1; and also $f(t)$, $t_1 \cdot t_2$, and $t_1 + t_2$, if f is a function symbol and t, t_1, and t_2 are terms. The terms are names of real numbers.

The *atomic formulas* are $t_1 = t_2$ or $t_1 \leqslant t_2$, where t_1 and t_2 are terms. The set of *formulas* is defined inductively exactly as in languages of $\mathfrak{L}_1$, with one addition: $\forall f(Q)$ and $\exists f(Q)$ are formulas if Q is a formula and f is the symbol for a variable function. The notions of a free occurrence of a variable (x or f), of a closed formula, and so on carry over to L_2Real in the obvious way. We shall use the same type of abbreviated notation here as in Chapter I. The standard interpretation of formulas which is implicit in the language should be obvious from the definitions and from the following examples.

2.3. *The formula* $Z(y)$: "*y is an integer*." It is perhaps not completely obvious how to write this formula. We can write, "y can be obtained from 0 by repeatedly adding or subtracting 1," or else "any function f which has period 1 and vanishes at 0 must also vanish at y," i.e.,

$$Z(y):\quad \forall f\big((f(0)=0 \wedge \forall x(f(x)=f(x+1)))\Rightarrow f(y)=0\big).$$

2.4. *The formula* CH: "*Any subset of* **R** *either has the same cardinality as* **R**, *or else is countable or finite*."

We first restate the formula in different words: "Given a set of zeros of any function h, either there exists a function g mapping it onto all **R**, or else there exists a function f mapping the integers onto all of this set." We then have:

$$\text{CH:}\quad \forall h\big(\exists g\, \forall y\, \exists x(h(x)= 0 \wedge y=g(x)) \vee \exists f\, \forall y(h(y)=0 \Rightarrow \exists x(Z(x) \wedge y=f(x)))\big).$$

Notice that the formula $Z(x)$ occurs as part of the CH.

We further write the completeness axiom C:

2.5. *The formula* C: "*Any subset of* **R** (*the set of values of a function f*) *which is bounded from above has a least upper bound z.*" We write:

$$\text{C:} \quad \forall f(\exists y \, \forall x(f(x) \leqslant y) \Rightarrow \exists z \, \forall y(\forall x(f(x) \leqslant y) \Leftrightarrow z \leqslant y)).$$

All the other formulas we are interested in are simpler and do not require any special comment.

We now give a precise definition of the property of "truth" for closed formulas in L_2Real; this property was described informally in §1. We emphasize that it is not an absolute property, but rather depends on the choice of the probability space Ω which is used to construct the "model" of the real numbers.

2.6. *The algebra of truth values.* As in §1, we set

I = a set;

$\Omega = [0, 1]^I$ with Lebesgue measure;

B = the algebra of measurable sets in Ω modulo sets of measure zero;

0 = the class of the empty set in B;

1 = the class of Ω in B.

We have the following operations in B:

a',	the "complement" of the element $a \in B$;
$a \wedge b$,	the "intersection" of two elements $a, b \in B$;
$a \vee b$,	the "union" of two elements $a, b \in B$.

These operations satisfy the usual identities and give a Boolean algebra structure on B. We write $a \leqslant b$ if $a \wedge b = a$.

Moreover, the operations of intersection and union extend uniquely to *infinite families* of elements, and continue to satisfy the usual identities which hold in the algebra of all subsets of any given set. We shall omit the verification of all this. We only note that sets here are identified "modulo sets of measure zero," and that identities of the type $(A \bmod 0) \wedge (B \bmod 0) = (A \cap B) \bmod 0$ do not carry over to infinite families.

Finally, B satisfies the following *countable chain condition*: if $a_\alpha \wedge a_\beta = 0$ for all distinct indices α and β then $a_\alpha \neq 0$ for at most countably many indices α. This follows because Lebesgue measure is positive and additive. Technically speaking, B is a *complete Boolean algebra with the countable chain condition*. The precise origin of B and the fact that it has a measure play a less important role.

2.7. *The interpretation set.* We now introduce a large set $\overline{M}$, each point ξ of which corresponds to the assignment of certain values to all the symbols in the alphabet of L_2Real. If ξ is fixed, each formula becomes a concrete statement about measurable functions (random variables) on Ω and about functionals on them (compare with §2 of Chapter II).

More precisely, we set

$\bar{R}$ = the set of measurable real-valued functions on Ω;
$\bar{R}^{(1)}$ = the set of all possible maps $\bar{f}: \bar{R} \Rightarrow \bar{R}$ which satisfy the condition:

$\forall \bar{x}, \bar{y} \in \bar{R}$

$$\left(\text{the set } \{\omega \in \Omega | \bar{x}(\omega) = \bar{y}(\omega)\} \leqslant \{\omega \in \Omega | \bar{f}(\bar{x})(\omega) = \bar{f}(\bar{y})(\omega)\} \bmod 0\right).$$

The definition of $\bar{R}^{(1)}$ has the following intuitive meaning. If we ignore the "mod 0," the condition simply means that the value of the random variable $\bar{f}(\bar{x})$ at each trial (each point in Ω) must be determined by the value of $\bar{x}$ *at this trial*. Of course, this is a very natural requirement if we want functions $\bar{f}$ to be adequate reflections of properties of ordinary real-valued functions in the sense of §1. The addition of "mod 0" weakens this requirement by saying "with conditional probability one."

We now return to the set $\bar{M}$. A point $\xi \in \bar{M}$ consists of a choice of

$x^{\xi} \in \bar{R}$, for each variable symbol x;

$f^{\xi} \in \bar{R}^{(1)}$, for each symbol f for a variable function.

Here is the interpretation of the expressions in the language which corresponds to a given choice of ξ:

(a) *Terms*. Let t be a term, and let $\xi \in \bar{M}$. Then $t^{\xi} \in \bar{R}$ is the random variable which is defined inductively in the obvious way.

(b) *The truth function* $\|\ \|$ *on atomic formulas*. Let P be the atomic formula $t_1 \leqslant t_2$ or $t_1 = t_2$. Its truth value at a point $\xi \in \bar{M}$ is the element of the algebra B which is defined as follows:

$$\|t_1 \leqslant t_2\|(\xi) = \{\omega \in \Omega | t_1^{\xi}(\omega) \leqslant t_2^{\xi}(\omega)\} \bmod 0,$$

and similarly for $t_1 = t_2$.

(c) *The truth function* $\|P\|(\xi)$ *in the general case*. The general definition proceeds by induction. The rules when formulas are joined by connectives are the same as in subsection 5.7 of Chapter II:

$$\begin{aligned}
\|\neg P\| &= \|P\|', \\
\|P \vee Q\| &= \|P\| \vee \|Q\|, \\
\|P \wedge Q\| &= \|P\| \wedge \|Q\|, \\
\|P \Rightarrow Q\| &= \|P\|' \vee \|Q\|, \\
\|P \Leftrightarrow Q\| &= (\|P\| \wedge \|Q\|) \vee (\|P\|' \wedge \|Q\|').
\end{aligned}$$

Here, for brevity, we have omitted the ξ. Finally,

$$\|\forall x P\|(\xi) = \bigwedge_{\xi'} \|P\|(\xi') \qquad \text{(over all } \xi' \text{ which only differ from } \xi \text{ by a variation of } x);$$

$$\|\exists x P\|(\xi) = \bigvee_{\xi'} \|P\|(\xi') \qquad \text{(over the same } \xi');$$

and similarly when we quantify over variable functions. Intuitively, the value of the truth function of an assertion about random variables is the set of trials mod 0 for which this assertion becomes true as a fact about real numbers.

2.8. **Lemma.** *If P is a closed formula, then $\|P\|(\xi)$ does not depend on the choice of $\xi \in \overline{M}$ and only takes the value* 0 *or* 1.

This is proved by a simple induction on the length of P. It is just as easy to prove a more general fact: if P is any formula and ξ and ξ' do not differ on variables which occur freely in P, then $\|P\|(\xi) = \|P\|(\xi')$. Compare with Proposition 2.10 in Chapter II.

This value of $\|P\|(\xi)$ which is common for all ξ if P is closed can be denoted simply $\|P\|$. We are now ready to formulate the basic definition of this section:

2.9. **Definition.** A formula P in $\mathrm{L_2Real}$ is said to be "true" if $\|P\|(\xi) = 1$ for all $\xi \in \overline{M}$.

3 The Continuum Hypothesis is not deducible in $\mathrm{L_2Real}$

3.1. **Fundamental Lemma**

(a) *"Truth" is preserved under the rules of deduction.*
(b) *The first-order logical axioms and the versions of them in* $\mathrm{L_2Real}$ *are "true."*
(c) *The special axioms of* $\mathrm{L_2Real}$ *are "true."*
(d) *The* CH *is not "true" if* card $I > \omega_1$.

This lemma implies

3.2. **Theorem.** *The* CH *is not deducible from the axioms in* $\mathrm{L_2Real}$.

In this section we give those parts of the proof of the Fundamental lemma which are also essential for the "real" Cohen theorem, as well as for our simplified problem. We note that Theorem 3.2 is weaker than Cohen's theorem because the language $\mathrm{L_2Real}$ contains fewer means of expression than the language of set theory. Although the Continuum Hypothesis can be stated in L_2Real, because of Gödel's general results we have no basis for expecting, even if the CH were deducible, that the proof could also be given in this language. For example, the deduction could require us to introduce functionals of functions, functionals of functionals, and so on. The language of set theory, which we shall return to in §4, contains the means for considering all of these finite and even transfinite levels at once.

3.3. PROOF OF 3.1(a). If $\|P\| = 1$ and $\|P \Rightarrow Q\| = 1$, then $\|P\|' = 0$ and $\|P\|' \vee \|Q\| = 1$, so that $\|Q\| = 1$. Secondly, if $\|P\| = 1$, then $\|P\|(\xi) = 1$ for all $\xi \in \overline{M}$; but then (here ξ' runs through all variations of ξ along x)

$$\|\forall x P\|(\xi) = \bigwedge_{\xi'} \|P\|(\xi') = \bigwedge_{\xi'} 1 = 1. \qquad \square$$

We similarly prove this for Gen over functions.

3.4. PROOF OF 3.1(b) (SKETCH).

Tautologies. Their "truth" is proved in §5 of Chapter II.

Quantifier axioms. The proof proceeds by induction on the length of the formulas in the axiom schemes. Since it is completely straightforward, we shall omit it.

3.5. PROOF OF 3.1(c) (SKETCH). We shall list the axioms and make some brief comments.

The special axioms of set theory: The axioms of equality and the axiom (schema) of choice

$$\text{AC:} \quad \forall x\, \exists y P(x, y) \Rightarrow \exists f\, \forall x\; P(x, f(x)),$$

where P is any formula which does not have any free variables except x and y, and where f is free for y in P.

The special axioms of field theory: The axioms of the additive group, the axioms of the multiplicative group, and the distributivity of addition with respect to multiplication.

The special order axioms:

$$x \leqslant y \vee y \leqslant x,$$
$$(x \leqslant y \wedge y \leqslant x) \Leftrightarrow x = y,$$
$$x \leqslant y \Rightarrow (x + z \leqslant y + z),$$
$$(x \leqslant y \wedge 0 \leqslant z) \Rightarrow xz \leqslant yz.$$

The completeness axiom (see 2.5).

Among these axioms, the greatest effort is needed to verify that the axiom of choice and the completeness axiom are "true." But these computations resemble those in the proof that the CH is false, which will be given in detail below. Hence, the verification of these two axioms will be omitted.

The first axiom of equality is trivial. The second axiom is first verified for atomic formulas P, and then we use induction on the length of P. The argument is rather tedious, but simple.

The axioms of an ordered field are verified without difficulty. We shall limit ourselves to one example: "every nonzero number has an inverse," i.e.

$$\|\forall x(\,\square(x=0)\Rightarrow\exists y(xy=1))\|=\bigwedge_{\bar{x}\in\bar{R}}\Big(\|\bar{x}=0\|\vee\bigvee_{\bar{y}\in\bar{R}}\|\bar{x}\bar{y}=1\|\Big).$$

To verify that this truth value equals 1, it suffices to prove this for each term on the right, i.e., for each fixed $\bar{x}\in\bar{R}$. Then, in turn, for that $\bar{x}$ it suffices to construct a random variable $\bar{y}\in\bar{R}$ such that $\|\bar{x}=0\|\vee\|\bar{x}\bar{y}=1\|=1$. We set

$$\bar{y}(\omega)=\begin{cases}\bar{x}(\omega)^{-1}, & \text{if } \bar{x}(\omega)\neq 0,\\ 0, & \text{if } \bar{x}(\omega)=0.\end{cases}$$

□

3.6. Proof of 3.1(d). We first recall the formula for the CH:

$$\forall h\big(\exists g\,\forall y\,\exists x(h(x)=0\wedge y=g(x))\vee$$
$$\exists f\,\forall y\big(h(y)=0\Rightarrow\exists x(Z(x)\wedge y=f(x))\big)\big).$$

We let P_1 and P_2 denote the first and the second alternatives in this formula. Thus, the CH has the form $\forall h(P_1\vee P_2)$. We must prove that $\|\forall h(P_1\vee P_2)\|(\xi)=0$ for any point $\xi\in\bar{M}$. By the definition in 2.7,

$$\|\forall h(P_1\vee P_2)\|(\xi)=\bigwedge_{\xi'}(\|P_1\|(\xi')\vee\|P_2\|(\xi')),$$

where ξ' runs through all variations of ξ along h. To show that this value is 0, it suffices to find a point ξ' such that $\|P_1\|(\xi')=\|P_2\|(\xi')=0$. Since all the variables except h are bound in P_1 and P_2, choosing ξ' is equivalent to choosing $h^{\xi'}=\bar{h}\in\bar{R}^{(1)}$. We shall give $\bar{h}$ explicitly; this will be a function "whose set of zeros has intermediate cardinality."

To do this, as in §1 we fix a subset $\mathcal{J}\subset I$ having cardinality strictly between ω_0 and card I. Recall that for each $i\in I$, $\bar{x}_i\in\bar{R}$ is the "ith coordinate" function. Further, for each random variable $\bar{x}\in\bar{R}$, we choose a subset $\Omega(\bar{x})\subset\Omega$ such that

$$\bigvee_{j\in\mathcal{J}}\|\bar{x}=\bar{x}_j\|=\Omega(\bar{x})\bmod 0$$

(here we use the completeness of B). Finally, we define $\bar{h}\in\bar{R}^{(1)}$ as follows for every $\bar{x}\in\bar{R}$ and $\omega\in\Omega$:

$$\bar{h}(\bar{x})(\omega)=\begin{cases}0, & \text{if } \omega\in\Omega(\bar{x}),\\ 1, & \text{otherwise}.\end{cases}$$

3.7. Correctness Lemma

(a) *For fixed* $\bar{x}$, $\bar{h}(\bar{x})$ *is measurable as a function of* ω, *so that* $\bar{h}$ *maps* $\bar{R}$ *to* $\bar{R}$.

(b) *For every* $\bar{x} \in \bar{R}$ *we have*

$$\|\bar{h}(\bar{x}) = 0\| = \bigvee_{j \in \mathcal{J}} \|\bar{x} = \bar{x}_j\|.$$

(c) $\bar{h} \in \bar{R}^{(1)}$ (*see* 2.7), *so that there exists a point* $\xi' \in M$ *for which* $h^{\xi'} = \bar{h}$.

PROOF.

(a) $\bar{h}(\bar{x})$ only takes the values 0 and 1 on Ω, and the set where it takes each of these two values is measurable by the definition and by the completeness of B.

(b) is obvious from the definition.

(c) We must verify that for all $\bar{x}, \bar{y} \in \bar{R}$ we have

$$\{\omega \in \Omega | \bar{x}(\omega) = \bar{y}(\omega)\} \leqslant \{\omega \in \Omega | \bar{h}(\bar{x})(\omega) = \bar{h}(\bar{y})(\omega)\} \bmod 0.$$

We shall show that the set of points $\omega \in \Omega$ for which both $\bar{x}(\omega) = \bar{y}(\omega)$ and $\bar{h}(\bar{x})(\omega) \neq \bar{h}(\bar{y})(\omega)$ has measure zero.

It suffices to consider the case $\bar{h}(\bar{x})(\omega = 0, \bar{h}(\bar{y})(\omega) = 1$, i.e., to show that

$$\|\bar{x} = \bar{y}\| \wedge \|\bar{h}(\bar{x}) = 0\| \wedge \|\bar{h}(\bar{y}) = 1\| = 0.$$

We write the second term in the form $\bigvee_{j \in \mathcal{J}} \|\bar{x} = \bar{x}_j\|$ (by 3.7(b)) and apply the distributive axiom to the first and second terms (where we use the completeness of B). We further use the fact that $\|\bar{x} = \bar{y}\| \wedge \|\bar{x} = \bar{x}_j\| \leqslant \|\bar{y} = \bar{x}_j\|$. We then obtain:

$$\|\bar{x} = \bar{y}\| \wedge \|\bar{h}(\bar{x}) = 0\| \leqslant \bigvee_{j \in \mathcal{J}} \|\bar{y} = \bar{x}_j\| = \|\bar{h}(\bar{y}) = 0\|,$$

which immediately gives us the required result. □

Explanation. Since the choice of $\bar{h}$ is the essential step in the proof, we would like to give some motivation for this choice. Recall that h is the name of the function the cardinality of whose set of zeros interests us. We choose a concrete $\bar{h}$ to "disprove" the CH in such a way that the "almost everywhere zeros" of $\bar{h}$ include the elements of the set $\{x_j | j \in \mathcal{J}\}$, which has intermediate cardinality in the naive sense of the word (compare with §1). However, $\bar{h}$ cannot be an arbitrary map from $\bar{R}$ to $\bar{R}$; it must satisfy the strong condition $\bar{h} \in \bar{R}^{(1)}$. Hence, along with all the $\bar{x}_j$, the almost everywhere zeros of $\bar{h}$ might also have to include various other $\bar{y} \in \bar{R}$, and might have to "partly include" still other $\bar{z} \in \bar{R}$. We say "partly include" to convey the possibility that $\|\bar{h}(\bar{z}) = 0\|$ is neither 0 nor 1, so that $\bar{z}$ has a "certain probability" of being a zero of $\bar{h}$.

Thus, the "set of zeros" of $\bar{h}$ might be bigger than we want, and we might expect to encounter difficulties in proving that this set cannot be mapped onto all of $\bar{R}$ (the alternative P_1). On the other hand, it would seem that this situation would make it trivial to disprove the alternative P_2 (mapping Z onto the entire set of zeros). But even this is wrong! As we noted before, we can have $\|Z(\bar{x})\| = 1$ for many $\bar{x}$ which are not constant integer functions on Ω. Moreover, for still other $\bar{x}$ we have $\|Z(\bar{x})\| \neq 0, 1$, so that the "set of integers" in our model has grown considerably.

A final remark: in this discussion we have been essentially dealing with the concept of a "B-random set," which will be a central idea in what follows (see §4). That is, the "set of zeros of $\bar{h}$" is random in the sense that, for each $\bar{z} \in \bar{R}$, the assertion "$\bar{z} \in$ (zeros of $\bar{h}$)" is naturally assigned the Boolean truth value $\|\bar{h}(\bar{z}) = 0\|$.

We now return to the proof that $\|\mathrm{CH}\| = 0$.

3.8. PROOF THAT $\|P_1\|(\xi') = 0$. By the rules for computing truth functions, we find:

$$\|P_1\|(\xi') = \bigvee_{\bar{g}} \bigwedge_{\bar{y}} \bigvee_{\bar{x}} \left\{ \|\bar{h}(\bar{x}) = 0\| \wedge \|\bar{y} = \bar{g}(\bar{x})\| \right\},$$

where $\bar{h}$ was defined above, $\bar{g}$ runs through all elements of $\bar{R}^{(1)}$, and $\bar{x}$ and $\bar{y}$ run through all elements of $\bar{R}$. We suppose that $\|P_1\|(\xi') \neq 0$, and show that this leads to a contradiction. We write the above formula for $\|P_1\|(\xi')$ as $\bigvee_{\bar{g}} a(\bar{g})$.

If $\|P_1\|(\xi') \neq 0$, then $a(\bar{g}) \neq 0$ for some concrete function $\bar{g} \in \bar{R}^{(1)}$. We take this function $\bar{g}$ and set

$$a = \bigwedge_{\bar{y}} \bigvee_{\bar{x}} \Big(\bigvee_{j \in \mathcal{J}} \|\bar{x} = \bar{x}_j\| \wedge \|\bar{y} = \bar{g}(\bar{x})\| \Big).$$

Here we have substituted $\bigvee_{j \in \mathcal{J}} \|\bar{x} = \bar{x}_j\|$ for $\|\bar{h}(\bar{x}) = 0\|$ using 3.7(b). Furthermore, we have $\|\bar{x} = \bar{x}_j\| \wedge \|\bar{y} = \bar{g}(\bar{x})\| \leqslant \|\bar{y} = \bar{g}(\bar{x}_j)\|$. Using this and distributivity, we find

$$a \leqslant \bigwedge_{\bar{y}} \bigvee_{j \in \mathcal{J}} \|\bar{y} = \bar{g}(\bar{x}_j)\|.$$

In particular, for each $\bar{x}_i$ in place of $\bar{y}$, we have

$$a \leqslant \bigvee_{j \in \mathcal{J}} \|\bar{x}_i = \bar{g}(\bar{x}_j)\|.$$

If, as we have supposed, $a \neq 0$, then for each i there exists a $j(i) \in \mathcal{J}$ such that

$$\|\bar{x}_i = \bar{g}(\bar{x}_{j(i)})\| \neq 0.$$

Since I is uncountable and card $\mathcal{J} <$ card I, it follows that there exists a $j_0 \in \mathcal{J}$ such that $j_0 = j(i)$ for all i in an uncountable subset $I_0 \subset I$. But this contradicts the countable chain condition on B, because the terms in the family $\|\bar{x}_i = \bar{g}(\bar{x}_{j_0})\|$ $(i \in I_0)$ are pairwise disjoint. In fact,

$$\|\bar{x}_{i_1} = \bar{g}(\bar{x}_{j_0})\| \wedge \|\bar{x}_{i_2} = \bar{g}(\bar{x}_{j_0})\| \leqslant \|\bar{x}_{i_1} = \bar{x}_{i_2}\| = 0,$$

if $i_1 \neq i_2$. □

Notice to what extent this proof parallels the "naive" argument in §1. By assumption, the function $\bar{y}$ maps the zeros of $\bar{h}$ onto $\bar{R}$ "with nonzero probability." But the exact meaning of the computations cannot readily be stated in words.

Computation of $\|Z(y)\|$. The formula for $Z(y)$, "y is an integer," was given in 2.3. Since this formula occurs in P_2, we must compute $\|Z(y)\|$ in order to compute $\|P_2\|$.

3.9. **Lemma.** *Let* $\eta \in \bar{M}$ *and* $\bar{y}^\eta = \bar{y} \in \bar{R}$. *Then*

$$\|Z(y)\|(\eta) = \bigvee_{n \in Z} \|\bar{y} = n\| = \{\omega \in \Omega | \bar{y}(\omega) \in Z\} \bmod 0.$$

Proof. We must show that

$$\bigwedge_{\bar{f}} \Big(\|\bar{f}(0) = 0\|' \vee \Big(\bigvee_{\bar{x}} \|\bar{f}(\bar{x}) = \bar{f}(\bar{x}+1)\|' \Big) \vee \|\bar{f}(\bar{y}) = 0\| \Big) = \bigvee_{n \in Z} \|\bar{y} = n\|.$$

We prove this equality by proving inequality in both directions.

The inequality $\leqslant$. It suffices to find a concrete function $\bar{f} \in \bar{R}^{(1)}$ for which the corresponding term on the left is contained in the right-hand side. We define $\bar{f}$ by setting $\bar{f}(\bar{x})(\omega) = \sin^2 \pi\bar{x}(\omega)$ (here, instead of $\sin^2 \pi z$, we could take any measurable function with period 1 and zeros only at the integers). It is easy to see that $\bar{f}(\bar{x}) \in \bar{R}$ and $\bar{f} \in \bar{R}^{(1)}$. Then $\|\bar{f}(0) = 0\|' = 0$ and $\|\bar{f}(\bar{x}) = \bar{f}(\bar{x}+1)\|' = 0$. Hence we need only verify that

$$\|\sin^2 \pi\bar{y} = 0\| \leqslant \bigvee_{n \in Z} \|\bar{y} = n\|,$$

and this is obvious.

The inequality $\geqslant$. It suffices to show that, for any fixed values of $n \in Z$, $\bar{f} \in \bar{R}^{(1)}$ and $\bar{y} \in \bar{R}$, we have:

$$\|\bar{y} = n\| \leqslant b \vee c,$$

where

$$b = \|\bar{f}(0) = 0\|' \vee \Big(\bigvee_{\bar{x}} \|\bar{f}(\bar{x}) = \bar{f}(\bar{x}+1)\|' \Big); \qquad c = \|\bar{f}(\bar{y}) = 0\|.$$

But the inclusion $a \leqslant b \vee c$ is equivalent to $a \wedge c' \leqslant b$. Furthermore, in our situation we have

$$a \wedge c' = \|\bar{y} = n\| \wedge \|\bar{f}(\bar{y}) = 0\|' \leqslant \|\bar{f}(n) = 0\|'.$$

(Here n in $\bar{f}(n)$ is the constant random variable which is everywhere equal to n.)

It is thus sufficient to see that

$$\|\bar{f}(n)=0\|' \leqslant \|\bar{f}(0)=0\|' \vee \Big(\bigvee_{\bar{x}} \|\bar{f}(\bar{x})=\bar{f}(\bar{x}+1)\|' \Big),$$

or, taking complements, that

$$\|\bar{f}(n)=0\| \geqslant \|\bar{f}(0)=0\| \wedge \Big(\bigwedge_{\bar{x}} \|\bar{f}(\bar{x})=\bar{f}(\bar{x}+1)\| \Big).$$

The right side can only become larger if we only take the intersection over the terms with $\bar{x}=0, 1, 2, \ldots, n-1$. But this obviously gives

$$\|\bar{f}(0)=0\| \wedge \|\bar{f}(0)=\bar{f}(1)= \ldots =\bar{f}(n)\| \leqslant \|\bar{f}(n)=0\|. \qquad \square$$

3.10. Proof that $\|P_2\|(\xi')=0$. Using Lemma 3.9 and the rules for computing truth functions, we find:

$$\|P_2\|(\xi')=\bigvee_{\bar{f}} \bigwedge_{\bar{y}} \Big(\|\bar{h}(\bar{y})=0\|' \vee \bigvee_{\bar{x}} \Big(\bigvee_{n} \|\bar{x}=n\| \wedge \|\bar{y}=\bar{f}(\bar{x})\| \Big)\Big).$$

Since $\bar{f} \in \bar{R}^{(1)}$, we have $\|\bar{x}=n\| \leqslant \|\bar{f}(\bar{x})=\bar{f}(n)\|$, so that $\|\bar{x}=n\| \wedge \|\bar{y}=\bar{f}(\bar{x})\| \leqslant \|\bar{y}=\bar{f}(n)\|$.

Now it suffices to prove that the term corresponding to any concrete choice of $\bar{f}$ is equal to 0. We suppose that this is not the case, and show that we obtain a contradiction. Let $a \neq 0$ be the term corresponding to $\bar{f}$. By the previous paragraph, we have

$$a \leqslant \bigwedge_{\bar{y}} \Big(\|\bar{h}(\bar{y})=0\|' \vee \bigvee_{n} \|\bar{y}=\bar{f}(n)\| \Big).$$

In particular, for every $j \in \mathcal{J}$ we must have (with $\bar{x}_j$ in place of $\bar{y}$).

$$a \leqslant \bigvee_{n} \|\bar{x}_j=\bar{f}(n)\|$$

(where we have $\|\bar{h}(\bar{x}_j)=0\|'=0$ by 3.7(b)). Hence, for every j there exists an integer $n(j)$ such that $0 \neq \|\bar{x}_j=\bar{f}(n(j))\|$. Since $\mathcal{J}$ is uncountable, there exists an n_0 and an uncountable subset $\mathcal{J}_0 \subset \mathcal{J}$ such that $n(j_0)=n_0$ for all $j_0 \in \mathcal{J}_0$. Then the $\|\bar{x}_j=\bar{f}(n_0)\|$ for $j \in \mathcal{J}_0$ form an uncountable set of pairwise disjoint nonzero elements of B. This contradicts the countable chain condition on B. $\square$

4 Boolean-valued universes

4.1. In this section we fix a complete Boolean algebra B (see 2.6) and construct the universe V^B of "B-random sets." It will be a model for the Zermelo–Fraenkel axioms in the same generalized sense in which the random variables $\bar{R}$ were a model for the real numbers $\mathbf{R}$ in §3. In §§5–7 we verify that all the axioms of L_1Set are "true," and then in §8 we verify that the Continuum Hypothesis is "false" for a suitable choice of B.

The objects of V^B will be denoted by capital letters $X, Y, Z, \cdots$. Any two objects determine elements $\|X \in Y\| \in B$ and $\|X = Y\| \in B$. The intuitive meaning, say, of the first of these is as follows: if B is the algebra of measurable sets in a probability space, then $\|X \in Y\|$ is the maximal set on which "X is an element of Y with probability one." Since we do not deal with probability measures in the general case, we shall simply call the elements of B "probabilities," and then $\|X \in Y\|$ is simply the probability that X belongs to Y.

It is not trivial to construct precise definitions, because we want the axiom of extensionality to be "true." If a random set must be uniquely determined by its elements (which are also random), even in a generalized sense, then this random set cannot be "too" random (see 4.3).

We shall assume that as a set B is an element of the von Neumann universe V. Then all the objects of V^B will also be elements of V, and all our constructions can be expressed in L_1Set. In principle, this allows us to take a more formalistic point of view than we shall in fact take. The proof given below of the independence of the CH could then be used as a guide for constructing a much more syntactic version, based on an "internal interpretation" of the language L_1Set in itself. In this context the assumption that the Zermelo–Fraenkel axioms are consistent in the statement of Theorem 1.6 becomes a necessary precaution, since (by Gödel's result) this consistency cannot be established using only the language L_1Set itself. However, in our treatment this condition is pure hypocrisy, since by assuming the "existence" of the universe V, which is a model for the axioms, we automatically "prove" that those axioms are consistent (see subsection 18 of the Appendix to Chapter II).

4.2. *Construction of V^B*. For every ordinal α we construct the set V_α^B by transfinite recursion, and then set $V^B = \bigcup_\alpha V_\alpha^B$. The first step is: $V_0^B = \varnothing$.

Inductive assumption. The set V_α^B is defined for the ordinal $\alpha \geqslant 0$; for every element $X \in V_\alpha^B$ the set $D(X) \subset V_\alpha^B$ is defined (its intuitive meaning will be explained below); for every pair of elements $X, Y \in V_\alpha^B$ the "Boolean truth functions"

$$\|X \in Y\| \in B, \qquad \|X = Y\| \in B$$

are defined (intuitively, they should be thought of as the "probability that X is an element of Y" and the "probability that X coincides with Y," respectively).

By assumption, this data satisfies the following conditions:

(a) If $\beta_1 \leqslant \beta_2 \leqslant \alpha_1$, then $V_{\beta_1}^B \leqslant V_{\beta_2}^B$.

(b) If $\beta < \alpha$ and $X \in V_{\beta+1}^B \setminus V_\beta^B$, then $D(X) = V_\beta^B$.

(c$_1$) $\|X \in Y\| = \bigvee_{Z \in D(Y)} (\|X = Z\| \wedge \|Z \in Y\|)$ $\qquad (1)_\alpha$

(the condition $(1)_\alpha$ expresses the requirement that the formula $x \in y \Leftrightarrow \exists z(x = z \wedge z \in y)$, which is easily deduced from the Zermelo–Fraenkel axioms, must be "true").

$$(\mathrm{c}_2)\ \|X = Y\| = \Big(\bigwedge_{Z \in D(X)} \|Z \in X\|' \vee \|Z \in Y\|\Big) \wedge \Big(\bigwedge_{Z \in D(Y)} \|Z \in Y\|' \vee \|Z \in X\|\Big) \quad (2)_\alpha$$

(this condition expresses the "truth" of the formula $x = y \Leftrightarrow (\forall z(z \in x \rightarrow z \in y) \wedge \forall z(z \in y \Rightarrow z \in x))$. We note that it is not completely clear at this point why, for example, in $(1)_\alpha$ we only took the union over Z in $D(Y)$; it would seem natural to take all Z. Later we shall see that the formula remains true if we take the Boolean union over all Z.

This completes the description of the data for V_α^B. We now give explicitly the recursive construction of $V_{\alpha+1}^B$ and the corresponding data.

Definition of $V_{\alpha+1}^B$ and D. We set $V_{\alpha+1}^B = V_\alpha^B \cup V_{\alpha+1}^{B*}$, where $V_{\alpha+1}^{B*}$ consists of all possible functions Z with domain of definition V_α^B and range of values $\subset B$ which satisfy the following "extensionality condition":

$$\|X = Y\| \wedge Z(X) = \|X = Y\| \wedge Z(Y), \quad \text{for all } X, Y \in V_\alpha^B. \qquad (3)$$

A little later we shall define $\|X \in Z\| = Z(X)$ for $X \in V_\alpha^B$ and $Z \in V_{\alpha+1}^B \setminus V_\alpha^B$. Thus, as before, (3) can be thought of as reflecting the formula

$$(x = y \wedge x \in z) \Leftrightarrow (x = y \wedge y \in z).$$

Compare also with the comment in 2.7 concerning the definition of $\bar{R}^{(1)}$.

We shall call the elements of $V_{\alpha+1}^B \setminus V_\alpha^B$ *new* elements (of rank $\alpha + 1$), and we shall call the elements of V_α^B *old* elements. We set $D(Z) = V_\alpha^B$ if Z is a new element.

Definition of the Boolean truth functions. These functions have already been defined for pairs of old elements. We further set:

$$\|X \in Y\| = Y(X), \quad \text{if } X \text{ is old and } Y \text{ is new}; \qquad (4)$$

$$\|X = Y\| = \Big(\bigwedge_{Z \in D(X)} \|Z \in X\|' \vee \|Z \in Y\|\Big) \wedge \Big(\bigwedge_{Z \in D(Y)} \|Z \in Y\|' \vee \|Z \in X\|\Big). \qquad (5)$$

Because of $(2)_\alpha$, (5) automatically holds if X and Y are both old elements; in the other cases, (5) uniquely determines $\|X = Y\|$ if we use (4) and the fact that Z only runs through old elements in (5). Finally, we set:

$$\|X \in Y\| = \bigvee_{Z \in D(Y)} \|X = Z\| \wedge \|Z \in Y\|, \tag{6}$$

if X is a new element and Y is either new or old. The right side is uniquely determined using (4) and (5), since $D(Y) \subset V_\alpha^B$.

Formulas (4) and (6) show the following. As a first approximation we might say that a random set Y of rank α "consists" of sets Z of lower rank which occur in Y with probability $Y(Z)$; these probabilities can be chosen rather arbitrarily, subject only to the extensionality condition (3).

However, we then find (in formula (6) for new X and old Y) that we must automatically "include" more and more elements X in Y with probabilities already assigned by formula (6). It is conditions (3) and (6) which prevent our sets from being completely random.

Definition of V_α^B and other data for limiting ordinals α. We simply set $V_\alpha^B = \cup_{\beta<\alpha} V_\beta^B$, and then all the other data has already been determined.

4.3. *Verification that the definitions are correct*. Properties 4.2(a) and (b) are obviously preserved in going from α to $\alpha + 1$; we must verify $(1)_{\alpha+1}$ and $(2)_{\alpha+1}$. Now the only identity here which is not completely obvious is obtained by taking X old and Y new in $(1)_{\alpha+1}$:

$$Y(X) = \bigvee_{Z \in V_\alpha^B} \|X = Z\| \wedge Y(Z).$$

This is verified as follows. We obtain $\geqslant$ by writing the right-hand side in the form $\bigvee_Z \|X = Z\| \wedge Y(X)$ using (3). We obtain $\leqslant$ by considering the term with $Z = X$ and taking into account that $\|X = X\| = 1$ for all X (as follows immediately from (5)).

This completes the construction of the Boolean-valued universe.

4.4. Examples and remarks. We examine some special cases of these constructions in order to clarify their structure.

(a) Obviously $V_1^B = \{\varnothing\}$, since there exists a unique "empty" function, whose domain of definition is the subset $V_0^B = \varnothing$. We compute $V_2^B = V_1^B \cup V_2^{B*}$. We let $\{\varnothing\}_b \in V_2^{B*}$ denote the function of the one-element set V_1^B which takes the value $b \in B$. All these functions are extensional, so that

$$V_2^B = \{\varnothing, \{\varnothing\}_b, \quad \text{for all } b \in B\}.$$

It follows from (4) that

$$\|\varnothing \in \{\varnothing\}_b\| = b.$$

It is clear from (5) that

$$\|\varnothing = \{\varnothing\}_b\| = b'.$$

Intuitively, these formulas mean that $\{\varnothing\}_b$ consists of one element $\varnothing$ "over b" and is empty away from b. Again applying (5), we find:

$$\|\{\varnothing\}_a = \{\varnothing\}_b\| = (a' \vee b) \wedge (a \vee b') = (a \wedge b) \vee (a' \wedge b').$$

Thus, $\{\varnothing\}_a$ and $\{\varnothing\}_b$ coincide when either they are both empty or they both consist of one element $\varnothing$: this agrees with intuition. Now applying (6), we find:

$$\|\{\varnothing\}_a \in \{\varnothing\}_b\| = \|\{\varnothing\}_a = \varnothing\| \wedge \|\varnothing \in \{\varnothing\}_b\| = a' \wedge b$$

(i.e., the only possible inclusion, which has the form $\varnothing \in \{\varnothing\}$, holds when $\{\varnothing\}_a$ is empty and $\{\varnothing\}_b$ is nonempty).

Finally, let $X \in V_3^{B*}$ be an extensional function on the subset V_2^B with values in B. Then, by (6),

$$\|X \in \{\varnothing\}_b\| = \|X = \varnothing\| \wedge \|\varnothing \in \{\varnothing\}_b\| = \|X = \varnothing\| \wedge b,$$

and by (5)

$$\|X = \varnothing\| = \Big(\bigwedge_{a \in B} \|\{\varnothing\}_a \in X\|'\Big) \wedge \|\varnothing \in X\|'$$

$$= \Big(\bigvee_{a \in B} \|\{\varnothing\}_a \in X\| \vee \|\varnothing \in X\|\Big)'.$$

Thus, intuitively, $\|X = \varnothing\|$ means the complement of the support of X in B, and $\|X \in \{\varnothing\}_b\|$ is the set where both X is empty and $\{\varnothing\}_b$ is nonempty, which again agrees with the usual formula $\varnothing \in \{\varnothing\}$. This shows how new objects X can be random elements of old objects with nonzero probabilities.

(b) We consider the case $B = \{0, 1\}$. The corresponding probability space consists of one point, so our random sets become completely determined. What happens is: the universe V^B maps naturally onto the von Neumann universe V in such a way that, if $\tilde{X}$ denotes the image of $X \in V^B$, then all X and Y satisfy the conditions:

$$\|X \in Y\| = 1 \Leftrightarrow \tilde{X} \in \tilde{Y},$$
$$\|X = Y\| = 1 \Leftrightarrow \tilde{X} = \tilde{Y}.$$

To construct this map we first set $\tilde{\varnothing} = \varnothing$. We now suppose that the map $V_\alpha^{\{0,1\}} \to V_\alpha$ has already been constructed with the required properties, and we extend the map to $\alpha + 1$. To do this, for any new element $X \in V_{\alpha+1}^{\{0,1\}}$ we first find the subset of $V_\alpha^{\{0,1\}}$ on which X takes the value 1, and we then take the image of this subset in V_α, which is an element $\tilde{X}$ of $\mathcal{P}(V_\alpha) = V_{\alpha+1}$; by definition, our map takes X to this $\tilde{X}$. We leave the verification of the properties of this map to the reader.

(c) *Boolean truth functions for the formulas in* $\mathrm{L_1Set}$.

We define these truth functions in an analogous manner to §2. We introduce the interpretation class $\overline{M}$: each point $\xi \in \overline{M}$ assigns to every variable symbol x in $\mathrm{L_1Set}$ some object $x^\xi = X$ of the universe V^B. We further assume that every point ξ maps the symbol $\varnothing$ in L_1Set to the empty set.

If P is the atomic formula $x \in y$ or $x = y$ in L_1Set, then $\|P\|(\xi)$ is defined to be $\|x^\xi \in y^\xi\| \in B$ or $\|x^\xi = y^\xi\| \in B$, respectively. The value of $\|P\|(\xi)$ for all other P is defined inductively using exactly the same formulas as in subsection 2.7. We need only note that, although the expressions $\bigvee_\xi a_\xi$ and $\bigwedge_\xi a_\xi$ must be taken over families indexed by the *class* $\overline{M}$ when we compute with quantifiers, all the *different* elements of such a family form a *subset* of B, so that such an expression makes sense. We shall call a formula P "true" (in the model V^B) if $\|P\|(\xi) = 1$ for all ξ, and we shall call P "false" if $\|P\|(\xi) = 0$ for all ξ.

As in §3 of Chapter II, it can be verified that all the tautologies and logical quantifier axioms are "true" and that the rules of deduction preserve "truth." Hence, it remains for us to show that the Zermelo–Fraenkel axioms are "true" (for any B) and that the Continuum Hypothesis is "false" (for suitable B).

5 The axiom of extensionality is "true"

We begin by proving some relations between the truth functions. First of all, it is clear from formula (5) in §4 that $\|X = Y\| = \|Y = X\|$ and $\|X = X\| = 1$. The following lemma is a less immediate consequence of the formulas.

5.1. **Lemma.** *For any $X, Y, Z \in V^B$ we have:*

$$\|X = Y\| \wedge \|Y = Z\| \leqslant \|X = Z\|, \tag{I}$$

$$\|X = Y\| \wedge \|Y \in Z\| \leqslant \|X \in Z\|, \tag{II}$$

$$\|X \in Y\| \wedge \|Y = Z\| \leqslant \|X \in Z\|. \tag{III}$$

PROOF.

(a) (III) *holds if $X \in D(Y)$*. In fact, then by formula (5) in §4

$$\|Y = Z\| \leqslant \|X \in Y\|' \vee \|X \in Z\|,$$

so that, if we intersect both sides with $\|X \in Y\|$, we obtain (III).

(b) (III) *holds if $X, Y \in V_\alpha^B$ and Z is a new element of $V_{\alpha+1}^B$*. In fact, we choose $U \in D(Y)$ and apply the special case of (III) proved in (a):

$$\|U \in Y\| \wedge \|Y = Z\| \leqslant \|U \in Z\|.$$

We take the Boolean intersection of both sides with $\|X = U\|$ and then the Boolean sum over all $U \in D(Y)$. Now applying formula (6) in §4 to the left-hand side and using distributivity, we obtain:

$$\|X \in Y\| \wedge \|Y = Z\| \leqslant \bigvee_{U \in D(Y)} \|X = U\| \wedge \|U \in Z\|$$

$$\leqslant \bigvee_{U \in D(Z) = V_\alpha^B} \|X = U\| \wedge \|U \in Z\| = \|X \in Z\|.$$

(c) (I) *holds in* $V^B_{\alpha+1}$ *if* (III) *holds in* V^B_α. We consider an element $U \in D(X) \in V^B_\alpha$. By (a), we have

$$\|U \in X\| \wedge \|X = Y\| \leqslant \|U \in Y\|.$$

We take the Boolean intersection with $\|Y = Z\|$

$$\|U \in X\| \wedge \|X = Y\| \wedge \|Y = Z\| \leqslant \|U \in Y\| \wedge \|Y = Z\|.$$

Here the right side is always $\leqslant \|U \in Z\|$. In fact, if $Y \in V^B_\alpha$ this follows by part (b) or by the induction assumption, and if Y is a new element of $V^B_{\alpha+1}$ then it follows by part (a).

We have thus shown that for all $X, Y, Z \in V^B_{\alpha+1}$ and all $U \in D(X)$:

$$\|U \in X\| \wedge \|X = Y\| \wedge \|Y = Z\| \leqslant \|U \in Z\|.$$

Because $a \wedge b \leqslant c$ implies $b \leqslant a' \vee c$ in any Boolean algebra, we then obtain

$$\|X = Y\| \wedge \|Y = Z\| \leqslant \|U \in X\|' \vee \|U \in Z\|,$$

and hence

$$\|X = Y\| \wedge \|Y = Z\| \leqslant \bigwedge_{U \in D(X)} \|U \in X\|' \vee \|U \in Z\|.$$

Interchanging X and Z, we find that for all $U \in D(Z)$

$$\|Z = Y\| \wedge \|Y = X\| \leqslant \bigwedge_{U \in D(Z)} \|U \in Z\|' \vee \|U \in X\|.$$

These last two formulas, together with (5), clearly imply (I).

(d) (II) holds in $V^B_{\alpha+1}$ if (I) holds in $V^B_{\alpha+1}$. In fact, let $U \in D(Z)$. By (I), we have

$$\|X = Y\| \wedge \|Y = U\| \leqslant \|X = U\|.$$

We take the Boolean intersection with $\|U \in Z\|$ and then the Boolean sum over all $U \in D(Z)$:

$$\|X = Y\| \wedge \Big(\bigvee_{U \in D(Z)} \|U \in Z\| \wedge \|Y = U\| \Big)$$
$$\leqslant \bigvee_{U \in D(Z)} \|Z = U\| \wedge \|U \in Z\|.$$

Applying $(1)_{\alpha+1}$ in §4, we obtain (II).

(e) (III) holds in $V^B_{\alpha+1}$ if (II) holds in $V^B_{\alpha+1}$. In fact, let $U \in D(Y)$. By part (a), we have

$$\|U \in Y\| \wedge \|Y = Z\| \leqslant \|U \in Z\|.$$

Intersecting with $\|X = U\|$ and applying (II) to the right-hand side, we obtain

$$\|X = U\| \wedge \|U \in Y\| \wedge \|Y = Z\| \leqslant \|X \in Z\|.$$

Finally, if we take the Boolean sum over all $U \in D(Y)$ and use formula (1) in §4, we obtain (III). □

Obviously, parts (a)–(e) prove the inductive step for α to $\alpha + 1$. We are now in a position to establish the basic result of this section.

5.2. **Proposition.** *The axiom of extensionality*

$$x = y \Leftrightarrow \forall z (z \in x \Leftrightarrow z \in y)$$

is "true."

PROOF. The formula $\|P \Leftrightarrow Q\|(\xi) = 1$ is equivalent to $\|P\|(\xi) = \|Q\|(\xi)$. It is therefore sufficient to prove that for all $X, Y \in V^B$

$$\|X = Y\| = \bigwedge_{Z \in V^B} (\|Z \in X\| \vee \|Z \in Y\|') \wedge (\|Z \in X\|' \vee \|Z \in Y\|).$$

The inequality $\geqslant$ follows immediately from formula (2) in §4. To obtain the opposite inequality, we write two obvious corollaries of formula (III) in Lemma 5.1:

$$\|X = Y\| \leqslant \|Z \in X\| \vee \|Z \in Y\|',$$
$$\|X = Y\| \leqslant \|Z \in X\|' \vee \|Z \in Y\|,$$

and we take the intersection over all Z. The proposition is proved. □

We note that formula (2) implies the following general extensionality property: for all $X, Y, Z \in V^B$

$$\|X = Y\| \wedge \|Y \in Z\| = \|X = Y\| \wedge \|X \in Z\|.$$

5.3. **Corollary.** *The axioms of equality in* $\mathrm{L_1Set}$ *are "true."*

In fact (see Proposition 4.6 in Chapter II), the axioms of equality in our case consist of: the "true" formula $x = x$, the axiom of extensionality (in the form $x = y \Rightarrow (P(x) \Rightarrow P(y))$ with $P(x) = z \in x$), and the "true" formula $x = y \Rightarrow (x \in z \Rightarrow y \in z)$ (in which $P(x) = x \in z$), since the only atomic formulas $P(x)$ in $\mathrm{L_1Set}$ are $z \in x$ and $x \in z$. □

5.4. *Remark.* In most computations, we shall only need to know the values of $\|X \in Y\|$ and $\|X = Y\|$, and not the precise definition of the objects X and Y. In this connection, we note that the following two binary relations on V^B coincide (as easily follows from (III) and the axiom of extensionality):

$$\text{(a) } \|X = Y\| = 1,$$
$$\text{(b) } \forall Z \in V^B, \qquad \|Z \in X\| = \|Z \in Y\|.$$

We shall call such X and Y *equivalent* and write $X \sim Y$.

6 The axioms of pairing, union, power set, and regularity are "true"

6.1. The computations in the previous section show that the basic work in ensuring that the axiom of extensionality is "true" was already incorporated into the definition of the universe V^B. The explicit formulas for recursively computing $\|X \in Y\|$ and $\|X = Y\|$ reflected so many special properties of inclusion and equality that together they guaranteed that the general axiom must hold.

In order to verify several of the other axioms, we must essentially define in V^B analogues of certain operations in V, such as forming the unordered pair, the set of subsets, and so on. These operations can be defined by means of formulas in $\mathrm{L}_1\mathrm{Set}$. However, recall that, if $P(x)$ is a formula with one free variable x, then the $x^\xi \in V$ for which $P(x)(\xi)$ is true generally form a class and not a set.

It will be convenient to introduce the auxiliary notion of a "random class" in V^B. Using this concept, we shall often construct the operations in V^B in two stages: the value of the operation will at first be a random class, which we then "identify" with a random set using a separate argument.

6.2. **Definition.**

(a) A random class is any function W on V^B with values in B which satisfies the following extensionality condition:

$$W(X) \wedge \|X = Y\| = W(Y) \wedge \|X = Y\|, \quad \text{for all } X, Y \in V^B.$$

(b) A random class W is said to be equivalent to a random set $Z \in V^B$ (written $W \sim Z$) if

$$W(X) = \|X \in Z\|, \quad \text{for all } X \in V^B.$$

6.3. Examples and remarks

(a) For any random set Z the function $X \mapsto \|X \in Z\|$ is extensional by (II), §5, and so is a random class. By analogy, we often write $\|X \in W\|$ instead of $W(X)$ if W is any random class.

(b) There exist random classes which are not equivalent to random sets. One such example is the "universal" random class $W(X) = 1$ for all X. (If W were a set, we would have $\|W \in W\| = 1$, contradicting the regularity axiom, which will be shown to be "true" below.)

(c) Let W be a random class, and let α be any ordinal. We define the element $W_\alpha \in V^B_{\alpha+1}$ as follows:

$D(W_\alpha) = V^B_\alpha$, $W_\alpha =$ the restriction of W to V^B_α (as a function; see 4.2). It is easy to see that for all $X \in V^B$ we have:

$$\|X \in W_\alpha\| \leqslant \|X \in W\|. \tag{1}$$

In fact, let $U \in V_\alpha^B$ and $X \in V^B$. We then have

$$\|X = U\| \wedge W_\alpha(U) = \|X = U\| \wedge W(U) = \|X = U\| \wedge W(X) \leqslant W(X),$$

so that, by (6), §4:

$$\|X \in W_\alpha\| = \bigvee_{U \in V_\alpha^B} \|X = U\| \wedge W_\alpha(U) \leqslant W(X) = \|X \in W\|.$$

We shall often show that some class W which we are interested in is equivalent to a set by finding an ordinal α such that $W \sim W_\alpha$. It is clear from (1) that this follows if $\|X \in W\| \leqslant \|X \in W_\alpha\|$ for all X.

(d) Let W, W_1, and W_2 be random classes. Then W', $W_1 \wedge W_2$, and $W_1 \vee W_2$ are also random classes, since the extensionality condition is trivially verified for these functions. We shall write $W_1 \cap W_2$ and $W_1 \cup W_2$ instead of $W_1 \wedge W_2$ and $W_1 \vee W_2$, respectively.

(e) Let W be a random class, and let X be a random set. We show that $W \cap X$ is equivalent to a random set. More precisely, if $D(X) = V_\alpha^B$, then $W \cap X \sim (W \cap X)_\alpha$. In fact, for any $Y \in V^B$ it follows by (6), §4 that:

$$\begin{aligned}
\|Y \in (W \cap X)_\alpha\| &= \bigvee_{U \in V_\alpha^B} \|U = Y\| \wedge \|U \in (W \cap X)_\alpha\| \\
&= \bigvee_{U \in V_\alpha^B} (\|U = Y\| \wedge \|U \in W\|) \wedge \|U \in X\| \\
&= \bigvee_{U \in V_\alpha^B} \|U = Y\| \wedge \|Y \in W\| \wedge \|U \in X\| \\
&= \|Y \in W\| \wedge \|Y \in X\| = \|Y \in W \cap X\|.
\end{aligned}$$

This result implies that the separation axioms are "true" (see subsection 4.9(b) of Chapter II).

The following proposition gives a general method for constructing random classes.

6.4. **Proposition.** *Let $P(x, y_1, \ldots, y_n)$ be a formula which does not contain any free variables besides $x, y_1, \ldots, y_n$. Let $Y_1, \ldots, Y_n \in V^B$ be fixed. Then the function*

$$X \mapsto W(X) = \|P(X, Y_1, \ldots, Y_n)\|$$

is a random class.

Intuitively, W contains every set X with probability equal to the probability that $P(X, \ldots, Y_n)$ is true. $Y_1, \ldots, Y_n$ play the role of "constants."

Proof. We use the "truth" of the following axiom of equality:

$$\|\forall x\, \forall y_1 \cdots \forall y_n (x = y \Rightarrow (P(x, y_1, \ldots, y_n) \Rightarrow P(y, y_1, \ldots, y_n)))\| = 1.$$

If we take a point ξ in the interpretation class which assigns to x, y,

$y_1, \ldots, y_n$ the values $X, Y, Y_1, \ldots, Y_n$, respectively, then we find that

$$\|X = Y\| \leqslant \|P(X, Y_1, \ldots, Y_n)\|' \vee \|P(Y, Y_1, \ldots, Y_n)\|,$$

or

$$\|X = Y\| \wedge W(X) \leqslant W(Y),$$

so that W is extensional. □

We are now ready to verify the axioms.

6.5. **Proposition.** *The axiom of pairing*

$$\forall u\, \forall w\, \exists x\, \forall z (z \in x \Leftrightarrow z = u \vee z = w)$$

is "true."

Proof. By definition we have

$$\|\forall u\, \forall w\, \exists x\, \forall z (z \in x \Leftrightarrow z = u \vee z = w)\|$$
$$= \bigwedge_U \bigwedge_W \bigvee_X \bigwedge_Z \|Z \in X \Leftrightarrow Z = U \vee Z = W\|.$$

Hence it suffices if, for any $U, W \in V^B$, we find an $X \in V^B$ such that for all $Z \in V^B$

$$\|Z \in X\| = \|Z = U\| \vee \|Z = W\|. \tag{2}$$

For fixed U and W we consider the right side of (2) as a function of Z. This function is a random class X by Proposition 6.4, since it corresponds to the formula $z = U \vee z = W$. We show that it is equivalent to a random set; more precisely, if $U, W \in V_\alpha^B$, then $X \sim X_\alpha$. By the remark at the end of 6.3(c), it suffices to verify that for all Z

$$\|Z \in X\| \leqslant \|Z \in X_\alpha\|.$$

But since $\|U \in X_\alpha\| = 1$, it follows by formula (II) in §5 that

$$\|Z = U\| \leqslant \|Z \in X_\alpha\|,$$

and similarly

$$\|Z = W\| \leqslant \|Z \in X_\alpha\|,$$

which gives the required inequality. □

6.6. **Proposition.** *The axiom of union*

$$\forall x\, \exists y\, \forall u (\exists z (u \in z \wedge z \in x) \Leftrightarrow u \in y)$$

is "true."

Proof. We fix $X \in V^B$ and construct a random set Y such that for all $U \in V^B$

$$\|U \in Y\| = \|\exists z (U \in z \wedge z \in X)\| = \bigvee_{Z \in V^B} \|U \in Z\| \wedge \|Z \in X\|.$$

By Proposition 6.4, there exists a random class Y with this property. We show that if $D(X) = V_\alpha^B$, then $Y \sim Y_\alpha$. Since $D(Y_\alpha) = D(X)$, we have

$$\|U \in Y_\alpha\| = \bigvee_{Z \in D(X)} \|U = Z\| \wedge \|Z \in Y_\alpha\|$$

$$= \bigvee_{Z \in D(X)} \|U = Z\| \wedge \Big(\bigvee_{Z_1 \in V^B} \|Z \in Z_1\| \wedge \|Z_1 \in X\| \Big). \quad (3)$$

We show that the inner sum in (3) may be taken only over $Z_1 \in D(X)$. In fact, for any Z_1

$$\|Z_1 \in X\| = \bigvee_{Z_2 \in D(X)} \|Z_1 = Z_2\| \wedge \|Z_2 \in X\|,$$

so that

$$\|Z \in Z_1\| \wedge \|Z_1 \in X\| = \bigvee_{Z_2 \in D(X)} \|Z \in Z_1\| \wedge \|Z_1 = Z_2\| \wedge \|Z_2 \in X\|$$

$$\leqslant \bigvee_{Z_2 \in D(X)} \|Z \in Z_2\| \wedge \|Z_2 \in X\|. \quad (4)$$

Taking this into account, in (3) we first sum over Z for fixed $Z_1 \in D(X)$. Since $D(Z_1) \leqslant D(X)$, the sum over $Z \in D(X)$ coincides with the sum over $Z \in D(Z_1)$, and is equal to $\|U \in Z_1\|$. Thus,

$$\|U \in Y_\alpha\| = \bigvee_{Z_1 \in D(X)} \|U \in Z_1\| \wedge \|Z_1 \in X\|$$

$$= \bigvee_{Z_1 \in V^B} \|U \in Z_1\| \wedge \|Z_1 \in X\| = \|U \in Y\|,$$

by (4). □

6.7. **Proposition.** *The power set axiom*

$$\forall x\ \exists y\ \forall z (z \subset x \Leftrightarrow z \in y)$$

is "true." (Recall that $z \subset x$ is abbreviated notation for $\forall u (u \in z \Rightarrow u \in x)$.)

Proof. We fix $X \in V^B$ and construct a $Y \in V^B$ such that for all $Z \in V^B$

$$\|Z \in Y\| = \|Z \subset X\| = \bigwedge_{U \in V^B} \|U \in Z\|' \vee \|U \in X\|.$$

By Proposition 6.4, the right side defines Y as a random class. We show that, if $D(X) = V_\alpha^B$, then $Y \sim Y_{\alpha+1}$.

We first construct the element $Z_\alpha \in V_{\alpha+1}^B$ by considering Z as a random class. By (1) we have $\|U \in Z_\alpha\|' \geqslant \|U \in Z\|$; so that

$$\|Z \in Y\| \leqslant \|Z_\alpha \in Y\| = \|Z_\alpha \in Y_{\alpha+1}\|. \quad (5)$$

If we prove the inequality

$$\|Z \in Y\| \leqslant \|Z_\alpha = Z\|, \quad (6)$$

it will immediately follow from (5) and (6) that $Y \sim Y_{\alpha+1}$, since, by (II), §5,

$$\|Z \in Y\| \leqslant \|Z_\alpha \in Y_{\alpha+1}\| \wedge \|Z_\alpha = Z\| \leqslant \|Z \in Y_{\alpha+1}\|.$$

It remains to verify (6).

First let $U \in D(X) = V_\alpha^B$. Then $\|U \in Z_\alpha\| = \|U \in Z\|$, so that $\|U \in Z_\alpha \Leftrightarrow U \in Z\|' = 0$, and *a fortiori*

$$\|U \in X\| \wedge \|U \in Z_\alpha \Leftrightarrow U \in Z\|' = 0. \tag{7}$$

As U varies, the left side of (7) determines a random class of the form $X \cap W$, where W corresponds to the formula $\neg(u \in Z_\alpha \Leftrightarrow u \in Z)$. Since $D(X) = V_\alpha^B$, it follows by 6.3(c) that $X \cap W \sim (X \cap W)_\alpha$. But, according to (7), $(X \cap W)_\alpha$ is the zero function on V_α^B. Thus, $\|U \in X \cap W\| = 0$ for all $U \in V^B$. Consequently,

$$\|U \in X\| \leqslant \|U \in Z_\alpha \Leftrightarrow U \in Z\| \quad \text{for all } U. \tag{8}$$

To prove (6), we now write the left and right-hand sides separately (using the "truth" of the formula $Z_\alpha = Z \Leftrightarrow \forall u(u \in Z_\alpha \Leftrightarrow u \in Z)$):

$$\|Z \in Y\| = \bigwedge_{U \in V^B} \|U \in Z\|' \vee \|U \in X\|,$$

$$\|Z_\alpha = Z\| = \bigwedge_{U \in V^B} \|U \in Z_\alpha \Leftrightarrow U \in Z\|.$$

It is now clear that the inequality in (6) holds term by term. In fact, for $\|U \in X\|$ this follows from (8), and for $\|U \in Z\|'$ it follows because

$$\|U \in Z_\alpha \Leftrightarrow U \in Z\| = (\|U \in Z_\alpha\|' \vee \|U \in Z\|) \wedge (\|U \in Z_\alpha\| \vee \|U \in Z\|'),$$

and $\|U \in Z\|' \leqslant \|U \in Z_\alpha\|'$ for all U. □

6.8. **Proposition.** *The regularity axiom*

$$\forall x(\exists y(y \in x) \Rightarrow \exists y(y \in x \wedge y \cap x = \varnothing))$$

is "true."

Proof. We fix $X \in V^B$. The axiom with the "constant" X in place of x has the form $R \Rightarrow S$. We must show that $\|R \Rightarrow S\| = 1$. It suffices to prove that $\|R\| \wedge \|S\|' = 0$, where

$$\|R\| = \bigvee_{Y \in V^B} \|Y \in X\|, \tag{9}$$

$$\|S\|' = \bigwedge_{Y \in V^B} \|Y \in X\|' \vee \Big(\bigvee_{Z \in V^B} \|Z \in Y\| \wedge \|Z \in X\|\Big). \tag{10}$$

We suppose that $\|R\| \wedge \|S\|' = a \neq 0$, and show that this leads to a contradiction. It follows from (9) and (10) that there exists a $Y \in V^B$ such that $\|Y \in X\| \wedge a \neq 0$. We choose Y to have the least rank of any element with this property.

It is again clear from (9) and (10) that

$$\|Y \in X\| \wedge a \leqslant \bigvee_{Z \in V^B} \|Z \in Y\| \wedge \|Z \in X\|.$$

On the right we may sum only over $Z \in D(Y)$, without changing the value of the sum. Hence, there must exist a $Z \in D(Y)$ such that

$$\|Z \in X\| \wedge \|Y \in X\| \wedge a \neq 0,$$

so that $\|Z \in X\| \wedge a \neq 0$. But the rank of Z is less than the rank of Y, contradicting the choice of Y. □

7 The axioms of infinity, replacement, and choice are "true"

7.1. We begin this section by describing two more methods for constructing random sets. The first of them, which is very widely used, solves the following problem. Suppose we are given a set of objects $X_i \in V^B$, $i \in I$, and a set of elements $a_i \in B$. We would like to construct a random set X which contains each X_i with probability a_i, but such an X might not exist. However, it turns out that there always exists an X with $\|X_i \in X\| \geqslant a_i$ for all $i \in I$; moreover, there exists a least X with this property.

7.2. **Lemma.**

(a) *Under the conditions in* 7.1, *the function X of Y*

$$\|Y \in X\| = \bigvee_{i \in I} a_i \wedge \|Y = X_i\| \tag{1}$$

is a random class X which is equivalent to a random set. In addition, $\|X_i \in X\| \geqslant a_i$, and, if X' is any random class such that $\|X_i \in X'\| \geqslant a_i$ for each i, then $\|Y \in X'\| \geqslant \|Y \in X\|$ for all Y.

We shall say that X (or the equivalent random set) *collects the X_i with probabilities a_i*.

(b) *Under the same conditions, the function Z of Y*

$$\|Y \in Z\| = \bigvee_i a_i \wedge \|Y \in X_i\| \tag{2}$$

is a random class Z which is equivalent to a random set. If we also have $a_i \wedge a_j = 0$ for all $i \neq j$, then $\|Z = X_i\| \geqslant a_i$, and, for any random class Z' such that $\|Z' = X_i\| \geqslant a_i$ for each i, we have $\|Y \in Z'\| \geqslant \|Y \in Z\|$ for all Y.

We shall say that Z *glues together the X_i with probabilities a_i*.

Proof. It is easily verified that the functions Z and X defined by formulas (1) and (2) are extensional.

There exists an ordinal α such that $X_i \in V_\alpha^B$ for all i. We show that $X \sim X_\alpha$ and $Z \sim Z_\alpha$. For any $Y \in V^B$ we have:

$$\begin{aligned}
\|Y \in X_\alpha\| &= \bigvee_{U \in V_\alpha^B} \|Y = U\| \wedge \|U \in X_\alpha\| \\
&= \bigvee_{U \in V_\alpha^B} \bigvee_i \|Y = U\| \wedge a_i \wedge \|U = X_i\| \\
&= \bigvee_{U \in V_\alpha^B} \bigvee_i a_i \wedge \|Y = X_i\| \wedge \|U = X_i\|.
\end{aligned}$$

If we consider the term with $U = X_i$ on the right, we obtain $a_i \wedge \|Y = X_i\| \leqslant \|Y \in X_\alpha\|$, so that $\|Y \in X\| \leqslant \|Y \in X_\alpha\|$ by (1), and the assertion follows by 6.3(c).

Similarly, for any $Y \in V^B$ we have

$$\begin{aligned}
\|Y \in Z_\alpha\| &= \bigvee_{U \in V_\alpha^B} \bigvee_i \|Y = U\| \wedge a_i \wedge \|U \in X_i\| \\
&= \bigvee_{U \in V_\alpha^B} \bigvee_i a_i \wedge \|Y \in X_i\| \wedge \|Y = U\|.
\end{aligned}$$

Since $\|Y \in X_i\| = \bigvee_{U \in V_\alpha^B} \|Y = U\| \wedge \|Y \in X_i\|$, it follows that $a_i \wedge \|Y \in X_i\| \leqslant \|Y \in Z_\alpha\|$, and $\|Y \in Z\| \leqslant \|Y \in Z_\alpha\|$ by (2).

Now let X' and Z' be any random sets with the properties in (a) and (b). It is clear from (1) that $\|X_i \in X\| \geqslant a_i$. If $\|X_i \in X'\| \geqslant a_i$ for each i, then $\|Y \in X'\| = \bigvee_U \|Y = U\| \wedge \|U \in X'\| \geqslant \bigvee_i \|Y = X_i\| \wedge \|X_i \in X'\| \geqslant \|Y \in X\|$ by (1).

Similarly, if $a_i \wedge a_j = 0$ for $i \neq j$, then it is clear from (2) that $a_i \wedge \|Y \in Z\| = a_i \wedge \|Y \in X_i\|$, so that

$$a_i \wedge \|X_i = Z\| = \bigvee_Y a_i \wedge \|Y \in X_i \Leftrightarrow Y \in Z\| = a_i$$

and $\|X_i = Z\| \geqslant a_i$. Now if $\|X_i = Z'\| \geqslant a_i$ for each i, then

$$\begin{aligned}
\|Y \in Z'\| &\geqslant \|Y \in Z'\| \wedge \|Z' = X_i\| \\
&= \|Y \in X_i\| \wedge \|Z' = X_i\| \geqslant a_i \wedge \|Y \in X_i\|,
\end{aligned}$$

so that $\|Y \in Z'\| \geqslant \|Y \in Z\|$. □

Here is our first application of Lemma 7.2(a):

7.3. Proposition. *The axiom of infinity*

$$\exists x(\varnothing \in x \wedge \forall u(u \in x \Rightarrow \{u\} \in x))$$

is "true."

Proof. When we proved that the axiom of pairing is "true," we constructed for any U, $W \in V^B$ an element $Z \in V^B$ (unique up to equivalence) with the property that $\|Y \in Z\| = \|Y = U \vee Y = W\|$ for all Y. It is natural to let $\{U, W\}^B$ denote this element Z, and let $\{U\}^B = \{U, U\}^B$.

We now verify the axiom of infinity. We set $X_0 = \varnothing$, $X_1 = \{\varnothing\}^B, \ldots, X_n = \{X_{n-1}\}^B, \ldots$. Further, we let $X \in V^B$ be the element which collects all the X_i with probabilities 1. We show that

$$\|\varnothing \in X \wedge \forall u(u \in X \Rightarrow \{u\} \in X)\| = 1.$$

It is obviously sufficient to prove that for all $U \in V^B$ we have $\|U \in X\| \leqslant \|\{U\}^B \in X\|$, that is, by (1):

$$\bigvee_{i=0}^{\infty}\|U = X_i\| \leqslant \bigvee_{i=0}^{\infty}\|\{u\}^B = X_i\|.$$

In fact, since the formula $u = x \Leftrightarrow \{u\} = \{x\}$ is "true," and since $X_{i+1} = \{X_i\}^B$, it immediately follows that

$$\|U = X_i\| = \|\{U\}^B = X_{i+1}\|. \qquad \square$$

7.4. **Lemma.** *Let W be a random class. Then there exists an element $X \in V^B$ such that*

$$\bigvee_{U \in V^B} W(U) = W(X).$$

The left-hand side may be represented in the form $\|\exists x(x \in W)\| = \|W \neq \varnothing\|$. Hence, intuitively, the lemma says that the probability that a given class is nonempty coincides with the probability that a suitable element occurs in it.

Proof. We first show that there exists an ordinal β such that $\bigvee_{U \in V^B} W(U) = \bigvee_{U \in V^B_\beta} W(U)$. In fact, let $a_\gamma = \bigvee_{U \in V^B_\gamma} W(U)$, and for any $a \in B$ set $\gamma(a) = \min(\gamma | a_\gamma > a)$ (or $\gamma(a) = 0$ if $a_\gamma \ngtr a$ for all γ). Finally, set $\beta = \sup_{a \in B} \gamma(a)$. This is an ordinal, because B is a set. If $\gamma > \beta$, then $a_\gamma \geqslant a_\beta$ by monotonicity, but we cannot have $a_\gamma > a_\beta$ because of the choice of β.

Thus, let $\bigvee_U W(U) = \bigvee_{U \in V^B_\beta} W(U)$. We index all the elements in V^B_β by an initial segment of ordinals (by the axiom of choice!): $V^B_\beta = \{U_\alpha\}_{\alpha \in I}$. We set

$$a_\alpha = W(U_\alpha) \wedge \Big(\bigvee_{\gamma < \alpha} W(U_\gamma)\Big)', \quad \alpha \in I.$$

Obviously $a_\alpha \wedge a_\gamma = 0$ for $\alpha \neq \gamma$. Using Lemma 7.2(b), we glue together the sets U_α with probabilities $a_\alpha (\alpha \in I)$. We obtain a set X satisfying the conditions $\|X = U_\alpha\| \geqslant a_\alpha \geqslant W(U_\alpha)$. Using the extensionality of W, we find:

$$W(X) \geqslant \bigvee_{\alpha \in I} \|X = U_\alpha\| \wedge W(U_\alpha) = \bigvee_{\alpha \in I} W(U_\alpha) = \bigvee_{U \in V^B} W(U). \quad \square$$

7.5. **Proposition.** *The replacement axiom*

$$\forall \bar{z}\, \forall u\big(\forall x(x \in u \Rightarrow \exists! y\, P(x, y, \bar{z}))$$

$$\Rightarrow \exists w\, \forall y\big(y \in w \Leftrightarrow \exists x(x \in u \wedge P(x, y, \bar{z}))\big)\big)$$

is "true" (*here* $\bar{z} = \langle z_1, \ldots, z_n \rangle$).

Proof. We fix a "vector" $\bar{Z} = \langle Z_1, \ldots, Z_n \rangle$ with $Z_i \in V^B$ and an element $U \in V^B$. We shall write $P(x, y)$ instead of $P(x, y, \bar{Z})$. If we write the axiom with the "constants" Z_i and U in the form $R \Rightarrow S$, then we must prove that $\|R \Rightarrow S\| = 1$.

7.6. *The special case*: If $\|R\| = 1$, then $\|S\| = 1$.

We first show how the general case follows from this special case. Let $a \in B$, and let B_a denote the set $\{b \in B \mid b \leqslant a\}$. The operations on B induce a Boolean algebra structure on B_a with unit element $1_a = a$. The natural mapping $B \to B_a$: $b \mapsto b \wedge a$ is a homomorphism. An easy induction on a allows us to construct a surjective map of universes $V^B \to V^{B_a}$: $X \mapsto X_a$ such that for all $X, Y \in V^B$ we have

$$\|X_a \in Y_a\| = \|X \in Y\| \wedge a,$$
$$\|X_a = Y_a\| = \|X = Y\| \wedge a.$$

Now, to prove Proposition 7.5 from the special case 7.6, we choose $a = \|R\|$. Then $\|R\|_a = 1_a$, so that 7.6 implies that $\|S\|_a = 1_a$. This means that $\|S\| \geqslant a$, and hence $\|R \Rightarrow S\| = 1$. (Here we have used 7.6 in V^{B_a}; clearly $\|R\|_a = \|R_a\|$, where R_a is the obvious image of R in V^{B_a}.)

7.7. Proof of 7.6. The condition $\|R\| = 1$ means that for any $X \in V^B$

$$\|X \in U\| \leqslant \|\exists! y\, P(X, y)\|. \tag{3}$$

To show that $\|S\| = 1$, it is sufficient if, given $U \in V^B$, we find a $W \in V^B$ such that for all $Y \in V^B$

$$\|Y \in W\| = \bigvee_{X \in V^B} \|X \in U\| \wedge \|P(X, Y)\|. \tag{4}$$

It follows from 6.5 that the formula (4) defines W as a random class. We find an ordinal α such that $W \sim W_\alpha$.

To do this, we first note that in (4) we may take the sum only over

$$\|Y \in W\| = \bigvee_{X \in D(U)} \|X \in U\| \wedge \|P(X, Y)\| \tag{5}$$

(the argument here is the same as after formula (3) in §6). We now apply Lemma 7.4 to the class $W_X(Y) = \|P(X, Y)\|$. It follows that for every $X \in D(U)$ there exists an element $Y_X \in V^B$ such that

$$\|\exists y\, P(X, y)\| = \|P(X, Y_X)\|. \tag{6}$$

(Because $\|\exists! y\, P(X, y)\| \leqslant \|\exists y\, P(X, y)\|$, we can use these Y_X to estimate $\|X \in U\|$ with the help of (9) below.)

We set $\alpha_X = \min(\alpha | Y_X \in V_\alpha^B)$, and

$$\alpha = \sup(\alpha_X | X \in D(U)),$$

and then show that $W \sim W_\alpha$ for this α. We must verify that $\|Y \in W\| \leqslant \|Y \in W_\alpha\|$ for every Y. By (5) and by formula (II) in §5, this follows if for any $X \in D(U)$ we have

$$\|X \in U\| \wedge \|P(X, Y)\| \leqslant \|Y = Y_X\| \wedge \|Y_X \in W_\alpha\|. \tag{7}$$

In the first place, by (3), (6), (5), and the definition of α, we have:

$$\begin{aligned} &\|X \in U\| \leqslant \|P(X, Y_X)\|, \\ &\|X \in U\| \leqslant \|Y_X \in W\| = \|Y_X \in W_\alpha\|. \end{aligned} \tag{8}$$

Further, we consider the following formula, which is "true" because it is deducible from the logical axioms and the axioms of equality:

$$\forall x(\exists! y\ P(x, y) \wedge P(x, y_1) \wedge P(x, y_2) \Rightarrow y_1 = y_2).$$

We thereby obtain:

$$\|\exists! y\ P(X, y)\| \wedge \|P(X, Y)\| \wedge \|P(X, Y_X)\| \leqslant \|Y = Y_X\|. \tag{9}$$

Finally, it follows from (3), (8), and (9) that

$$\|X \in U\| \wedge \|P(X, Y)\| \leqslant \|Y = Y_X\| \wedge \|Y_X \in W_\alpha\|,$$

i.e., we have (7). □

7.8. **Proposition.** *The axiom of choice is "true."*

PROOF. Recall that the axiom of choice has the form $\forall x\ \exists y(Q \wedge R \wedge S \wedge T)$, where

Q denotes: $\forall z(z \in y \Rightarrow \exists u\ \exists w(z = \langle u, w\rangle))$ ("y is a binary relation");
R denotes: $\forall u\ \forall w_1\ \forall w_2(\langle u, w_1\rangle \in y \wedge \langle u, w_2\rangle \in y \Rightarrow w_1 = w_2)$ ("y is a function");
S denotes: $\forall u(\exists w(\langle u, w\rangle \in y) \Rightarrow u \in x)$ ("the domain of definition of y is contained in x");
T denotes: $\forall u(u \neq \varnothing \wedge u \in x \Rightarrow \exists w(w \in u \wedge \langle u, w\rangle \in y))$ ("the domain of definition of y coincides with x, and y chooses one element from each nonempty element of x").

We fix $X \in V^B$ and construct the corresponding "choosing function" Y. To do this:

(a) We index $D(X)$ by an initial segment of ordinals:

$$D(X) = \{U_0, U_1, \ldots, U_\alpha, \ldots\}, \quad \alpha \in I.$$

(b) For each $U_\alpha \in D(X)$ we use Lemma 7.4 to find an element $W_\alpha \in V^B$ such that

$$\|W_\alpha \in U_\alpha\| = \bigvee_{W \in V^B} \|W \in U_\alpha\|.$$

(c) For each $\alpha \in I$ we set

$$a_\alpha = \|U_\alpha \in X\| \wedge \Big(\bigvee_{\beta<\alpha} \|U_\beta \in X\|' \vee \|U_\beta = U_\alpha\|' \Big).$$

(d) Finally, we let Y denote the set which collects the "ordered pairs" $\langle U_\alpha, W_\alpha \rangle^B$ with probabilities a_α, $\alpha \in I$. Here, of course, $\langle U, W\rangle^B = \{\{U\}^B, \{U, W\}^B\}$.

The idea of this construction is as follows. In each U_α we choose the element W_α which belongs to U_α "with the largest possible probability." We then put together the graph of the choice function Y from the "pairs" $\langle U_\alpha, W_\alpha\rangle^B$, where we take the pairs in the order they are indexed, but only include a given $\langle U_\alpha, W_\alpha\rangle^B$ to the extent that U_α "was not already considered earlier as belonging to X."

We now substitute X and Y in place of x and y in the axiom of choice, and, letting Q, R, S, and T now denote the corresponding formulas with these constants, we show that $\|Q\| = \|R\| = \|S\| = \|T\| = 1$. We shall constantly be using the following formula, which follows from (1) and the definition of Y;

$$\|Z \in Y\| = \bigvee_a \|Z = \langle U_\alpha, W_\alpha\rangle^B\| \wedge a_\alpha. \tag{10}$$

7.9. $\|Q\| = 1$. By the definition of Q, this means that for all $Z \in V^B$ we must have

$$\|Z \in Y\| \leqslant \bigvee_{U, W} \|Z = \langle U, W\rangle^B\|,$$

but this is obvious from (10).

7.10. $\|R\| = 1$. By the definition of R, for any U, W^1, $W^2 \in V^B$ we must prove the inequality

$$\|\langle U, W^1\rangle^B \in Y\| \wedge \|\langle U, W^2\rangle^B \in Y\| \leqslant \|W^1 = W^2\|.$$

Using (10), we rewrite the left-hand side in the form

$$\bigvee_{\alpha,\beta} \|U = U_\alpha\| \wedge \|W^1 = W_\alpha\| \wedge a_\alpha \wedge \|U = U_\beta\| \wedge \|W^2 = W_\beta\| \wedge a_\beta.$$

Since $\|U = U_\alpha\| \wedge \|U = U_\beta\| \leqslant \|U_\alpha = U_\beta\|$ and $\|U_\alpha = U_\beta\| \wedge a_\alpha \wedge a_\beta = 0$ for $\alpha \neq \beta$ (see the definition of a_α), it follows that in this sum we need only consider the terms with $\alpha = \beta$. But such a term is $\leqslant \|W^1 = W_\alpha\| \wedge \|W^2 = W_\alpha\| \leqslant \|W^1 = W^2\|$, as required.

7.11. $\|S\| = 1$. This is equivalent to the inequality

$$\|\langle U, W\rangle^B \in Y\| \leqslant \|U \in X\|.$$

But, by (10), the left-hand side equals

$$\bigvee_\alpha \|U = U_\alpha\| \wedge \|W = W_\alpha\| \wedge a_\alpha \leqslant \bigvee_\alpha \|U = U_\alpha\| \wedge \|W = W_\alpha\| \wedge \|U_\alpha \in X\|$$

$$\leqslant \bigvee_\alpha \|U = U_\alpha\| \wedge \|U_\alpha \in X\| = \|U \in X\|.$$

7.12. $\|T\| = 1$. We must prove that for any $U \in V^B$

$$\|U \in X\| \wedge \|U \neq \varnothing\| \leqslant \bigvee_{W \in V^B} \|W \in U\| \wedge \|\langle U, W\rangle^B \in Y\|. \quad (11)$$

We first show that it suffices to prove (11) for $U \in D(X)$, i.e., for all U_α, $\alpha \in I$. In fact, suppose (11) holds for all U_α. Then for $U \in V^B$ we have

$$\|U \in X\| = \bigvee_\alpha \|U = U_\alpha\| \wedge \|U_\alpha \in X\|,$$

$$\|U \neq \varnothing\| = \bigvee_{U_1 \in V^B} \|U_1 \in U\|,$$

and hence

$$\|U \in X\| \wedge \|U \neq \varnothing\| = \bigvee_{\alpha, U_1} \|U_1 \in U\| \wedge \|U = U_\alpha\| \wedge \|U_\alpha \in X\|$$

$$\leqslant \bigvee_{\alpha, U_1} \|U_1 \in U_\alpha\| \wedge \|U = U_\alpha\| \wedge \|U_\alpha \in X\| \qquad \text{(by (III) in §5)}$$

$$= \bigvee_\alpha \|U_\alpha \neq \varnothing\| \wedge \|U = U_\alpha\| \wedge \|U_\alpha \in X\|$$

$$\leqslant \bigvee_{\alpha, W \in V^B} \|W \in U_\alpha\| \wedge \|\langle U_\alpha, W\rangle^B \in Y\| \wedge \|U = U_\alpha\| \qquad \text{(by (11) for } U_\alpha)$$

$$\leqslant \bigvee_W \|W \in U\| \wedge \|\langle U, W\rangle^B \in Y\|.$$

(Here we used the fact that

$$\|\langle U_\alpha, W\rangle^B \in Y\| \wedge \|U = U_\alpha\|$$
$$= \bigvee_\beta \|U_\alpha = U_\beta\| \wedge \|W = W_\beta\| \wedge \alpha_\beta \wedge \|U = U_\alpha\|$$
$$\leqslant \bigvee_\beta \|U = U_\beta\| \wedge \|W = W_\beta\| \wedge \alpha_\beta$$
$$= \|\langle U, W\rangle^B \in Y\|.)$$

Thus, it remains to prove (11) for U_α, $\alpha \in I$. Now

$$\|U_\alpha \neq \varnothing\| = \|\exists w(w \in U_\alpha)\| = \bigvee_W \|W \in U_\alpha\| = \|W_\alpha \in U_\alpha\|.$$

Hence (11) can be rewritten

$$\|U_\alpha \in X\| \wedge \|W_\alpha \in U_\alpha\| \leqslant \bigvee_W \|W \in U_\alpha\| \wedge \|\langle U_\alpha, W\rangle^B \in Y\|. \quad (12)$$

We prove this by induction on α. (12) is obvious for $\alpha = 0$, since the term on the right with $W = W_0$ coincides with the left-hand side. Suppose (12) holds for $\beta < \alpha$.

By the definition of a_α, we have

$$\|U_\alpha \in X\| = a_\alpha \vee \Big(\bigvee_{\beta < \alpha} \|U_\beta \in X\| \wedge \|U_\beta = U_\alpha\| \Big).$$

If we substitute this formula in the left-hand side of (12), we find that we must prove two inequalities:

$$a_\alpha \wedge \|W_\alpha \in U_\alpha\| \leqslant \bigvee_W \|W \in U_\alpha\| \wedge \|\langle U_\alpha, W\rangle^B \in Y\|, \tag{13}$$

$$\|U_\beta \in X\| \wedge \|U_\beta = U_\alpha\| \wedge \|W_\alpha \in U_\alpha\|$$
$$\leqslant \bigvee_W \|\langle U_\alpha, W\rangle^B \in Y\| \wedge \|W \in U_\alpha\|, \quad \textit{for all } \beta < \alpha. \tag{14}$$

The inequality (13) is obvious if we look at the term on the right with $W = W_\alpha$. The inequality (14) reduces to the induction assumption as follows. The left-hand side of (14) is:

$$\leqslant \|U_\beta \in X\| \wedge \|U_\beta = U_\alpha\| \wedge \|W_\alpha \in U_\beta\|$$
$$\leqslant \|U_\beta \in X\| \wedge \|U_\beta = U_\alpha\| \wedge \|W_\beta \in U_\beta\|$$

by the definition of W_β. Further, using the induction assumption and extensionality, we have

$$\|U_\beta = U_\alpha\| \wedge \|U_\beta \in X\| \wedge \|W_\beta \in U_\beta\|$$
$$\leqslant \bigvee_W \|W \in U_\beta\| \wedge \|\langle U_\beta, W\rangle^B \in Y\| \wedge \|U_\beta = U_\alpha\|$$
$$\leqslant \bigvee_W \|W \in U_\alpha\| \wedge \|\langle U_\alpha, W\rangle^B \in Y\|,$$

which completes the verification of the axiom of choice. □

8 The Continuum Hypothesis is "false" for suitable B

8.1. We recall (Lemma 7.2(a)) that the set $X \in V^B$ collects the sets $\{X_i\}$ with probabilities $a_i \in B (i \in I)$ if $\|Y \in X\| = \bigvee_i \|Y = X_i\| \wedge a_i$ for all Y. Using this definition, we can introduce a useful canonical mapping $t \mapsto \hat{t}$ from the von Neumann universe V to the universe V^B. Let $\hat{\varnothing} = \varnothing$ (recall that $\|Y \in \varnothing\| = 0$ for all Y), and, if $\hat{s}$ has already been defined for all $s \in V_\alpha$, then for $t \in V_{\alpha+1}$ we *let $\hat{t}$ collect all the $\hat{s}$ for $s \in t$ with probabilities* 1. In other words, for any $Y \in V^B$

$$\|Y \in \hat{t}\| = \bigvee_{s \in t} \|Y = \hat{s}\|. \tag{1}$$

(Here the collecting set $\hat{t}$ is not uniquely defined, i.e., it is only defined modulo equivalence, so that, strictly speaking, we should also specify the rank of $\hat{t}$, for example by saying that it equals the rank of t. This is not essential for us, however, since we shall only be interested in the truth functions, which do not change if we replace an object by an equivalent object.)

We now formulate some additional conditions (besides completeness) which must be imposed on the Boolean algebra B for the purposes of this section. Recall that ω_0 is the first infinite ordinal, ω_1 is the first ordinal having cardinality $> \omega_0$, and ω_2 is the first ordinal having cardinality $> \omega_1$.

8.2. *Conditions on B.*

(a) The countable chain condition, which, we recall, says that if we have a family of elements $\{a_i\}$, $i \in I$, such that $a_i \neq 0$ and $a_i \wedge a_j = 0$ for $i \neq j$, then I is at most a countable set.

(b) There exists a family of elements $b(n, \alpha) \in B$, indexed by the set $\omega_0 \times \omega_2$, with the following property: if $Z(\alpha)$ collects the elements $\hat{n}$, $n \in \omega_0$, with probabilities $b(n, \alpha)$, then $\|Z(\alpha) = Z(\beta)\| = 0$ for $\alpha \neq \beta$, $\alpha, \beta \in \omega_2$.

The second condition has the following intuitive meaning. It is easy to see that $\|Z(\alpha) \subset \hat{\omega}_0\| = 1$. In fact, this equality is equivalent to $\|\forall x(x \in Z(\alpha) \Rightarrow x \in \hat{\omega}_0)\| = 1$, i.e., to

$$\forall X \in V^B, \qquad \|X \in Z(\alpha)\| \leqslant \|X \in \hat{\omega}_0\|,$$

and this is obvious from (1), since $\hat{\omega}_0$ collects the $\hat{n}$ with probabilities 1, and $Z(\alpha)$ collects the $\hat{n}$ with probabilities $b(n, \alpha) \leqslant 1$.

Thus, condition (b) means that we can find ω_2 distinct subsets $Z(\alpha) \subset \hat{\omega}_0$, so that, in the naive sense, we have card $\mathcal{P}(\hat{\omega}_0) > \omega_1$. This is precisely the negation of the Continuum Hypothesis. Of course, it is still necessary to show that this intuitive idea can be made into a proof.

8.3. *The existence of B with the required properties.* We could use measurable sets, as in §3. However, in order to vary our approach, and to prepare for §9, we give another construction. Let $\{0, 1\}$ be the discrete two-point space, let $I = \omega_0 \times \omega_2$, and let $S = \{0, 1\}^I$ be the space of vectors whose coordinates are indexed by I and take the values 0 or 1. We introduce the direct product topology on S. It has a standard basis of open sets consisting of all vectors whose coordinates indexed by a finite subset $\mathcal{J} \subset I$ are fixed.

If $a \subset S$, we set

$$a' = \text{the complement of the closure of } a \text{ in } S,$$

and we set $a'' = (a')'$. Sets $a \subset S$ with $a'' = a$ are called regular open sets in S.

8.4. **Theorem.** *Let*

$$B = \{a \subset S \mid a'' = a\},$$
$$a \wedge b = a \cap b,$$
$$a \vee b = (a \cup b)''.$$

Then B with the operations $\wedge$, $\vee$, and $'$ is a complete Boolean algebra with the countable chain condition, and $\bigvee_i a_i = (\bigcup_i a_i)''$ for any family of $a_i \in B$.

We omit the proof (see J. B. Rosser, *Simplified Independence Proofs*, Academic Press, New York, 1969, Chapter 2).

8.5. **Lemma.** *Under the conditions in* 8.4, *let*

$$b(n, \alpha) = \textit{the set of vectors with } 1 \textit{ in the } (n, \alpha) \textit{ place},$$

and let $Z(\alpha)$ *be defined as in* 8.2(b). *Then*

$$\|Z(\alpha) = Z(\beta)\| = 0, \quad \textit{for } \alpha \neq \beta.$$

Proof. By formula (5) in §4, we have

$$\|Z(\alpha) = Z(\beta)\| = \bigwedge_{n \in \omega_0} (b(n, \alpha) \vee b(n), \beta) \wedge (b(n, \alpha)' \wedge b(n, \beta)').$$

The right side can only become larger if we replace $\wedge$ by $\cap$ and $\vee$ by $\cup$; here the primes $'$ coincide with the ordinary complements. If we had $\|Z(\alpha) = Z(\beta)\| \neq 0$, then there would exist an element X in the standard basis of the topology (see the beginning of 8.3) which is contained in

$$\bigcap_{n \in \omega_0} (b(n, \alpha) \cap b(n, \beta)) \cup (b(n, \alpha)' \cap b(n, \beta)').$$

But this intersection consists of all vectors having the same (n, α)-coordinate and (n, β)-coordinate for all n, while all coordinates except for a finite number range freely in any element X of the standard basis of the topology. □

8.6. *Formulation of the negation of the Continuum Hypothesis.* We shall prove that the following is "true":

$\neg$CH: $\forall x$(("x is an ordinal"$\wedge$"x is not finite"$\wedge\forall y(y \in x \Rightarrow$"$y$ is finite"))$\Rightarrow\exists w$("there is no function from x onto all of w"$\wedge$"there is no function from w onto $\mathcal{P}(x)$")).

Here:

x is finite: $\forall y(y \subset x \wedge y \neq x \Rightarrow$"there is no function from y onto all of x").

We leave the translation of the other abbreviated notation to the reader.

The premise in $\neg$CH says that "x is the first infinite ordinal," and the conclusion says that "w is a set having cardinality intermediate between that of x and that of $\mathcal{P}(x)$." We shall abbreviate $\neg$CH as follows:

$$\forall x(P(x) \Rightarrow \exists w(Q_1(x, w) \wedge Q_2(x, w))). \tag{2}$$

8.7. **Reduction Lemma.** *Let* $P(x)$ *and* $Q(x)$ *be two formulas in the Zermelo–Fraenkel language having one free variable* x *and satisfying the properties*:

The formula $\exists! x\ P(x)$ *is deducible from the axioms, and*

$X_0 \in V^B$ *is an element such that* $\|P(X_0)\| = 1$.

Then $\|P(X)\| = \|X = X_0\|$ *for all* X, *and if* $\|Q(X_0)\| = 1$, *it follows that* $\|\forall x(P(x) \Rightarrow Q(x))\| = 1$.

PROOF. We first note that $\|\exists x\, P(x)\| \geqslant \|\exists!\, x\, P(x)\| = 1$, since all the axioms are "true" in V^B, and the rules of deduction preserve "truth." It hence follows from Lemma 7.4 that there exists an object $X_0 \in V^B$ with $\|P(X_0)\| = 1$.

Further, $P(x) \wedge P(y) \Rightarrow x = y$ is also deducible, so that, if we apply this with X in place of x and X_0 in place of y, we find that

$$\|P(X)\| \leqslant \|X = X_0\|. \tag{3}$$

But $\|P(X)\| \wedge \|X = X_0\| = P(X_0) \wedge \|X = X_0\| = \|X = X_0\|$. Hence the inequality in (3) may be replaced by equality.

Finally, we suppose that $\|Q(X_0)\| = 1$. Then, by what was just proved,

$$\begin{aligned}\|P(X)\| = \|Q(X_0)\| \wedge \|X = X_0\| &= \|Q(X)\| \wedge \|X = X_0\| \\ &= \|Q(X)\| \wedge \|P(X)\|,\end{aligned}$$

so that $\|P(X)\| \leqslant \|Q(X)\|$, and $\|\forall x(P(x) \Rightarrow Q(x))\| = 1$. □

This lemma can be applied to $\neg$CH in the form (2), since the formula $\exists!\, x\, P(x)$, where $P(x)$ is the premise "x is the first infinite ordinal," is deducible from the axioms. We shall not give this formal deduction, and shall consider the uniqueness of ω_0 to be common knowledge. Now, by Lemma 8.7, to verify $\neg$CH it suffices to show the following facts:

8.8. $\|P(\hat{\omega}_0)\| = 1$. (In other words, $\hat{\omega}_0$ plays the role of X_0 in our situation.)

8.9. $\|Q_1(\hat{\omega}_0, \hat{\omega}_1)\| = 1$.

8.10. $\|Q_2(\hat{\omega}_0, \hat{\omega}_1)\| = 1$. (This then implies that $\|\exists\omega(Q_1(\hat{\omega}_0, \omega) \wedge Q_2(\hat{\omega}_0, \omega))\| = 1$, and completes the verification of the conditions of the lemma.)

8.8. is verified almost mechanically, and we leave it as an exercise.

8.11. VERIFICATION OF 8.9. We must show that, if B satisfies the countable chain condition, then

$$\|\exists \text{ a function from } \hat{\omega}_0 \text{ onto all of } \hat{\omega}_1\| = 0.$$

The proof that follows carries over word for word to the more general case, when instead of ω_0 and ω_1 we take any pair $s, t \in V$ such that card $s <$ card t and card s is infinite.

We suppose that

$$0 \neq a = \|\exists f(f \text{ is a function} \wedge \forall y(y \in \hat{\omega}_1 \Rightarrow \exists x(x \in \hat{\omega}_0 \wedge \langle x, y\rangle \in f)))\|,$$

and we show that this leads to a contradiction. There must exist an $F \in V^B$ such that

$$a \leqslant \|F \text{ is a function}\| \wedge \Big(\bigwedge_Y \cdots\Big).$$

For every $\alpha \in \omega_1$ we consider the term in $\bigwedge_Y \cdots$ corresponding to

$Y = \hat{\alpha}$ and use the fact that $\|\hat{\alpha} \in \hat{\omega}_1\| = 1$. We obtain:

$$a \leqslant \|F \text{ is a function}\| \wedge \Big(\bigvee_X \|X \in \hat{\omega}_0\| \wedge \|\langle X, \hat{\alpha}\rangle^B \in F\|\Big). \tag{4}$$

By (1), we have

$$\begin{aligned}\|X \in \hat{\omega}_0\| \wedge \|\langle X, \hat{\alpha}\rangle^B \in F\| &= \bigvee_{n<\omega_0} \|X = \hat{n}\| \wedge \|\langle X, \hat{\alpha}\rangle^B \in F\| \\ &= \bigvee_{n<\omega_0} \|X = \hat{n}\| \wedge \|\langle \hat{n}, \hat{\alpha}\rangle^B \in F\|,\end{aligned}$$

so that, if we sum first over X and then over n, we may write (4) in the form

$$a \leqslant \|F \text{ is a function}\| \wedge \Big(\bigvee_{n<\omega_0} \|\langle \hat{n}, \hat{\alpha}\rangle^B \in F\|\Big).$$

Hence, for every $\alpha < \omega_1$ there is an $n(\alpha) < \omega_0$ such that

$$\|F \text{ is a function}\| \wedge \|\langle n(\hat{\alpha}), \hat{\alpha}\rangle^B \in F\| \neq o.$$

Then there exists an n_0 and a subset $\mathcal{J} \subseteq \omega_1$ of cardinality ω_1 such that

$$0 \neq a_\alpha = \|F \text{ is a function}\| \wedge \|\langle \hat{n}_0, \hat{\alpha}\rangle^B \in F\|, \quad \text{for all } \alpha \in \mathcal{J}.$$

It remains to show that $a_\alpha \wedge a_\beta = 0$ for $\alpha \neq \beta$, which contradicts the countable chain condition on B. Now by the definition of a function

$$a_\alpha \wedge a_\beta = \|F \text{ is a function}\| \wedge \|\langle \hat{n}_0, \hat{\alpha}\rangle^B \in F\| \wedge \|\langle \hat{n}_0, \hat{\beta}\rangle^B \in F\| \leqslant \|\hat{\alpha} = \hat{\beta}\|,$$

so that it suffices to show that $\alpha \neq \beta$ implies $\|\hat{\alpha} = \hat{\beta}\| = 0$.

In fact, if, say, $\gamma \in \alpha$ but $\gamma \notin \beta$, then the formula (5) in §4 for $\|\hat{\alpha} = \hat{\beta}\|$ has a zero term, namely $\|\hat{\gamma} \in \hat{\alpha}\|' \vee \|\hat{\gamma} \in \hat{\beta}\|$. (To check that $\|\hat{\gamma} \in \hat{\beta}\| = 0$ if $\gamma \notin \beta$ we have to know that $\|\hat{\gamma} = \hat{\delta}\| = 0$ if $\gamma \neq \delta$, but we only have to know this for γ and δ of lower rank than α and β, so that the detailed proof uses induction on the rank.) □

8.12. Verification of 8.10. We must show that

$$\|\exists \text{ a function from } \hat{\omega}_1 \text{ onto } \mathcal{P}(\hat{\omega}_0)\| = 0,$$

that is, that

$$\|\exists g(g \text{ is a function} \wedge \forall z(z \subset \hat{\omega}_0 \Rightarrow \exists y(y \in \hat{\omega}_1 \wedge \langle y, z\rangle \in g)))\| = 0.$$

Suppose that for some $G \in V^B$ we have

$$0 \neq a = \|G \text{ is a function}\| \wedge \Big(\bigwedge_Z \cdots \Big).$$

For every $\alpha < \omega_2$ we consider the term corresponding to $Z = Z(\alpha)$ (see the definition in 8.2 and 8.5), and we use the fact that

$$0 \neq a \leqslant \|G \text{ is a function}\| \wedge \Big(\bigvee_Y \|Y \in \hat{\omega}_1\| \wedge \|\langle Y, Z(\alpha)\rangle^B \in G\|\Big). \tag{5}$$

By (1), we have

$$\|Y \in \hat{\omega}_1\| \wedge \|\langle Y, Z(\alpha)\rangle^B \in G\| = \bigvee_{\beta<\omega_1} \|Y = \hat{\beta}\| \wedge \|\langle Y, Z(\alpha)\rangle^B \in G\|$$
$$= \bigvee_{\beta<\omega_1} \|Y = \hat{\beta}\| \wedge \|\langle \hat{\beta}, Z(\alpha)\rangle^B \in G\|.$$

Summing first over Y, we rewrite (5) in the form

$$0 \neq a \leqslant \|G \text{ is a function}\| \wedge \bigvee_{\beta<\omega_1} \|\langle \hat{\beta}, Z(\alpha)\rangle^B \in G\|.$$

Hence, for every $\alpha < \omega_2$ there is a $\beta(\alpha) < \omega_1$ such that

$$0 \neq a_\alpha = \|G \text{ is a function}\| \wedge \|\langle \beta(\alpha), Z(\alpha)\rangle^B \in G\|.$$

Then there exists a $\beta_0 < \omega_1$ and a subset $\mathcal{J} \subset \omega_2$ of cardinality ω_2 such that

$$0 \neq a_\alpha = \|G \text{ is a function}\| \wedge \|\langle \hat{\beta}_0, Z(\alpha)\rangle^B \in G\|, \quad \text{for all } \alpha \in \mathcal{J}.$$

As in 8.11, we obtain a contradiction to the countable chain condition if we show that $a_\alpha \wedge a_\beta = 0$ for $\alpha \neq \beta$. But this follows from

$$a_\alpha \wedge a_\beta \leqslant \|Z(\alpha) = Z(\beta)\| = 0$$

by Lemma 8.5. □

9 Forcing

9.1. By choosing the Boolean algebra B in various ways, one can use the corresponding models V^B to show that many different assertions P are consistent with the Zermelo–Fraenkel axioms. But each choice of B for a given P such that $\|P\| = 1$ in V^B presents a separate problem.

There is another interpretation of this method which is closer to Cohen's original idea. From this point of view we start not with a universe V and a Boolean algebra B, but with an (often countable) transitive model M and an ordered set C of "forcing conditions." It is usually more obvious how to choose a suitable C than how to choose a suitable B for proving that a given proposition P is consistent. One might say that B embodies the "physical meaning" of the problem, while C expresses its "logical meaning." Anyway, it is not difficult to go from one version to the other, and in either case it takes about the same amount of work to verify the "truth" of the axioms.

In this section we discuss the second version, using forcing, with most of the proofs omitted. The details can be found in Cohen's original article, and also in Jech's book *Lectures in Set Theory with Particular Emphasis on the Method of Forcing*, Springer-Verlag Lecture Notes in Mathematics 217, 1971, and in J. R. Shoenfield's article, "Unramified forcing," *Proc. Symp. in Pure Math.*, vol. 13, 1, 357–381 (American Math. Soc., Providence, 1972).

9.2. Before introducing the general concept of forcing, we consider a special case which arises in a typical problem.

Let X and Y be two sets, for example $\mathcal{P}(\omega_0)$ and ω_2. We consider the proposition P : "card $X \geqslant$ card Y," which in this special case is the negation of the CH. One possible approach to constructing a model (in the usual rather than Boolean sense) of $\mathrm{L_1Set}$ in which P is true, is as follows.

We take our original countable transitive model M of set theory (i.e., of the special axioms of $\mathrm{L_1Set}$), which was shown to exist in §7 of Chapter II. Let X_M and Y_M be the "representatives" of X and Y in M. (This means that if, say, X is defined by the formula $\exists!\, x\, P(x)$, then $X_M = x^{\xi}$, where ξ is a point of the interpretation class for which $|P(x)|_M(\xi) = 1$; see §7 of Chapter II.) We assume that X_M is infinite and Y_M is nonempty. Then "from an external point of view" X_M is countable and Y_M is at most countable, so there automatically exists a function F which maps X_M onto all of Y_M. A natural idea would be to add (the graph of) F to M, i.e., to consider the least countable model N of the axioms which contains M and F. Then N has a map from X_M onto Y_M, but it is very likely that $X_N \neq X_M$ and $Y_N \neq Y_M$. What we need in N is a map from X_N onto Y_N.

As we have shown when discussing Skolem's paradox in Chapter II, at least for certain pairs (such as $X = \omega_0$, $Y = \mathcal{P}(\omega_0)$), we cannot obtain a map from X onto Y in this way. In those cases when we can construct such a map, we must choose F very carefully. Cohen's idea was that F, rather than being chosen so as to satisfy some conditions, should be chosen so as to avoid reflecting any specific properties of M, i.e., F should be "generic." We shall formulate this more precisely.

It turns out to be important to start not by choosing F directly, but by choosing the set

$$G = \{\text{restrictions of } F \text{ to finite subsets of } X_M\}.$$

Clearly, F is uniquely determined from G: $F = \bigcup_{g \in G} g$ (recall that a function is the same as its graph). Hence F is contained in any model which contains G. But now we must give an axiomatic characterization of the suitable G without using F explicitly. Here are the properties which G must satisfy:

9.3.

(a) $G \subseteq C$, where C is the set of maps from finite subsets of X_M to Y_M.

It is important that $C \in M$, because the formula in $\mathrm{L_1Set}$ which defines C is (M, V)-absolute. We need this remark in order to motivate the general definitions later.

(b) $\varnothing \in G$; if $p \in G$ and $q \in C$, where $q \subseteq p$, then $q \in G$; for any $p_1, p_2 \in G$ there is a $p \in G$ such that $p \supseteq p_1 \cup p_2$.

Suppose we have chosen such a set G of maps from finite subsets of X_M to Y_M. Then $\bigcup_{g \in G} g$ is also a map from some subset of X_M to Y_M. In order for this map to be defined on all of X_M and to be surjective, it is necessary and sufficient for the following additional conditions to hold:

$$\forall Z \in X_M, \qquad G \cap \{p \in C \mid p \text{ is defined at } Z\} \neq \varnothing,$$
$$\forall Z \in Y_M, \qquad G \cap \{q \in C \mid q \text{ takes the value } Z\} \neq \varnothing.$$

We call a subset $D \subseteq C$ *dense in* C if for all $p \in C$ there is a $q \in D$ with $p \subseteq q$. The set of maps p defined at Z and the set of maps q taking the value Z are dense, and, moreover, are elements of M by the same consideration of (M, V)-absoluteness. Hence the two requirements at the end of the last paragraph are included in the last condition, that G be *generic*:

(c) $G \cap W \neq \varnothing$ for all dense subsets $D \subseteq C$ which are elements of M.

Although it is not yet evident, it is precisely the condition that G be generic which ensures that the properties of the sets X_M and Y_M will be preserved as much as possible after we add G to the model.

We now define the general concept of "forcing conditions."

9.4. *Forcing conditions*. These are the elements in any partially ordered set $(C, <)$ which has a maximal element 1. Usually C and $<$ lie in the original model M.

A set G is called *generic over* M (relative to C) if the following conditions hold:

(a) $G \subseteq C$;
(b) $1 \in G$; if $p \in G$ and $q \in C$, where $q \geqslant p$, then $q \in G$; for any $p_1, p_2 \in G$ there is a $p \in G$ such that $p \leqslant p_1$ and $p \leqslant p_2$;
(c) $G \cap D \neq \varnothing$ for all dense subsets $D \subseteq C$ with $D \in M$ (D is dense if, for all $p \in C$, there is a $q \in D$ with $q \leqslant p$).

If the reader compares this definition with the special case in 9.3, he or she will notice that we have replaced $\subseteq$ by $\geqslant$ and $\varnothing$ by 1. This is in keeping with Cohen's original point of view, according to which $p \geqslant q$ if, when p is considered as a "condition" imposed, say, on F, more F's satisfy p than q. (Each p fixes the restriction of F to some finite subset of X_M.)

9.5. *The existence of generic sets. Let M and C be fixed. If $M \cap \mathcal{P}(C)$ is countable, then for every $p \in C$ there exists a generic set G containing p.*

In fact, we index the elements of $M \cap \mathcal{P}(X)$ as $X_1, X_2, X_3, \ldots,$ and then set

$$p_1 = p, \qquad p_{n+1} = \begin{cases} p_n, & \text{if } p_n \leqslant q \text{ for all } q \in X_n; \\ \text{any } q \in X_n \text{ such that } q < p_n, & \text{otherwise.} \end{cases}$$

Finally, we set $G = \{q \in C \mid \exists n (p_n \leqslant q)\}$.

Conditions (a) and (b) for G to be generic are trivial to verify. Condition (c) follows because, if $D \in M$ and D is dense, then there exist n and q for which $D = X_n$, $q \in X_n$, and $q \leqslant p_n$, so that $p_{n+1} \in D \cap G$.

9.6. *The connection with Boolean models.* As mentioned before, we have considerable freedom in our choice of the set C of forcing conditions and the generic subset $G \subseteq C$. Exactly how one "forces" a given proposition P was explained briefly in 9.2. We now show how to construct an axiom model $M[G]$ which contains M and G, once C and G have already been chosen.

The article by Shoenfield gives a direct construction, but we shall make use of an analogy with V^B, as in Jech's presentation. In this approach $M[G]$ is constructed in three basic steps:

(a) Corresponding to the set C we construct a canonical complete Boolean algebra B.
(b) We construct a Boolean universe M^B over B which is "relativized" by means of M.
(c) We construct a canonical maximal ideal $I_G \subseteq B$ determined by G and the "fibre" of the universe M^B over the quotient algebra $B/I_G \cong \{0, 1\}$. It is this fibre which will be the model $M[G]$.

We now discuss these steps separately and in more detail.

9.7. *Ordered sets and Boolean algebras.* Every Boolean algebra B has a canonical partial ordering: $a \leqslant b$ if $a \wedge b = a$. All elements of the structure of B are uniquely determined by this partial ordering. The induced ordering on $B - \{0\}$ is *separable*. By definition, this means that, if $a, b \neq 0$ and $a \nleqslant b$, then there exists $c \leqslant a$, $c \neq 0$ such that there is no $d \neq 0$ for which $d \leqslant b$ and $d \leqslant c$. (It suffices to take $c = a \wedge b'$.) Such b and c are called *disjoint*.

Now let C be a fixed partially ordered set. We consider the class of (nonstrictly) order-preserving maps of C into different *complete* Boolean algebras B such that 0 is not contained in the image.

9.8. **Proposition.** *In this class of maps there exists a unique universal map $e : C \to B$ with the following properties*:

(a) *$e(c)$ is the maximal separable ordered quotient set of C such that $c_1, c_2 \in C$ are disjoint $\Leftrightarrow e(c_1), e(c_2) \in B$ are disjoint*;
(b) *$e(c)$ is dense in $B - \{0\}$.*

B can be realized as the algebra of regular open sets in the space C with the topology defined by the basis $U_c = \{x \in C \mid x \leqslant c\}$, $c \in C$.

Now we can indicate how I_G is constructed from the generic subset $G \subseteq C$:

$$G_1 = \{b \in B \mid \exists p \in G, e(p) \leqslant b\},$$
$$I_G = B \setminus G_1.$$

It is not hard to prove that I_G is a maximal ideal in B, i.e., the kernel of a Boolean homomorphism $B \to \{0, 1\}$. The set G_1 is precisely the preimage of

1 under this homomorphism. Since G is generic in C, we have the following property of G_1: *for any subset $A \subseteq B$ such that $\bigvee_{a \in A} a = 1$ and $a_1 \wedge a_2 = 0$ whenever $a_1 \neq a_2 \in A$, there exists a unique element $a \in A \cap G_1$.*

9.9. *The universe M^B.* This universe is constructed from M and B in exactly the same way as V^B was constructed from V and B, with one essential difference: *all constructions are relativized with respect to M.* This means that, instead of B, we take the algebra B_M which "represents" B in M (see 9.2); only ordinals $\alpha \in M$ are used in the construction of M_α^B, and so on. A rigorous presentation of these constructions would require much more formalization using the expressive means in L_1Set than seems desirable in this section. In such a presentation both the general plan and the details of the work would remain essentially the same as before.

The basic result of these constructions is that to every closed formula P in L_1Set with constants in M corresponds a Boolean truth value $\|P\| \in B_M$. Here the value 1 corresponds to the axioms, and deductions preserve "truth."

The next step cuts down the size of M^B, again giving a transitive standard submodel.

9.10. *Construction of $M[G]$.* For brevity, we shall write B instead of B_M, and so on. The construction essentially consists in going from "random" sets $X, Y \in M^B$ to "determined" sets $\bar{X}, \bar{Y}$, where we say that $\bar{X} \in \bar{Y}$ if the truth value $\|X \in Y\|$ goes to 1 under the homomorphism $B \to B/I_G = \{0, 1\}$, i.e., if $\|X \in Y\| \in G_1$ (see 9.8). More precisely, we inductively define

$$i(\varnothing) = \varnothing,$$

and let $M[G]$ denote the image of the map $i : M^B \to V$. This notation is justified by the following result. Suppose that C and $<$ belong to M and that the subset $G \subseteq C$ is generic.

9.11 **Proposition.** *$M[G]$ is a model for the Zermelo–Fraenkel axioms which contains M and G. If M is countable, then $M[G]$ is the least such model.*

$M[G]$ contains M for the following reason. If we let $X \mapsto \hat{X}$ denote the map $M \to M^B$ which is constructed as in 8.1, then it is easy to show that $\bar{\hat{X}} = X$.

$M[G]$ contains G because $G = \bar{G'}$, where G' is the object in M^B which collects all the $\hat{b}$, $b \in B$, with probabilities b.

$M[G]$ is an axiom model basically because M^B is a Boolean axiom model. However, here we use in an essential way the assumption that G is generic. (Shoenfield verifies this result directly, without using M^B.)

9.12. Example. We return to the assertion "card $\mathcal{P}(\omega_0) \geqslant \omega_2$" in 9.2. By the above discussion, to prove that it is consistent with the axioms we choose a countable model M and then set:

$$C = \{\text{maps of finite subsets of } \mathcal{P}(\omega_0) \text{ to } \omega_2\},$$
$$G \subseteq C = \text{a generic subset of } C.$$

If we consider a map from a subset of $\mathcal{P}(\omega_0)$ to ω_2 as a function from $\omega_0 \times \omega_2$ to $\{0, 1\}$, and if, instead of "relative" constructions in M, we consider "absolute" constructions in V, then the Boolean algebra B that we obtain from C turns out to be the same algebra which was constructed in 8.3 and 8.4. This explains the appearance of B. The ideal I_G did not play any role in §8 because we were not trying to construct a standard model.

9.13. We conclude with a very general theorem of Easton, which shows how little we understand the behavior of the function 2^k (k a cardinal).

Let α be a limit ordinal. Its *cofinality* cf (α) is the least ordinal β such that α is the union of β ordinals less than α. An infinite cardinal k is called *regular* if cf $(k) = k$ and is called *singular* if cf $(k) < k$. König (1905) proved that cf $(2^k) > k$.

9.14. **Theorem** (Easton, 1965). *Let F be any (nonstrictly) monotonic function on a subclass of the regular cardinals which takes values in the class of cardinals and which satisfies*: cf $(\aleph_{F(K)}) > \aleph_K$. *Then the assertion* "$\forall$ regular $k \in$ dom F, $2^{\aleph_K} = \aleph_{F(K)}$" *does not contradict the Zermelo–Fraenkel axioms.*

If the domain of F is a set, Easton's theorem can be obtained using a model of the form $M[G]$, where M is a model in which the generalized continuum hypothesis holds (Gödel proved that such an M exists; see the next chapter). If the domain of F is a class (for example, the class of all regular cardinals), the concept of forcing must be generalized to the case when C is a class.

The question of the behavior of 2^k for singular k has not yet been "solved" in this weak sense.

CHAPTER IV

The continuum problem and constructible sets

1 Gödel's constructible universe

1.1. In this section we introduce the subclass $L \subset V$—"Gödel's constructible universe"—and establish its fundamental properties. Perhaps the shortest description of L is that it is the smallest transitive model of the axioms of L_1Set which contains all the ordinals. But the working definition of L, from which the name "constructible universe" is derived, is rather different.

We consider the following operations $F_1, \ldots, F_8$ on sets:

$$F_1(X, Y) = \{X, Y\},$$
$$F_2(X, Y) = X \setminus Y,$$
$$F_3(X, Y) = X \times Y,$$
$$F_4(X) = \{U \mid \exists W(\langle U, W\rangle \in X)\} = \text{dom } X$$
$$F_5(X) = \{\langle U, W\rangle \mid U, W \in X;\ U \in W\}$$
$$F_6(X) = \{\langle U_1, U_2, U_3\rangle \mid \langle U_2, U_3, U_1\rangle \in X\},$$
$$F_7(X) = \{\langle U_1, U_2, U_3\rangle \mid \langle U_3, U_2, U_1\rangle \in X\},$$
$$F_8(X) = \{\langle U_1, U_2, U_3\rangle \mid \langle U_1, U_3, U_2\rangle \in X\}.$$

We say that a set (or class) Y is closed with respect to an operation F of degree r if we have $F(Z_1, \ldots, Z_r) \in Y$ for all $Z_1, \ldots, Z_r \in Y$ such that $F(Z_1, \ldots, Z_r)$ is defined. For every $X \in V$ we let $\mathcal{G}(X)$ denote the smallest set $Y \supset X$ which is closed with respect to the operations $F_1, \ldots, F_8$. It will later be shown (subsection 1.4) that $\mathcal{G}(X)$ actually is a set. The following construction is analogous to the definition of V.

1.2. **Definition.**

$$L_0 = \varnothing;$$
$$L_{\alpha+1} = \mathcal{P}(L_\alpha) \cap \mathcal{G}(L_\alpha \cup \{L_\alpha\});$$
$$L_\alpha = \bigcup_{\beta<\alpha} L_\beta, \text{ if } \alpha \text{ is a limit ordinal};$$
$$L = \cup L_\alpha.$$

The elements of L are called *constructible sets*.

The operations $F_1, \ldots, F_8$ and simple combinations of them, together with the transfinite recursion in the definition of L, exhaust the arsenal of primitive set theoretic constructions used in mathematics. This can be seen by looking at Bourbaki's "compendium of the results of set theory," upon which all subsequent material in their voluminous treatise on the foundations of mathematics is based. The only way we could possibly (but not necessarily) leave L would be to apply the axiom of choice. This could happen provided that L is strictly less than V; but, as mentioned before, this question is undecidable in the Zermelo–Fraenkel axiom system (see also 5.16 below). Gödel was of the opinion that L does not exhaust V, as are most specialists who accept the semantics of L_1Set.

Of course, the constructibility of the elements of L should not be understood in a finitistic sense. The sets we construct at the $(\alpha + 1)$th stage are only the subsets of L_α which are obtained from the elements of the sets L_α and $\{L_\alpha\}$ using the explicit constructions F_i. But when we consider all the ordinals indexing the stages, we see that L is hopelessly infinite. Nevertheless, in many respects the construction of L is simpler than that of V, and L seems to provide a convenient framework for mathematics.

We now list some properties of L which follow easily from the definitions. The specific nature of the operations F_i plays a very secondary role in these properties.

1.3. $L_n = V_n$ *for all* $n \leqslant \omega_0$. This is true for L_0. Suppose it is true for L_n. It is clear from the definition that $L_n \in L_{n+1}$ and $\{X\} \in L_{n+1}$ for all $X \in L_n$. Moreover, any subset of L_n can be represented as a finite difference $(\cdots(L_n \setminus \{X_1\}) \setminus \{X_2\}) \setminus \cdots \setminus \{X_k\}$ where the $X_i \in L_n$ are the elements not in the given subset.

1.4. card L_α = card α *for all infinite ordinals* α. In fact, for $X \in V$ let

$$\Phi(X) = X \cup \bigcup_{i=1}^{3} F_i''(X \times Y) \cup \bigcup_{j=4}^{8} F_i''(X),$$

where $F''(X) = \{F(Y) | Y \in X\}$ is the image of F restricted to the elements of X. Then $\mathcal{G}(X) = \bigcup_{n=0}^{\infty} \Phi^n(X)$. It is hence clear that card $\mathcal{G}(X)$ = card X if X is infinite. We now prove the assertion 1.4 by induction on α.

Obviously card $L_\alpha \geqslant$ card α. Suppose that α is the least infinite ordinal for which card $L_\alpha >$ card α. By 1.3, we have $\alpha > \omega_0$. α cannot be a limit ordinal, or we would have card $L_\alpha = \Sigma_{\beta<\alpha}$ card $\beta =$ card α. But the case $\alpha = \beta + 1$ is also impossible, since in that case card $L_\alpha \leqslant$ card $\mathcal{G}(L_\beta \cup \{L_\beta\}) =$ card $(L_\beta \cup \{L_\beta\}) =$ card $\beta =$ card α. □

In particular, the result 1.4 shows that, beginning with $\omega_0 + 1$, the inclusion $L_\alpha \subset V_\alpha$ becomes a strict inequality, since card $V_{\omega_0+1} = 2^{\omega_0}$. Of course, this does not in principle exclude the possibility that $\forall\alpha\ \exists\beta > \alpha$, $L_\beta \supset V_\alpha$, but it seems that there is no such β even for $\alpha = \omega_0 + 1$.

1.5. *L is transitive*: $Y \in X \in L_\alpha \Rightarrow Y \in L_\alpha$, *i.e.* $L_\alpha \subset L_{\alpha+1}$. See subsection 13 of the Appendix to Chapter II; the proof is no different for L.

1.6. *L is a big class*: by definition, this means that *for any* $X \in V$ *with* $X \subset L$ *there exists a* $Y \in L$ *such that* $X \subset Y$.

On L we consider the function $\phi(x)$ which is equal to the least α for which $x \in L_\alpha$. Let $X \in V$, $X \subset L$. We consider the map ϕ restricted to X. By the replacement axiom, the values of ϕ form some set Y. The elements of Y are ordinals. Let $\beta = \cup Y$. Then for each $x \in X$ we have $\beta \geqslant \phi(x)$, so that $X \subset L_\beta$.

Effective numbering of L by ordinals.

We order pairs of ordinals $\langle\alpha, \beta\rangle$ by the relation

$$\langle\alpha_1, \beta_1\rangle < \langle\alpha_2, \beta_2\rangle \Leftrightarrow \text{either } \max(\alpha_1, \beta_1) < \max(\alpha_2, \beta_2),$$

or else these maximums are equal and $\alpha_1 < \alpha_2$,

or else these maximums are equal and $\alpha_1 = \alpha_2$ and $\beta_1 < \beta_2$.

Further, we order triples $\langle i, \alpha, \beta\rangle$, where $i = 0, \ldots, 8$ by the relation

$$\langle i_1, \alpha_1, \beta_1\rangle < \langle i_2, \alpha_2, \beta_2\rangle \Leftrightarrow \text{either } \langle\alpha_1, \beta_1\rangle < \langle\alpha_2, \beta_2\rangle,$$

or else $\langle\alpha_1, \beta_1\rangle = \langle\alpha_2, \beta_2\rangle$ and $i_1 < i_2$.

We call these triples *important*.

1.7. **Lemma.** *The class of important triples is well-ordered by the relation* $<$. *In addition, the following assertions hold*:

(a) *The next triple after* $\langle i, \alpha, \beta\rangle$ *has the form*:

$\langle i + 1, \alpha, \beta\rangle$, *if* $i \leqslant 7$;

$\langle 0, \alpha + 1, \beta\rangle$, *if* $i = 8$ *and* $\alpha + 1 < \beta$;

$\langle 0, \alpha + 1, 0\rangle$, *if* $i = 8$ *and* $\alpha + 1 = \beta$;

$\langle 0, \alpha, \beta + 1\rangle$, *if* $i = 8$ *and* $\alpha > \beta$;

$\langle 0, 0, \beta + 1\rangle$, *if* $i = 8$ *and* $\alpha = \beta$.

(b) *Limit triples have the form*:

$\langle 0, \alpha, \beta\rangle$, *if* $\alpha + 1 \leqslant \beta$ *and* α *is a limit ordinal*: *this is the limit of* $\langle i, \gamma, \beta\rangle$, $\gamma < \alpha$;

$\langle 0, \alpha, 0\rangle$, *if* α *is a limit ordinal*: *this is the limit of* $\langle i, \gamma, \alpha\rangle$, $\gamma < \alpha$;

$\langle 0, \alpha, \beta\rangle$, *if* $\alpha \geqslant \beta$ *and* β *is a limit ordinal*: *this is the limit of* $\langle i, \alpha, \gamma\rangle$, $\gamma < \beta$;

$\langle 0, 0, \beta\rangle$, *if* β *is a limit ordinal*: *this is the limit of* $\langle i, \alpha, \gamma\rangle$, $\alpha < \beta$, $\gamma < \beta$.

PROOF. The proof follows immediately from the definitions. We shall illustrate this by showing explicitly how to find the least triple in any nonempty class C of triples. We set

$$\gamma = \min\{\max(\alpha, \beta) | \langle i, \alpha, \beta\rangle \in C\};$$
$$C_\gamma = \{\langle i, \alpha, \beta\rangle \in C | \max(\alpha, \beta) = \gamma\}.$$

If C_γ does not contain any triples of the form $\langle i, \alpha, \gamma\rangle$, then let β_0 be the minimum of the third coordinates of triples in C_γ, and let i_0 be the least i such that $\langle i, \gamma, \beta_0\rangle \in C_\gamma$. Then $\langle i_0, \gamma, \beta_0\rangle$ is the least triple in C. Otherwise, let C'_γ consist of triples of the form $\langle i, \alpha, \gamma\rangle \in C_\gamma$, let α_0 be the minimum of the second coordinates in C'_γ, and let i_0 be the least i such that $\langle i, \alpha_0, \gamma\rangle \in C_\gamma$. Then $\langle i_0, \alpha_0, \gamma\rangle$ is the least triple in C.

The exact form of assertions (a) and (b) will be needed only in §5. The lemma implies that there exists a unique order-preserving isomorphism

$$K : \{\text{ordinals}\} \Rightarrow \{\text{important triples}\}.$$

Using this isomorphism, we recursively define a numbering mapping

$$N : \{\text{ordinals}\} \Rightarrow L.$$

Since we have $\alpha < \gamma$ and $\beta < \gamma$ if $\gamma > 0$, $i > 0$, and $K(\gamma) = \langle i, \alpha, \beta\rangle$, we may set:

$$N(\gamma) = \begin{cases} L_\alpha, & \text{for } i = 0; \\ F_i(N(\alpha), N(\beta)), & \text{for } i = 1, 2, 3; \\ F_i(N(\alpha)), & \text{for } i = 4, 5, 6, 7, 8. \end{cases}$$

1.8. **Lemma.**

(a) *The mapping* N *is correctly defined.*
(b) *The image of* N *coincides with all of* L.

PROOF.

(a) To verify correctness, it suffices to show that $\{L_\alpha\} \in L$ and that the class L is closed with respect to the operations F_i. In fact, then induction on γ shows that $N(\gamma) \in L$ if $N(\alpha) \in L$ for all $\alpha < \gamma$.

Let $X, Y \in L_\alpha$. Since L is transitive (see 1.5), we easily find that $F_1(X, Y)$, $F_2(X, Y)$, and $F_4(X)$ belong to $\mathcal{P}(L_\alpha)$, and hence to $L_{\alpha+1}$. For example,

$$U \in F_4(X) \Rightarrow \exists W \langle U, W \rangle \in L_\alpha \Rightarrow \{U\} \in L_\alpha \Rightarrow U \in L_\alpha.$$

Further, $X \times Y$ is a subset of the ordered pairs of elements in L_α. We showed that the unordered pairs lie in $L_{\alpha+1}$, so that the ordered pairs lie in $L_{\alpha+2}$, and finally $X \times Y \in L_{\alpha+3}$ and $F_5(X) \in L_{\alpha+4}$. Analogously, the elements of $F_i(X)$ for $i = 6, 7, 8$ are ordered triples of elements in L_α, so that $F_i(X) \in L_{\alpha+6}$.

(b) Let Z be the image of N. We show by induction on α that $L_\alpha \subset Z$. If α is a limit ordinal and $L_\gamma \subset Z$ for each $\gamma < \alpha$ then also $L_\alpha = \bigcup_{\gamma<\alpha} L_\gamma \subset Z$. Suppose $\alpha = \beta + 1$ and $L_\beta \subset Z$, and let $X \in L_\alpha$. Then $X \in \Phi^n(L_\beta \cup \{L_\beta\})$ and we show $X \in Z$ by induction on n.

(b_1) $n = 0$. Then either $X \in L_\beta$ so $X \in Z$ by the induction hypothesis, or else $X = L_\beta$, in which case $X = N(\gamma)$ for γ such that $K(\gamma) = \langle 0, \beta, 0 \rangle$.

(b_2) $n > 0$. Let $X = F_i(Y, Z)$, $i = 1, 2, 3$; $Y, Z \in \Phi^{n-1}(L_\beta \cup \{L_\beta\})$. By the induction hypothesis $Y = N(\gamma_1)$ and $Z = N(\gamma_2)$ for some ordinals γ_1, γ_2. Therefore $X = N(\gamma)$ where $K(\gamma) = \langle i, \gamma_1, \gamma_2 \rangle$.

Let $X = F_i(Y)$, $i = 4, \ldots, 8$; $Y \in \Phi^{n-1}(L_\beta \cup \{L_\beta\})$. The verification is analogous.

The lemma is proved. □

In §3 the numbering N will allow us to prove that a strong form of the axiom of choice is L-true. The fundamental step in the proof is to choose the element with the least N-number in each constructible set.

2 Definability and absoluteness

2.1. Let $M \subset V$ be a non-empty class, and let P be a formula in L_1Set. As in §7 of Chapter II, we shall consider the truth values $|P|_M(\xi)$ for $\xi \in \overline{M}$, where we take the standard interpetation of L_1Set in V restricted to M. We then say that the formula P is M-true if $|P|_M = 1$ for all ξ.

We shall also consider formulas "with constants in M," where we assume that the language L_1Set has been extended so that its alphabet includes names for all the elements of M. We shall designate these elements by the same letters as in the metalanguage ($X, Y, \ldots$ for sets; $\alpha, \beta, \ldots$ for ordinals, etc.), which we hope will not lead to confusion. We extend the definition of $|P|_M(\xi)$ to formulas with constants in M in the obvious way: we take $X^\xi = X$ for any constant X and any point ξ.

2.2. **Definition.** Let $X_i \in M$, $i = 1, \ldots, n$. Sets of the form

$$\left\{ \langle y_1^\xi, \ldots, y_n^\xi \rangle | \xi \in \overline{M}, y_i^\xi \in X_i \text{ for } i = 1, \ldots, n; |P|_M(\xi) = 1 \right\} \subset X_1 \times \cdots \times X_n$$

are called M-definable sets. Here P runs through all formulas with constants in M and free variables in the set $\{y_1, \ldots, y_n\}$.

If $P(y_1, \ldots, y_n, Z_1, \ldots, Z_m)$ is such a formula (where the notation shows the constants and free variables) and if $y_i^\xi = Y_i$, we shall often write "$P(Y_1, \ldots, Y_n, Z_1, \ldots, Z_m)$ is M-true" instead of $|P|_M(\xi) = 1$.

The next proposition, which, in particular, is applicable to L, is a basic instrument for proving many assertions about L.

2.3. **Proposition.** *Let $M \subset V$ be a transitive big class* (*see* 1.6) *which is closed with respect to the operations $F_1, \ldots, F_8$. Then all M-definable sets are elements of M.*

Proof. The proof is by induction on the number of connectives and quantifiers in the defining formula P.

(a) $P(y_1, \ldots, y_n; Z_1, \ldots, Z_m)$ is an atomic formula. It can have one of eight possible forms: the predicate can be either $\in$ or $=$, and on each side of $\in$ or $=$ we can have either a constant or a variable. But all of these cases reduce to two: $y_i \in y_j$ and $y_i \in Z_j$, if we are willing to make the formula a little more complicated. For example, since M is transitive, we have

$$\text{"}y = Z\text{" defines the same set as } \forall z(z \in Z \Leftrightarrow z \in y),$$

$$\text{"}Z \in y\text{" defines the same set as } \exists z(z = Z \wedge z \in y),$$

and so on. We therefore analyze these two basic cases.

(a_1) $y_i \in Z$. We have $Z \cap X_i = Z \setminus (Z \setminus X_i) \in M$ since Z and $X_i \in M$, and M is F_2-closed; and we have $X_1 \times \cdots \times X_{i-1} \times Z \cap X_i \times \cdots \times X_n \in M$, since M is F_3-closed. This last set is M-definable by the formula $y_i \in Z$, because M is transitive.

(a_2) $y_i \in y_j$. We use induction on $n \geqslant 3$. Let

$$Y = \{\langle Y_1, \ldots, Y_n\rangle | Y_k \in X_k \quad \text{for } k = 1, \ldots, n;\ Y_i \in Y_j\}.$$

The case $\langle i, j\rangle = \langle n-1, n\rangle$. Let $X_{n-1} \cup X_n \subset X \in M$. Then

$$Y = \times F_6(F_5(X) \times (X_1 \times \cdots \times X_{n-2}) \cap (X_{n-1} \times X_n) \times (X_1 \times \cdots \times X_{n-2})).$$

The case $\langle i, j\rangle = \langle n, n-1\rangle$. Again let $X_{n-1} \cup X_n \subset X \in M$. Then

$$Y = \times F_7(F_5(X) \times (X_1 \times \cdots \times X_{n-2}) \cap (X_{n-1} \times X_n) \times (X_1 \times \cdots \times X_{n-2})).$$

The case $n \notin \{i, j\}$. By the induction assumption, the set Y' which is M-defined by the formula $y_i \in y_j$ in $X_1 \times \cdots \times X_{n-1}$ lies in M. But $Y = Y' \times X_n$.

The case $n - 1 \notin \{i, j\}$. Let Y' be M-defined by the formula $y_i \in y_j$ in $X_1 \times \cdots \times X_{n-2} \times X_n$. Then $Y = F_8(Y' \times X_{n-1})$.

The case $n=2$ reduces to the case $n=3$ by taking the direct product with $\{\varnothing\}$ and projecting. The projection of $X_1 \times \cdots \times X_n$ onto X_1 is $F_4 \circ \cdots \circ F_4$ ($n-1$ times).

(b) *Connectives.* $\wedge$ corresponds to intersection, and $\neg$ corresponds to taking the complement (relative to $X_1 \times \cdots \times X_n$). M is closed with respect to these operations, and the other connectives can be expressed in terms of these two.

(c) *Quantifiers.* It suffices to verify $\exists$. This corresponds to projecting, because M is a big class. More precisely, let Y be M-defined by the formula $\exists y_{n+1} P(y_1, \ldots, y_n, y_{n+1})$ in $X_1 \times \cdots \times X_n$. We have:

$$\langle Y_1, \ldots, Y_n \rangle \in Y \Leftrightarrow$$
$$\text{there exists a } Y_{n+1} \in M \text{ such that } P(Y_1, \ldots, Y_{n+1}) \text{ is } M\text{-true.}$$

To each $\langle Y_1, \ldots, Y_n \rangle \in X_1 \times \cdots \times X_n$ we associate the least ordinal α for which there exists $Y_{n+1} \in M \cap V_\alpha$ such that $P(Y_1, \ldots, Y_{n+1})$ is M-true, if there is such a Y_{n+1}. This gives rise to a function on $Y \subset X_1 \times \cdots \times X_n$. Let A be the set of its values, and let $\beta = \cup A$. Then $X = M \cap V_\beta$ is a set, and $X \subset M$. Since M is a big class, there exists $X_{n+1} \in M$ such that $X \subset X_{n+1}$. By the induction assumption, the M-definable subset $Y' \subset X_1 \times \cdots \times X_n \times X_{n+1}$ consisting of those points $\langle Y_1, \ldots, Y_{n+1} \rangle$ for which $P(Y_1, \ldots, Y_{n+1})$ is M-true, belongs to M. But $Y = F_4(Y')$, and M is closed under F_4.

The proposition is proved. □

In order to be able to use Proposition 2.3, we need criteria for verifying M-truth. As remarked in §7 of Chapter II, the basic technical tool for this is the notion of *absoluteness*. A formula P is called M-absolute ((M, V)-absolute in the terminology of Chapter II), if $|P|_M(\xi) = |P|_V(\xi)$ for all $\xi \in \overline{M} \subset \overline{V}$. The standard method of proving that a formula is M-true is to prove that it is V-true and M-absolute.

The following lemma provides us with a large class of M-absolute formulas.

2.4. Lemma.

(a) *Atomic formulas are M-absolute for all M.*

(b) *If the formulas P, P_1, and P_2 are M-absolute, then so are the formulas $\neg P$ and $P_1 * P_2$ (where $*$ is any connective).*

(c) *Suppose that the class M is transitive, and is closed with respect to an operation f of degree r. If the formula P is M-absolute, then the "restricted quantifier" formulas*

$$\forall x (x \in f(y_1, \ldots, y_r) \Rightarrow P),$$
$$\exists x (x \in f(y_1, \ldots, y_r) \wedge P)$$

are also M-absolute.

Proof. Part (c) is the only assertion that might not be completely obvious. Before proving it, we make one remark. The formula $x \in f(y_1, \ldots, y_r)$ is written in a suitable extension of $\mathrm{L_1Set}$, and may be assumed to be V-equivalent to some formula $P(x, y_1, \ldots, y_r)$ in $\mathrm{L_1Set}$ (with constants in M) for which $\forall y_1, \ldots, \forall y_r\ \exists! x\ P$ or a restricted version of this formula is deducible from the Zermelo–Fraenkel axioms. This P determines the operation f. We also allow the case $r = 0$; then f is simply a constant in M. We shall identify f with its standard interpretation, i.e., we shall denote terms by $f(Y_1, \ldots, Y_r) \in M$ for $Y_1, \ldots, Y_r \in M$.

Now let $\xi \in \overline{M}$, $y_i^{\xi} = Y_i \in M$, $Q = \exists x(x \in f(y_1, \ldots, y_r) \wedge P)$, $Y = f(Y_1, \ldots, Y_r) \in M$. Then

$$|Q|_M(\xi) = \sup_{X \in M} (|X \in Y|_M \cdot |P|_M(\xi')),$$

where the $\xi' \in \overline{M}$ are variations of ξ along x such that $x^{\xi'} = X$. Since P is absolute, it follows that $|P|_M(\xi') = |P|_V(\xi')$, and since M is transitive, it follows that, if $X \notin M$, then $|X \in Y|_M = |X \in Y|_V = 0$. Hence, on the right we can write V everywhere in place of M and can let ξ' run through all variations of ξ along x in V with $x^{\xi'} = X$. The resulting expression equals $|Q|_V(\xi)$.

The quantifier $\forall$ can be handled analogously, or else can be reduced to $\exists$. The lemma is proved. □

We shall abbreviate the restricted quantifier formulas in 2.4 (c) as

$$(\forall x \in f(y_1, \ldots, y_r))P, \qquad (\exists x \in f(y_1, \ldots, y_r))P,$$

respectively.

If all the quantifiers in a formula Q are restricted in this way, we say that Q is a Σ_0-formula.

As a first application of the results in 2.3 and 2.4, we prove the following fact.

2.5. **Proposition.** *All ordinals are constructible.*

Proof. Suppose that this is not the case, and that β is the least nonconstructible ordinal. All of the elements in β are contained in L_α. Since L is transitive, it follows that all $\gamma \geqslant \beta$ are nonconstructible. Hence,

$$\beta = \{x | (x \text{ is an ordinal} \wedge x \in L_\alpha) \text{ is } V\text{-true}\}.$$

If we show that "V-true" may be replaced by "L-true" here, we immediately have a contradiction, since then $\beta \in L$ by Proposition 2.3.

To do this, it suffices to verify that the formula "x is an ordinal" is L-absolute. Using the regularity axiom, from which $\neg(y \in y)$ is deducible, we can write this formula in the following Σ_0-form:

$$(\forall y \in x)(\forall z \in y)(z \in x)$$
$$\wedge (\forall y_1 \in x)(\forall y_2 \in x)(y_1 \in y_2 \vee y_2 \in y_1 \vee y_1 = y_2)$$

and then apply Lemma 2.4. □

3 The constructible universe as a model for set theory

3.1. **Theorem.** *The Zermelo–Fraenkel axioms are L-true.*

Proof. The general principle for verifying the axioms is to note that every set whose existence is stipulated in a given axiom can be represented as a set defined by a Σ_0-formula with constants in L. We only occasionally have to perform a direct verification that a subformula is L-absolute.

(a) *Empty set*. This axiom is equivalent to the Σ_0-formula $\neg\exists x(x \in \varnothing)$, which is V-true.

(b) *Extensionality*. This axiom can be represented in Σ_0-form. In addition, in subsection 4.8 of Chapter II we verified this axiom for any transitive class.

(c) *Pairing*. A direct computation of the L-truth function gives 1, since L is closed with respect to forming pairs.

(d) *Regularity*. This follows by a direct computation using the transitivity of L.

(e) *Union*. Here it is somewhat more complicated to reduce the axiom to a Σ_0-formula. The axiom is written in the form

$$\forall x\ \exists y\ \forall u(\exists z(u \in z \wedge z \in x) \Leftrightarrow u \in y).$$

Let $\xi \in \bar{L}$, let ξ' be any variation of ξ along x, and let $X = x^{\xi'} \in L$. We must show that

$$|\exists y\ \forall u(\exists z(u \in z \wedge z \in X) \Leftrightarrow u \in y)|_L(\xi') = 1.$$

It suffices to find a $Y \in L$ such that

$$|\forall u(\exists z(u \in z \wedge z \in X) \Leftrightarrow u \in Y)|_L = 1,$$

i.e., such that for all $U \in L$

$$|(\exists z \in X)(U \in z)|_L = |U \in Y|_L.$$

We can clearly take $Y = \bigcup_{Z \in X} Z$ if we show that Y is constructible. Since L is transitive, we know that all the elements of Y are constructible. Hence, there exists a constructible set Y' such that $Y' \supset Y$. Then Y can be represented as follows (where we replace V-truth by L-truth using Lemma 2.4):

$$Y = \{U | U \in Y';\ (\exists z \in X)(U \in z) \text{ is } L\text{-true}\}.$$

Now the required assertion follows by Proposition 2.3.

In what follows we shall usually omit explicit mention of the points $\xi \in \bar{L}$.

(f) *Power set axiom* $\forall x\ \exists y\ \forall z(z \subset x \Leftrightarrow z \in y)$. We fix $X \in L$, form the set $Y = \mathcal{P}(X) \cap L$ of constructible subsets of X, and show that Y is constructible. In fact, let $Y' \supset Y$, where Y' is constructible. Then by

Lemma 2.4

$$Y = \{ Z | Z \in Y'; (Z \subset X) \text{ is } L\text{-true} \},$$

because $Z \subset X$ has the Σ_0-form $(\forall z \in Z)(z \in X)$. Now a direct computation gives

$$|\forall z(z \subset X \Leftrightarrow z \in Y)|_L = 1.$$

(g) *Infinity*. This axiom is L-true because of the constructibility of the set $\{\varnothing, \{\varnothing\}, \{\{\varnothing\}\}, \ldots\}$, which can be represented in the form

$$\{ Y | Y \in L_{\omega_0}; [Y = \varnothing \vee (\exists y \in L_{\omega_0})(Y = \{y\})] \text{ is } L\text{-true} \}.$$

(h) *Replacement*. Let $\bar{z} = \langle z_1, \ldots, z_n \rangle$. This axiom is written in the form

$$\forall \bar{z}\, \forall u \big(\forall x (x \in u \Rightarrow \exists ! y P(x, y, \bar{z}))$$
$$\Rightarrow \exists w\, \forall y (y \in w \Leftrightarrow \exists x (x \in u \wedge P(x, y, \bar{z})))\big).$$

We fix $Z_1, \ldots, Z_n \in L$, $\bar{Z} = \langle Z_1, \ldots, Z_n \rangle$, and $U \in L$. It is sufficient to consider the case when the premise is L-true, i.e., when, for all $X \in L$,

$$|X \in U \Rightarrow \exists ! y P(X, y, \bar{Z})|_L = 1.$$

We must find a value $W \in L$ of w for which the conclusion is L-true. We set W' = a constructible set containing as elements all constructible Y for which

$$(\exists x \in U) P(x, Y, \bar{Z}) \text{ is } L\text{-true}.$$

This set exists because, since the premise of the axiom is L-true, it follows that each $X \in U$ corresponds to at most one constructible Y. We then set

$$W = \{ Y | Y \in W'; (\exists x \in U)(P(x, Y, \bar{Z})) \text{ is } L\text{-true} \}.$$

This set is constructible by Proposition 2.3, and it follows from the way it is defined that

$$|\forall y (y \in W \Leftrightarrow \exists x (x \in U \wedge P(x, y, \bar{Z})))|_L = 1.$$

(i) *Axiom of choice*. The main intuitive point in the verification is the numbering N of the universe L that was constructed in 1.8. But the formal verification is much more complicated here than in the previous cases. A fair amount of work is needed to give a formalization of the construction in 1.7–1.8 which is sufficiently detailed to prove the following fact:

3.2. **Proposition.** *There exists a formula $N(x, y)$ in L with two free variables such that*

(a) *For any X, $Y \in V$, the formula $N(X, Y)$ is V-true if and only if X is an ordinal and $Y = N(X)$.*
(b) *$N(x, y)$ is L-absolute.*

We shall postpone the proof until §5, and shall make use of this proposition to verify the axiom of choice. We divide this verification into two steps.

3.3. UNIVERSAL CHOICE FUNCTION. Let $X \in L$ be a nonempty set. We construct the function Y which for every nonempty $Z \in X$ chooses the element U in Z with the least N-number (see 1.8):

$$Y = \Big\{ \langle Z, U \rangle | Z \in X, U \in \bigcup_{X' \in X} X';\ U \in Z \wedge \exists w \big(N(w, U) \wedge \forall z \big(z \in Z \Rightarrow \big(z = U \vee \forall w' (N(w', z) \Rightarrow w \in w') \big)\big)\big) \text{ is } V\text{-true} \Big\}.$$

We want to prove that $Y \in L$. By Proposition 2.3, this holds if we can define Y by means of the L-truth of a formula. We are not allowed mechanically to replace V by L, since it is not immediately obvious from its external form that this formula is L-absolute. We proceed as follows: taking into account the constructibility of the ordinals, we take all ordinals which occur as the least N-numbers of the elements of the constructible set $\cup_{X' \in X} X' = \cup(X)$, and we find a constructible set W which contains these ordinals. Then we replace $\exists w$ by $\exists w \in W$ and $\forall w'$ by $\forall w' \in W$ in the formula. The set Y does not change, and now V-truth may be replaced by L-truth, as can be seen by using Proposition 3.2 and Lemma 2.4.

3.4. We now compute the L-truth value of the axiom of choice:

$$\forall x \big(x \neq \varnothing \Rightarrow \exists y \big(y \text{ is a function} \wedge \operatorname{dom} y = x \wedge (\forall z \in x)(z \neq \varnothing \Rightarrow y(z) \in z) \big)\big).$$

It suffices to show that, if we take a nonempty $X \in L$ and the constructible choice function $Y \in L$ in 3.3, then

$$|Y \text{ is a function}|_L = |\operatorname{dom} Y = X|_L = |(\forall z \in X)(z \neq \varnothing \Rightarrow Y(z) \in z)|_L = 1.$$

The third formula here is V-true, and is written in Σ_0-form except for the subformula $Y(z) \in z$, which can be replaced by $(\forall u \in U(Y))(\langle z, u \rangle \in Y \Rightarrow u \in z)$. Thus, the third formula is L-absolute and hence L-true.

We verify that the first two formulas are absolute in §5. They are V-true by construction. This completes the proof of Proposition 3.1. □

We note that the same argument shows the following: *all the axioms*, with the possible exception of the axiom of choice, *are M-true for any transitive big class M which is closed with respect to the operations* $F_1, \ldots, F_8$.

4 The Generalized Continuum Hypothesis is L-true

4.1. We wish to show that the assertion "card $\mathcal{P}(\omega_\alpha) = \omega_{\alpha+1}$" is L-true. A certain amount of caution is essential here, because cardinality is not an L-absolute notion. If Y is a constructible set, let $\text{card}_L(Y)$ be the least ordinal β for which there exists in L a one-to-one onto function $f: Y \to \beta$. Hence "card $(Y) =$ card (Z)" is L-true iff $\text{card}_L(Y) = \text{card}_L(Z)$. Note that although $\text{card}_L(Y) \geqslant \text{card}(Y)$, equality fails if there are one-to-one onto functions $Y \to \beta$ in V, but no such function lies in L. The cardinal ω_α in L is the αth ordinal $\beta > \omega_0$ such that $\text{card}_L(\beta) = \beta$. Thus ω_α in L may not coincide with the "real" ω_α, that is, with ω_α in V.

We shall show that for each ordinal β and each constructible $X \subset \beta$ there is an ordinal γ with $X \in L_\gamma$ and $\text{card}_L(\gamma) = \text{card}_L(\beta)$. Hence $\mathcal{P}(\beta) \cap L \subset L_{\beta^+}$, where β^+ is the least ordinal greater than β such that $\text{card}_L(\beta^+) \neq \text{card}_L(\beta)$. The L-truth of the Generalized Continuum Hypothesis will then follow if we show the L-truth of "card $(\beta^+) = \beta^+$."

Our proof exploits throughout a proposition that requires a good deal of work formalizing the construction of L within L_1Set.

4.2. **Proposition.** *There exists a formula $L(x, y)$ of* L_1Set *with two independent variables such that*

(a) *for any X and Y in V, $L(X, Y)$ is V-true $\Leftrightarrow Y$ is an ordinal and $X \in L_Y$;*

(b) *for any transitive model $M \subset V$ of the axioms (without the axiom of choice), the formula $L(x, y)$ is M-absolute. In particular, it is L-absolute.*

We again postpone the proof until §5.

4.3. **Lemma.** *Let $X \subseteq \beta$ be constructible. Then $X \in L_\gamma$ for some ordinal γ such that* $\text{card}_L(\gamma) = \text{card}_L(\beta)$.

Proof. In this deduction, in addition to Proposition 4.2 we use versions of Propositions 7.3 and 7.6 of Chapter II which apply to the constructible universe. They are formulated precisely and proved below, in subsections 4.5 and 4.6.

Suppose that $X \subset \beta$ is constructible. Let δ be an ordinal such that $X \in L_\delta$. We enlarge the alphabet of L_1Set by adding names $\bar{\delta}$ and $\bar{X}$ for δ and X. Let $\mathfrak{S}$ be the set of formulas

$$\{\text{axioms of } \text{L}_1\text{Set}\} \cup \{L(\bar{X}, \bar{\delta})\}.$$

Let $N_0 \subset L$ be the set $\beta \cup \{X\} \cup \{\delta\}$. By Proposition 4.5 there is a constructible set N such that $N_0 \subset N$, all formulas in $\mathfrak{S}$ are (N, L)-absolute, and $\operatorname{card}_L(N) = \operatorname{card}_L(\beta)$. Thus $(N, \in)$ is a model for the axioms and, by Proposition 4.2 (a), for $L(\bar{X}, \bar{\delta})$. Now N might not be transitive, but then by Proposition 4.6 there is a transitive axiom model (M, ε) and a constructible isomorphism $f: (N, \in) \tilde{\to} (M, \varepsilon)$. Hence $L(\bar{X}, \bar{\delta})$ is M-true and $\operatorname{card}_L(M) = \operatorname{card}_L(N)$. What are the interpretations of the constants $\bar{X}$ and $\bar{\delta}$ in M?

Since the set $\beta \subset N$ is transitive, it goes to itself under the isomorphism f; hence so does the set $X \subset \beta$. Let δ_M be the image of δ under f. Since by Proposition 4.2(b) the formula $L(x, y)$ is M-absolute, and $L(\bar{X}, \bar{\delta})$ is M-true, it follows that $L(X, \delta_M)$ is V-true, so that δ_M is an actual ordinal and $X \in L_{\delta_M}$. Moreover, since $\delta_M \in M$ and M is transitive, $\delta_M \subset M$; hence $\operatorname{card}_L(\delta_M) \leqslant \operatorname{card}_L(M)$. Letting γ be the larger of δ_M and β, we have $\operatorname{card}_L(\gamma) = \operatorname{card}_L(\beta)$ and $X \in L_\gamma$. The lemma is proved. □

4.4. Deduction that the GCH is L-true from the lemma. Let β^+ be the smallest ordinal greater than β such that $\operatorname{card}_L(\beta^+) \neq \operatorname{card}_L(\beta)$. Then Lemma 4.3 implies the V-truth of the formula

$$\forall z\big(z \in L \Rightarrow (z \subset \beta \Rightarrow z \in L_{\beta^+})\big).$$

Since "$z \in L_{\beta^+}$" (i.e., the formula $L(z, \beta^+)$) is L-absolute, it follows that

$$\forall z\big(z \subset \beta \Rightarrow z \in L_{\beta^+}\big)$$

is L-true. Now if β is the cardinal ω_α in L then β^+ is the cardinal $\omega_{\alpha+1}$ in L. Hence for each α we have shown the L-truth of

$$\mathcal{P}(\omega_\alpha) \subset L_{\omega_{\alpha+1}}.$$

We claim that the following formula is also L-true:

$$\operatorname{card}\big(L_{\omega_{\alpha+1}}\big) = \omega_{\alpha+1}.$$

Since "$\operatorname{card}(\mathcal{P}(\omega_\alpha)) \leqslant \omega_{\alpha+1}$" is formally deducible in $\mathrm{L_1Set}$ from the preceding two formulas, and since all the axioms are L-true, this will show that the GCH is L-true.

Our claim is verified thus: In subsection 1.4 we proved that $\operatorname{card}(L_\gamma) = \operatorname{card}(\gamma)$ for each ordinal γ. Indeed, that proof can be formalized in $\mathrm{L_1Set}$, using the formula $L(x, y)$ of Proposition 4.2. That is, the assertion "$\forall \gamma\big(\operatorname{card}(L_\gamma) = \operatorname{card}(\gamma)\big)$" is deducible from the axioms (see 5.17). Since the axioms are L-true, this assertion is then L-true. But since "$\operatorname{card}(\omega_{\alpha+1}) = \omega_{\alpha+1}$" is trivially L-true, the claim follows. This completes the proof. □

4.5. **Proposition.** *Let $\mathfrak{S}$ be a constructible countable set of L-true formulas in the language* $\mathrm{L_1Set}$, *and let M_0 be a constructible set. Then there exists a constructible set $M \supset M_0$, $\operatorname{card}_L(M) \leqslant \operatorname{card}_L(M_0) + \omega_0$, such that all of the formulas in $\mathfrak{S}$ are (L, M)-absolute.*

Proof. The general scheme is the same as in subsection 7.3 of Chapter II, but some additional precautions are required. The main point is to prove that, if $P(x,\bar{y}), \bar{y}=(y_1,\ldots,y_n)$, is a formula in $\mathfrak{S}$, then there exists a constructible set $M \supset M_0$ with $\operatorname{card}_L(M) \leqslant \operatorname{card}_L(M_0)+\omega_0$ which can be constructed constructibly from P and has the property that $\exists x(P(x,\bar{y}))$ is (L, M)-absolute. After this we must verify constructible closure over all $P \in \mathfrak{S}$.

We reproduce the construction in subsection 7.3 of Chapter II. We construct the set M_i by induction. Let $\bar{Y}=\langle Y_1,\ldots,Y_n\rangle \in M_i \times \cdots \times M_i$. We let $\hat{M}_i(\bar{Y})$ denote the class $\{X \mid P(X, Y_1,\ldots,Y_n)$ is L-true$\}$. We let $\tilde{M}_i(\bar{Y})$ denote $\varnothing$ if $\hat{M}_i(\bar{Y})$ is empty, and $\hat{M}_i(\bar{Y}) \cap L_\alpha$ for the least α for which this intersection is nonempty, otherwise. Since $L(x,y)$ is absolute (see §5), it is not hard to see that the function $\tilde{M}_i$, $\operatorname{dom} \tilde{M}_i = M_i \times \cdots \times M_i$, is constructible. Because the constructible axiom of choice holds in L, we can obtain a constructible function F_i by choosing one element from each nonempty $\tilde{M}_i(\bar{Y})$. Let N_i be the set of values of $\tilde{M}_i$. This set is constructible, since all of our constructions are absolute; and, if M_i is infinite, then $\operatorname{card}_L(N_i)=\operatorname{card}_L(M_i)$. We set $M_{i+1}=M_i \cup N_i$ and $M=\cup M_i$. The set M has the required properties; obviously, $\operatorname{card}_L(M)+\omega_0 = \operatorname{card}_L(M_0)+\omega_0$ in L. The formal transition from $\{M_i\}$ to M is realized by considering a function which "closes" M_0, as in subsection 5.11 below. □

4.6. **Proposition.** *For every constructible set N such that the extensionality axiom is N-true there exist a unique constructible transitive set M and isomorphism $f:(N,\in)\tilde{\to}(M,\varepsilon)$.*

Proof. The plan of proof is the same as in subsection 7.6 of Chapter II. First let "f is a continuous $(\alpha+1)$-sequence" be the formula "α is an ordinal"$\wedge$"f is a function"$\wedge \operatorname{dom} f=\alpha+1 \wedge (\forall \beta \in \alpha+1)(\beta$ a limit ordinal$\Rightarrow f(\beta)=\cup_{\gamma\in\beta} f(\gamma))$. This formula is shown to be L-absolute as in subsection 5.14 below. Now consider the L-absolute operation $\phi(Z)=\{X \mid X \in N \wedge X \cap N \subset Z\}$, and let $\varnothing_N$ be the unique member of N such that $\varnothing_N \cap N=\varnothing$. Finally, let $\psi(x,y)$ be the formula

$$(\exists f)\big(\text{“}f \text{ is a continuous } (x+1)\text{-sequence”} \wedge f(0)=\varnothing_N \wedge$$
$$\times(\forall \beta \in x)\big(f(\beta+1)=\phi(f(\beta))\big) \wedge y=f(x)\big).$$

Then ψ is L-absolute, as can be shown as in subsections 5.14 and 5.15 below, and $\psi(x,y)$ is L-true if and only if $y=N_x$ in the sense of Chapter II, subsection 7.6.

We now set $\hat{N}=\cup_\alpha N_\alpha=\{z \mid (\exists \alpha)(\exists y \subset N)(\psi(\alpha,y)\wedge z \in y)\}$. We show that $\hat{N}=N$. Clearly $\hat{N}\subset N$, and if $N\setminus\hat{N}=Y$ were nonempty, it would follow by the regularity axiom, which holds in L, that $\exists Z(Z\in Y \wedge Z\cap Y=\varnothing)$. For this Z we would have $Z\subset\hat{N}$, hence $Z\subset N_\alpha$ for a suitable α, so that $Z\in N_{\alpha+1}$, which is a contradiction.

The implication $Z \subset \hat{N} \Rightarrow \exists\alpha(Z \subset N_\alpha)$, which we have used here, follows because there exists an absolute function on $\hat{N}$ which associates to each X the least α for which $X \in N_\alpha$. The replacement axiom shows that there exists an ordinal α_0, namely, the least upper bound of the values of this function, for which $\hat{N} = N = N_{\alpha_0}$. This ordinal, which is fixed for N, occurs in our subsequent construction, which is verified to be absolute as in §5.

Let "h is a constructing $(\alpha + 1)$-sequence for N, M" be the formula "h is a continuous $(\alpha + 1)$-sequence"$\wedge h(0) = \{\langle \varnothing_N, \varnothing \rangle\} \wedge$"$(\forall \beta \in \alpha)$ $(h(\beta + 1)$ is a function $\wedge \operatorname{dom} h(\beta + 1) = N_{\beta+1} \wedge$ the value of $h(\beta + 1))$ on any $X \in N_{\beta+1}$ is the set of $h(\beta)$-images of elements of $X \cap N$)." Then for each α there is a unique such h; let M_α be the image of $h(\alpha)$. For $\alpha = \alpha_0$ we obtain a function $h : N \to M = M_{\alpha_0}$, where M is our desired constructible set and h is a constructible $\in$-isomorphism.

The proposition is proved. □

5 Constructibility formula

5.1. The purpose of this section is to prove Propositions 4.2 and 3.2. Both proofs are extremely straightforward, and simply consist in writing out explicitly the formulas $L(x, y)$ and $N(x, y)$ and verifying that the conditions in Lemma 2.4 apply. But since these formulas are very long, we perform the verifications in a series of "blocks," in order to improve their appearance and to make the interpretation and verification of the conditions in 2.4 easier. As soon as a block (subformula) is constructed and its absoluteness is verified, we replace it by an abbreviated notation in the next formula.

The material within each subsection is arranged in the following order: first the abbreviated notation for the formula which is being constructed and shown to be absolute in the subsection; then the complete form of the formula; and finally any remarks that may be needed regarding absoluteness. The "complete form" of the formula may contain abbreviated notation for subformulas. If such a subformula has not yet been interpreted in detail and shown to be absolute, this is done right after the complete form.

By absoluteness we mean "M-absoluteness for any transitive model M for the axioms without the axiom of choice."

Subsections 5.2–5.15 are devoted to the formula $L(x, y)$, and subsections 5.18–5.20 are devoted to the formula $N(x, y)$. As the material we are dealing with accumulates, we shall allow ourselves to omit more and more details and to rely on the reader's experience.

The formulas

$$z = \begin{cases} F_i(x, y), & i = 1, 2, 3; \\ F_j(y), & j = 4, 5, 6, 7, 8. \end{cases}$$

5.2. $z = \{x, y\}$: $(\forall u \in z)(u = x \vee u = y) \wedge x \in z \wedge y \in z$. This whole formula is clearly absolute by Lemma 2.4. From now on we shall not even comment on such simple cases.

5.3 $z = x \backslash y$: $(\forall u \in z)(u \in x \wedge u \notin y) \wedge (\forall u \in x)(u \notin y \Rightarrow u \in z)$.

5.4. $z = x \times y$: $(\forall u_1 \in x)(\forall u_2 \in y)(\langle u_1, u_2 \rangle \in z)$

$$\wedge (\forall u \in z)(\exists u_1 \in x)(\exists u_2 \in y)(u = \langle u_1, u_2 \rangle)$$

$\langle u_1, u_2 \rangle \in z$: $(\exists v \in z)(v = \langle u_1, u_2 \rangle)$;

$u = \langle u_1, u_2 \rangle$: $(\forall v \in u)(v = \{u_1\} \vee v = \{u_1, u_2\})$

$$\wedge \{u_1\} \in u \wedge \{u_1, u_2\} \in u;$$

$\{u_1, u_2\} \in u$: $(\exists v \in u)(v = \{u_1, u_2\})$.

5.5. $Z = F_4(y) = \operatorname{dom} y$: $(\forall u \in z)(\exists v \in \bigcup \bigcup (y))(\langle u, v \rangle \in y)$

$$\wedge (\forall u \in \bigcup \bigcup (y))(\forall v \in \bigcup \bigcup (y))(\langle u, v \rangle \in y \Rightarrow u \in z).$$

Here $\bigcup \bigcup$ appears because $\langle u, v \rangle = \{\{u\}, \{u, v\}\} \in y \Rightarrow u, v \in \bigcup \bigcup (y)$. This formula is absolute, since a transitive model is closed with respect to the operation $\bigcup$ (see 3.1(e)). We shall write $\bigcup^2 = \bigcup \bigcup$, and so on.

5.6. $z = F_5(y)$: $(\forall u \in z)(\exists v \in y)(\exists w \in y)(v \in w \wedge u = \langle v, w \rangle) \wedge (\forall v \in y)(\forall w \in y)(v \in w \Rightarrow \langle v, w \rangle \in z)$.

5.7. $z = F_6(y)$: $(\forall u \in z)(\exists u_1 \in \bigcup^4(y))(\exists u_2 \in \bigcup^4(y))(\exists u_3 \in \bigcup^2(y))(\langle u_1, u_2, u_3 \rangle \in y \wedge u = \langle u_3, u_1, u_2 \rangle) \wedge (\forall u_1 \in \bigcup^4(y))$ $(\forall u_2 \in \bigcup^4(y))$ $(\forall u_3 \in \bigcup^2(y))(\langle u_1, u_2, u_3 \rangle \in y \Rightarrow \langle u_3, u_1, u_2 \rangle \in z)$. Here $\bigcup^4$ appears for the same reason as $\bigcup^2$ in 5.5. The formulas $\langle u_1, u_2, u_3 \rangle \in y$, etc., are shown to be absolute in the same way as in 5.4.

The operations F_7 and F_8 are treated analogously to F_6.

The formulas

$$y = \begin{cases} F_i''(x \times x), & \text{for } i = 1, 2, 3; \\ F_j''(x), & \text{for } j = 4, 5, 6, 7, 8. \end{cases}$$

5.8. $y = F_i''(x \times x)$, $i = 1, 2, 3$:

$$(\forall u \in y)(\exists u_1 \in x)(\exists u_2 \in x)(u = F_i(u_1, u_2))$$
$$\wedge (\forall u_1 \in x)(\forall u_2 \in x)(F_i(u_1, u_2) \in y),$$

where $F_i(u_1, u_2) \in y$: $(\exists v \in y)(v = F_i(u_1, u_2))$.

5.9. $y = F_j''(x)$, $j = 4, \ldots, 8$: $(\forall u \in y)(\exists v \in x)(u = F_j''(v)) \wedge (\forall v \in x)$

$$(F_j''(v) \in y).$$

5.10. $y = \Phi(x)$ (see 1.4):

$$(\forall z \in y)(z \in x \vee z \in F_1''(x \times x) \vee \cdots \vee z \in F_8''(x)) \wedge (\forall z \in x)(z \in y)$$
$$\wedge (\forall z \in F_1''(x \times x))(z \in y) \wedge \cdots \wedge (\forall z \in F_8''y(x))(z \in y).$$

The class L is closed with respect to the operations F_i''. In fact suppose, for example, that $i \geqslant 4$, and let $X \in L$. Let $U \in L$ be a set containing all $F_i(Y)$ for $Y \in X$. Then

$$F_i''(X) = \{Z | Z \in U, (\exists y \in X)(Z = F_i(y)) \text{ is } V\text{-true}\}.$$

Since the formula $Z = F_i(y)$ has been shown to be absolute, we may replace "V-true" by "L-true" here, and then apply Proposition 2.3. Thus, the formula $y = \Phi(x)$ is L-absolute by Lemma 2.4.

If M is an arbitrary transitive model, then the verification that M is closed with respect to F_i'' is somewhat different. Namely, the formula $\forall x\, \exists! y(y = F_i''(x))$ is obviously V-true. The formal deduction of this formula does not use the axiom of choice. Hence, the formula is M-true for any transitive model M. We therefore have $Y \in M$ if $X \in M$, where $Y = F_i''(X)$. We shall use this device many times in what follows.

5.11. "g *closes* x," which is short for: "*g is a function on* ω_0, *and* $g(n) = \Phi^n(x)$ *for all* $n \in \omega_0$." We write the formula with the constant ω_0 and the free variables g and x:

$$\text{"}g \text{ is a function"} \wedge F_4(g) = \omega_0 \wedge g(0) = x$$
$$\wedge (\forall n \in \omega_0)(g(n+1) = \varnothing(g(n))).$$

Here:

(a) "g is a function":

$$(\forall u \in g)(\exists u_1 \in \bigcup{}^2(g))(\exists u_2 \in \bigcup{}^2(g))(u = \langle u_1, u_2 \rangle)$$
$$\wedge (\forall u_1 \in \bigcup{}^2(g))(\forall u_2 \in \bigcup{}^2(g))(\forall u_3 \in \bigcup{}^2(g))$$
$$(\langle u_1, u_2 \rangle \in g \wedge \langle u_1, u_3 \rangle \in g \Rightarrow u_2 = u_3).$$

(b) $g(0) = x$: $\langle \varnothing, x \rangle \in g$.

(c) $g(n+1) = \Phi(g(n))$:

$$(\exists y \in \bigcup{}^2(g))(\langle n, y \rangle \in g \wedge \langle n \cup \{n\}, \Phi(y) \rangle \in g),$$

where

$$\langle n \cup \{n\}, \Phi(y) \rangle \in g: \quad (\exists u \in \bigcup{}^2(g))(\exists v \in \bigcup{}^2(g))$$
$$(u = n \cup \{n\} \wedge v = \Phi(y) \wedge \langle u, v \rangle \in g).$$

Since $\omega_0 \in M$, the formula 5.11 is now easily seen to be absolute by the previous results.

In 5.11 we took the liberty of using g and n for variables of $\mathrm{L_1Set}$ in order to make the formulas intuitively clearer. In what follows we shall also use α, β, K, and N as variables, thereby temporarily ignoring our convention of only using small letters at the end of the Latin alphabet.

5.12. $y \in \mathcal{G}x$: $\exists g$("g closes x"$\wedge(\exists n \in \omega_0)(\langle n, y\rangle \in g)$). Here the quantifier over g is not restricted. Since the formula under the $\exists g$ sign is absolute, we may conclude directly from the definition $|\ |_L(\xi) = |\ |_V(\xi)$, $\xi \in \overline{M}$, that $y \in \mathcal{G}x$ is also absolute, provided we show that, for any $X \in M$, the function $G \in V$ which closes X lies in M. The formula $\forall x\ \exists!\ g$ ("g closes x") is obviously V-true. If we formalize the verification of this fact, we see that this formula is deducible from the axioms without the axiom of choice. Hence it is M-true. This implies that for any $X \in M$ we have $G \in M$.

5.13. $y \in \mathcal{P}(x) \cap \mathcal{G}(x \cup \{x\})$: $(\forall z \in y)(\forall v \in z)(v \in x) \wedge y \in \mathcal{G}(x \cup \{x\})$.

5.14. "f is the constructing $(\alpha+1)$-sequence," which is short for: "α is an ordinal"$\wedge$"f is a function"$\wedge \operatorname{dom} f = \alpha + 1 \wedge (\forall \beta \in \alpha+1)(f(\beta) = L_\beta)$.
Here:

(a) $(\forall \beta \in \alpha+1)(f(\beta) = L_\beta)$:

$$(\forall \beta \in \alpha+1)\big((\beta \text{ is a limit ordinal} \Rightarrow f(\beta) = \textstyle\bigcup_{\gamma\in\beta} f(\gamma))$$
$$\wedge \big(f(\beta+1) = \mathcal{P}(f(\beta)) \cap \mathcal{G}(f(\beta) \cup \{f(\beta)\})\big)\big).$$

(b) "β is a limit ordinal": "β is an ordinal"$\wedge(\forall \alpha \in \beta)(\beta \neq \alpha \cup \{\alpha\})$

(c) $f(\beta) = \bigcup_{\gamma\in\beta} f(\gamma)$: $(\exists v \in \bigcup^2(f))\big(v = \bigcup_{\gamma\in\beta} f(\gamma) \wedge \langle \beta, v\rangle \in f\big)$;

$$v = \textstyle\bigcup_{\gamma\in\beta} f(\gamma): \quad (\forall u \in v)(\exists \gamma \in \beta)(u \in f(\gamma))$$
$$\wedge \big(\forall u \in \textstyle\bigcup^3(f)\big)(u \in f(\gamma) \Rightarrow u \in v);$$
$$u \in f(\gamma): \quad \big(\exists w \in \textstyle\bigcup^2(f)\big)(\langle \gamma, w\rangle \in f \wedge u \in w).$$

(d) $f(\beta+1) = \mathcal{P}(f(\beta)) \cap \mathcal{G}(f(\beta) \cup \{f(\beta)\})$:

$$\big(\exists u \in \textstyle\bigcup^2(f)\big)\big(\langle \beta+1, u\rangle \in f \wedge (\forall v \in u)$$
$$\big(v \in \mathcal{P}(f(\beta)) \cap \mathcal{G}(f(\beta) \cup \{f(\beta)\})\big)$$
$$\wedge \forall v\big(v \in \mathcal{P}(f(\beta)) \cap \mathcal{G}(f(\beta) \cup \{f(\beta)\}) \Rightarrow v \in u\big)\big);$$
$$v \in \mathcal{P}(f(\beta)) \cap \mathcal{G}(f(\beta) \cup \{f(\beta)\}):$$
$$\big(\exists u \in \textstyle\bigcup^2(f)\big)(\langle \beta, u\rangle \in f \wedge v \in \mathcal{P}(u) \cap \mathcal{G}(u \cup \{u\})).$$

Finally, in order to verify directly that the subformula

$$\forall v\big(v \in \mathcal{P}(f(\beta)) \cap \mathcal{G}(f(\beta) \cup \{f(\beta)\}) \Rightarrow v \in u\big)$$

is M-absolute, it suffices to show that M is closed with respect to the operation $X \mapsto \mathcal{P}(X) \cap \mathcal{G}(X \cup \{X\})$. But M is closed with respect to both $\mathcal{G}$ and $X \mapsto \mathcal{P}(X) \cap M$, so the verification is complete.

5.15. $L(x, y)$: "y is an ordinal and $x \in L_y$": "y is an ordinal"$\wedge \exists f$("f is the constructing $(y+1)$-sequence"$\wedge(\exists z \in \bigcup^2(f))(\langle y, z\rangle \in f \wedge x \in z))$. Since the quantifier $\exists f$ is not bounded, in order to verify this last absoluteness statement we must show that the constructing $(Y+1)$-sequence F is an element of M for any ordinal Y in M. We use the same argument as in 5.12: the formula $\forall y$(y is an ordinal$\Rightarrow \exists! f$(f is the constructing $(y+1)$-sequence)) not only is V-true, but also is deducible from the axioms without the axiom of choice; therefore it is M-true.

This completes the proof of Proposition 4.2. □

5.16. *Remark*. The formula $\forall x\ \exists y\ L(x, y)$ is often written in the form: $V = L$, and is called the *axiom of constructibility*. The absoluteness of $L(x, y)$ implies that the following formula is L-true:

$$|\forall x\ \exists y\ L(x, y)|_L = \inf_{X \in L} \sup_{Y \in L} |L(X, Y)|_L = \inf_{X \in L} \sup_{Y \in L} |L(X, Y)|_V = 1.$$

Hence, this formula is consistent with the Zermelo–Fraenkel axioms. On the other hand, $V = L$ implies the Generalized Continuum Hypothesis (GCH), and, since the negation of the GCH is also consistent with the Zermelo–Fraenkel axioms, it follows that $\neg(V = L)$ is consistent with the axioms.

We now proceed to the proof of Proposition 3.2. This proof follows the same plan as the proof of Proposition 4.2. We return to the conventions and constructions in 1.7–1.8.

5.17. *Remark*. In subsection 4.4 we exploited the fact that the assertion "$\alpha \geqslant \omega_0 \Rightarrow \mathrm{card}(L_\alpha) = \mathrm{card}(\alpha)$" is formally deducible from the axioms of $\mathrm{L_1Set}$ (without the axiom of choice). We may now see that such a formal deduction can be obtained by exactly mimicking the proof in subsection 1.4. Indeed, from the definition of $L(x, y)$ we have the formal deducibility of "$L_{\alpha+1} = \mathcal{P}(L_\alpha) \cap \mathcal{G}(L_\alpha \cup \{L_\alpha\})$" and "$\beta$ a limit ordinal$\Rightarrow L_\beta = \bigcup_{\gamma \in \beta} L_\gamma$". Moreover, the following are deducible: "$\mathrm{card}(X) < \omega_0 \Rightarrow \mathrm{card}(X) < \mathrm{card}(\mathcal{P}(X)) < \omega_0$" and "$\mathrm{card}(X) \geqslant \omega_0 \Rightarrow \mathrm{card}(\mathcal{G}(X)) = \mathrm{card}(X)$". As a result, the assertions "$\mathrm{card}(L_{\omega_0}) = \omega_0$", "$\mathrm{card}(L_\alpha) \geqslant \omega_0 \Rightarrow \mathrm{card}(L_{\alpha+1}) = \mathrm{card}(L_\alpha)$" and "$\beta$ a limit ordinal $\Rightarrow \mathrm{card}(L_\beta) = \mathrm{card}(\bigcup_{\gamma \in \beta} L_\gamma)$" are all deducible. And from these and the axioms of $\mathrm{L_1Set}$ the desired assertion may be deduced (using, in particular, the deducibility of "$\mathrm{card}(\omega_0) = \omega_0$", "$\alpha \geqslant \omega_0 \Rightarrow \mathrm{card}(\alpha+1) = \mathrm{card}(\alpha)$", "$\beta$ is a limit ordinal$\Rightarrow \beta = \bigcup_{\gamma \in \beta} \gamma$" and in addition an instance of transfinite induction on the ordinals, which is of course also formally deducible in $\mathrm{L_1Set}$).

5.18. *The formula* $H(K, x)$: K is a function $\wedge\, x$ is an ordinal $\wedge$ dom $K = x + 1 \wedge K(0) = \langle 0, 0, 0\rangle \wedge (\forall y \in x + 1)(K(y)$ is an important triple $\wedge K(y + 1)$ is the next important triple after $K(y)) \wedge (y$ is a limit ordinal $\Rightarrow K(y) = \lim_{z \in y} K(z))$ *is absolute.*

We shall not analyze the subformulas which have been considered before. The following subformulas remain:

(a) "$K(y)$ is an important triple $\wedge\, K(y + 1)$ is the next important triple after $K(y)$"

(b) $K(y) = \lim_{z \in y} K(z)$.

We shall have to use the absoluteness of the auxiliary formula "$y = x_{(i)}$," which is short for: "x is an important triple (i.t.) and y is the ith coordinate of x," where $i = 1, 2$, or 3. That is:

$$(\exists u_1 \in \bigcup{}^3(x))(\exists u_2 \in \bigcup{}^3(x))(\exists u_3 \in \bigcup(x))$$
$$\times(x = \langle u_1, u_2, u_3\rangle \wedge u_1 \text{ is an ordinal} \wedge u_1 \leqslant 8$$
$$\wedge\, u_2 \text{ is an ordinal} \wedge u_3 \text{ is an ordinal} \wedge y = u_i).$$

The *complete form of* (a) is:

$$(\exists u \in \bigcup(K))(\exists v \in \bigcup(K))(\langle y, u\rangle \in K \wedge \langle y + 1, v\rangle \in K$$
$$\wedge\, u \text{ is an i.t.} \wedge v \text{ is the i.t. after } u).$$

According to Lemma 1.7(a), "u is an i.t.$\wedge v$ is the i.t. after u" can be written in the form $\bigvee_{i=1}^{5} C_i(u, v)$, where $C_i(u, v)$ is the formalization of the ith alternative in 1.7(a). For example,

$$C_1\colon\quad u \text{ is an i.t.} \wedge v \text{ is an i.t.} \wedge u_{(1)} \leqslant 7 \wedge v_{(1)} = u_{(1)} + 1$$
$$\wedge\, v_{(2)} = u_{(2)} \wedge v_{(3)} = u_{(3)};$$
$$C_2\colon\quad u \text{ is an i.t.} \wedge v \text{ is an i.t.} \wedge u_{(1)} = 8 \wedge u_{(2)} + 1 < u_{(3)}$$
$$\wedge\, v_{(1)} = 0 \wedge v_{(2)} = u_{(2)} + 1 \wedge v_{(3)} = 0.$$

The other C_i are analogous, and are absolute for the same reasons.

The *complete form of* (b). Here we need to know that the following auxiliary formulas are absolute:

$$u = \bigcup_{z \in y} K(z)_{(i)},\ i = 2 \text{ or } 3\colon\quad (\forall v \in u)(\exists z \in y)(v = K(z)_{(i)})$$
$$\wedge (\forall z \in y)(\exists v \in u)(v = K(z)_{(i)});$$

$$v = K(z)_{(i)}\colon\quad (\exists w \in \bigcup(K))(\langle z, w\rangle \in K \wedge w \text{ is an i.t.} \wedge v = w_{(i)}).$$

Then, using Lemma 1.7(b), we explain the formula $K(y) = \lim_{z \in y} K(z)$ as follows:

$$K(y)_{(1)} =$$

$$0 \wedge \exists u_2\, \exists u_3 \left(u_2 = \bigcup_{z \in y} K(z)_{(2)} \wedge u_3 = \bigcup_{z \in y} K(z)_{(3)} \wedge \bigvee_{i=1}^{4} D_i(u_2, u_3, y) \right),$$

where the alternatives D_i have the following structure, depending on "how $K(z)$ approaches $K(y)$";

$$D_1:\quad u_2 \in u_3 \wedge u_2 \text{ is a limit ordinal} \wedge \big((\exists z \in y)(K(z)_{(3)} = u_3)$$
$$\rightarrow K(y)_{(2)} = u_2 \wedge K(y)_{(3)} = u_3\big);$$

$$D_2:\quad u_2 = u_3 \wedge u_2 \text{ is a limit ordinal} \wedge \big((\exists z \in y)(K(z)_{(3)} = u_3)$$
$$\wedge (\forall z \in y)(K(z)_{(2)} \in u_2) \rightarrow K(y)_{(2)} = u_2 \wedge K(y)_{(3)} = 0\big);$$

$$D_3:\quad u_2 \geqslant u_3 \wedge u_3 \text{ is a limit ordinal} \wedge \big((\forall z \in y)(K(z)_{(3)} \in u_3)$$
$$\rightarrow K(y)_{(2)} = u_2 \in K(y)_{(3)} = u_3\big);$$

$$D_4:\quad u_2 = u_3 \wedge u_2 \text{ is a limit ordinal} \wedge \big((\forall z \in y)(K(z)_{(2)} \in u_2$$
$$\wedge K(z)_{(3)} \in u_3) \rightarrow K(y)_{(2)} = 0 \wedge K(y)_{(3)} = u_3\big).$$

It is therefore obvious that the D_i are absolute. Even though the quantifiers $\exists u_2$ and $\exists u_3$ are not restricted, there is no problem, since, when $K^\xi, y^\xi \in L$, this formula can only be V-true if $u_2^{\xi'}$ and $u_3^{\xi'}$ are uniquely determined ordinals, and lie in L, which gives us L-truth.

5.19. *The formula* $S(N, x)$: "x is an ordinal $\wedge$ N is a function $\wedge$ dom $N = x + 1 \wedge (\forall y \leqslant x + 1)(N(y)$ is a constructible set with N-number y)" *is absolute.*

We shall need to know that the following auxiliary formula is absolute:

$$y = (x)_i, \quad i = 1, 2, 3, \quad \text{where } K(x) = \langle (x)_1, (x_2), (x)_3 \rangle$$

(not to be confused with the formula $y = x_{(i)}$ in 5.16, which occurs here as a subformula): x is an ordinal $\wedge \exists K(H(K, x) \wedge (\exists u \in \cup(K))(\langle x, u\rangle \in K \wedge y = u_{(i)}))$. Even though $\exists K$ is not restricted, this does not cause any problem, because, for every ordinal $x^\xi \in L$, the value of K^ξ making $H(K^\xi, x^\xi)$ V-true lies in L. In fact, the V-true formula

$$\forall x\big(x \text{ is an ordinal} \Rightarrow \exists!K(H(K, x))\big)$$

is deducible from the axioms without the axiom of choice, and hence is L-true.

We now return to $S(N, x)$. We need only show that the subformula "$N(y)$ is a constructible set with N-number y" is absolute. By definition, this subformula can be written as $\bigvee_{i=0}^{8} Q_i(y, N)$, where the alternatives have the form:

$$Q_0\colon\quad (y)_1 = 0 \wedge \langle y, L_{(y)_2}\rangle \in N;$$

$$Q_i,\ 1 \leqslant i \leqslant 3\colon\quad (y)_1 = i \wedge \langle y, F_i(N((y)_2), N((y)_3))\rangle \in N;$$

$$Q_i,\ 4 \leqslant i \leqslant 8\colon\quad (y)_1 = i \wedge \langle y, F_i(N((y)_2))\rangle \in N.$$

The absoluteness of the subformulas that have not been analyzed is clear from the following complete forms of these formulas:

(a) $\langle y, L_{(y)_2}\rangle \in N$: $\quad (\exists z \in \cup(N))(\langle y, z\rangle \in N \wedge z \in L_{(y)_2})$;

$z = L_{(y)_2}$: $\quad (\exists u \in y + 1)(u = (y)_2 \wedge z = L_u)$;

$z = L_u$: $\quad (\forall v \in z)(v \in L_u) \wedge \forall v(v \in L_u \Rightarrow v \in z)$.

We can verify directly that the last subformula, with the unrestricted quantifier $\forall v$, *is absolute, since* $L_U \in L$ *for any ordinal* U, *and* L *is transitive.*

(b) $\langle y, F_i(N((y)_2))\rangle \in N, \qquad i = 4, \ldots, 8$:

$(\exists u, v, w \in \cup(N))(u = (y)_2 \wedge \langle u, v\rangle \in N \wedge w = F_i(v) \wedge \langle y, w\rangle \in N)$.

(c) $\langle y, F_i(N((y)_2), N((y)_3))\rangle \in N, \qquad i = 1, 2, 3$:

$(\exists u_2, u_3, v_2, v_3, w \in \cup(N))(u_2 = (y)_2 \wedge u_3 = (y)_3 \wedge \langle u_2, v_2\rangle \in N$
$\wedge \langle u_3, v_3\rangle \in N \wedge w = F_i(v_2, u_3) \wedge \langle y, w\rangle \in N)$.

5.20. *The formula* $N(x, y)$: "x is an ordinal $\wedge\, y = N(x)$" *is absolute.*
In fact, this formula is written in the form

$$\exists N(S(N, x + 1) \wedge \langle x, y\rangle \in N).$$

There is no problem with $\exists N$ being unrestricted, since we can apply the same type of argument as we have used many times before: for any ordinal x^ξ there is a unique $N^{\xi'}$ making this formula V-true, and then $N^{\xi'} \in L$, since the formula $\forall x$ (x is an ordinal $\Rightarrow \exists!\, N(S(N, x + 1))$) is deducible from the axioms without the axiom of choice, and hence is L-true.

This completes the proof of Proposition 3.2. □

6 Remarks on formalization

Gödel's theory, to which this chapter is devoted, is usually presented in a more syntactic version. We shall now briefly describe the system of basic ideas and the most important changes in the proofs in this version, in which the least possible appeal is made to the semantics.

6.1. Let $Q(x)$ be a formula in $\mathrm{L_1Set}$ with one free variable x. Let ZF be the set of all the (logical, special, and equality) axioms of $\mathrm{L_1Set}$ except for the axiom of choice. $Q(x)$ is said to be *transitive* if

$$\mathrm{ZF}\vdash (Q(x) \wedge y \in x) \Rightarrow Q(y).$$

6.2. The *relativization* P_Q of a formula P in $\mathrm{L_1Set}$ relative to Q is defined by induction on the number of connectives and quantifiers in P:

$$\begin{aligned}
(x \in y)_Q &\text{ is } Q(x) \wedge Q(y) \Rightarrow x \in y;\\
(x = y)_Q &\text{ is } Q(x) \wedge Q(y) \Rightarrow x = y;\\
(\neg P)_Q &\text{ is } \neg (P_Q);\\
(P_1 * P_2)_Q &\text{ is } (P_1)_Q * (P_2)_Q, \quad \text{for any connective } * ;\\
(\forall x P)_Q &\text{ is } \forall x (Q(x) \Rightarrow P);\\
(\exists x P)_Q &\text{ is } \exists x (Q(x) \wedge P).
\end{aligned}$$

6.3. $Q(x)$ is called an (*internal*) *model* of $\mathrm{L_1Set}$ if for any axiom $P \in \mathrm{ZF}$ we have

$$\mathrm{ZF}\vdash P_Q.$$

This model is *transitive* if Q is transitive.

A formula $P(y_1, \ldots, y_n)$ is called *Q-absolute* if

$$\mathrm{ZF}\vdash (Q(y_1) \wedge \cdots \wedge Q(y_n)) \Rightarrow (P \Leftrightarrow P_Q).$$

6.4. The connection between these concepts and our earlier ones is as follows. Every formula $Q(x)$ determines a class $M = \{X \in V \mid Q(X)$ is V-true$\}$. This class M has the property that

$$|P|_M(\xi) = |P_Q|_V(\xi), \qquad \forall \xi \in \overline{M},$$

for any formula P (as can easily be proved by induction on the number of connectives and quantifiers in P). Thus, to give a syntactic reformulation of our proofs we must make the following changes throughout;

(a) We only consider classes M which are defined by formulas Q, and all references to M are replaced by references to Q.
(b) We everywhere replace "P is V-true" by "P is deducible from ZF."
(c) We everywhere replace "P is M-true" by "P_Q is deducible from ZF."
(d) We everywhere replace "P is M-absolute" by "P is Q-absolute."

In order for the new assertions on deducibility from ZF to become sufficiently obvious, we must either do some additional work formalizing the proofs or else give more careful intuitive proofs. In particular, we must find finite subsets of ZF from which the various facts are deducible. The basic results are stated as follows in the new syntactic language:

6.5. $\exists y\ L(x, y)$ "*is*" *a transitive internal model of* $\mathrm{L_1Set}$.

6.6. $\mathrm{ZF}\vdash(\text{Axiom of choice})_{\exists y\, L(x,y)}$.

6.7. ZF$\vdash$(Generalized Continuum Hypothesis)$_{\exists y\, L(x,y)}$.

6.8. Thus, a completely syntactic version of Gödel's theory would consist of all the deductions implicit in 6.5–6.7, without any commentary. Of course, such a treatment has never been written. The formula $\exists y\, L(x, y)$ alone takes up several pages; without appealing to semantics, it would be impossible either to think up, or to explain, or even to copy down all this without making mistakes. The deductions of all the required relativized formulas $P_{\exists y\, L(x,y)}$ would also be extremely long. This situation gives us an instructive example of what was discussed in "Digression: Proof" in Chapter II.

7 What is the cardinality of the continuum?

After all we have learned about the Zermelo–Fraenkel language and axiom system, it might seem naive to return to this question. But we must do so if we consider mathematical meaning to be our primary concern.

Some specialists in the foundations of mathematics espouse a different point of view. Namely, they answer that the question itself is meaningless. It seems that Paul Cohen himself tends toward this viewpoint, at the same time admitting that "this is a hard decision" (P. Cohen, Comments on the foundations of set theory, Proc. Symp. Pure Math., vol. XIII, part I, American Math. Soc., Providence 1971, p. 12).

From this point of view it is natural to reject almost the entire semantics of $\mathrm{L_1Set}$, including all the V_α starting with $\alpha = \omega_0 + 1$ in the von Neumann universe. No half-way solutions can help matters, especially since questions concerning higher axioms of infinity or the so-called "measurable cardinals" are in an even worse position than the CH.

It thus becomes necessary to try to find alternative languages and semantics. Here the differences of opinion are wide and irreconcilable. The most clear-cut position is that of the constructivists, although even among them there are different shades of opinion. The constructivists do not recognize infinity as a usable concept, and reject noneffective existence proofs. (It turns out that in practice they often replace these noneffective proofs by a more carefully differentiated word usage—"there cannot not exist," or "there quasi-exists"—which is nearly synonymous with certain linguistic precautions adopted in classical texts.) In our opinion, the shortcoming in their point of view is that constructivism is in no sense "another mathematics." It is, rather, a sophisticated subsystem of classical mathematics, which rejects the extremes in classical mathematics and carefully nourishes its effective computational apparatus.

Unfortunately, it seems that it is these "extremes"—bold extrapolations, abstractions which are infinite and do not lend themselves to a constructivist interpretation—which make classical mathematics effective. One should try to imagine how much help mathematics could have provided twentieth century quantum physics if for the past hundred years it had developed using only abstractions from "constructive objects." Most likely, the standard calculations with infinite dimensional representations of Lie groups which today play an important role in understanding the microworld, would simply never have occurred to anyone.

It is not impossible that a new (or a completely forgotten old) conception of the continuum, in which the continuum has no "cardinality," could be found in the course of a deep investigation of the external world. The notion of a set consisting of elements may actually only be adequate for finite or countable sets, and "higher infinities" may turn out to be abstractions from objects of a completely different type.

Physics seems to point up a difference in principle between "counting" and the Eudoxos–Dedekind idealization of measurement. The counting procedure applies to regions of attraction—"attractors" (R. Thom)—which are units not having sharp boundaries. The parts of a unit, even if they have physical meaning, are nevertheless attractors of a different sort. But even these ideas apparently stop making sense in the microworld.

If nature has a fundamentally statistical aspect, it might be fruitful to consider mathematical models in which the statistical aspect appears as an undefined concept. The unexpected richness of the nonstandard interpretations of classical mathematics in Boolean-valued models agrees with the suggestion that all the words we say should be understood in a new way.

7.2. We now discuss a less radical point of view on the continuum problem, according to which this question of its cardinality is meaningful. Then the main problem once again becomes how to determine the place of the continuum on the scale of alephs.

Cohen concludes his book with the following opinion: "A point of view which the author feels may eventually come to be accepted is that CH is *obviously* false . . . . C is greater than $\aleph_n$, $\aleph_\omega$, $\aleph_\alpha$ where $\alpha = \aleph_\omega$ etc. This point of view regards C as an incredibly rich set given to us by one bold new axiom, which can never be approached by any piecemeal process of construction."

We thus have a conjectural estimate from below for C, and nothing more—not even a conjecture as to whether the cardinal C is regular or singular.

Of course, the real problem consists not only in guessing a plausible conjecture, but in supporting it with sufficiently convincing indirect evidence for it to become widely accepted, even if not proved. What sort of evidence could this be? In discussing new axioms for set theory, Gödel writes:

> "there may exist... other (hitherto unknown) axioms of set theory which a more profound understanding of the concepts underlying logic and mathematics would enable us to recognize as implied by these concepts.
> "Furthermore, however, even disregarding the intrinsic necessity of some new axiom, and even in case it had no intrinsic necessity at all, a

> decision about its truth is possible also in another way, namely, inductively by studying its "success," that is, its fruitfulness in consequences and in particular in "verifiable" consequences, i.e., consequences demonstrable without the new axiom, whose proofs by means of the new axiom, however, are considerably simpler and easier to discover, and make it possible to condense into one proof many different proofs. The axioms for the system of real numbers, rejected by the intuitionists, have in this sense been verified to some extent owing to the fact that analytic number theory frequently allows us to prove number theoretic theorems which can subsequently be verified by elementary methods. A much higher degree of verification than that, however, is conceivable. There might exist axioms so abundant in their verifiable consequences, shedding so much light upon a whole discipline, and furnishing such powerful methods for solving given problems (and even solving them, as far as that is possible, in a constructivistic way) that quite irrespective of their intrinsic necessity they would have to be assumed at least in the same sense as any well established physical theory." (K. Gödel, What is Cantor's continuum problem?, Amer. Math. Monthly, vol. 54, no. 9, 1947).

There is little to add here to this ardently expressed hope. But see §8 of Chapter VII, where it is shown using an idea of Gödel's own that *any* new independent axiom can shorten to an arbitrary extent the proofs of suitable assertions which are provable without the axiom. This result somewhat weakens our confidence in pragmatic criteria for truth.

II

COMPUTABILITY

CHAPTER V

Recursive functions and Church's thesis

1 Introduction. Intuitive computability

1.1. The first part of this book was primarily concerned with *mathematical proof*; we showed that the analogous concept in formal languages is that of formal deduction, after which the most interesting results were that certain intuitive mathematical assertions (such as the Continuum Hypothesis and its negation) are not deducible.

Our primary concern in the second part of the book is the notion of a *determinate computational process*, that is, the processing of information, or, briefly, the notion of an algorithm. In §2 we give a precise and presumably complete characterization of everything that can be obtained using computational algorithms. Then the most interesting results turn out to be assertions that certain intuitively defined functions cannot be computed by an algorithm (Chapter VI).

Both the theory of proof and the theory of computation can be presented in a large part independently of one another. This is the approach we have adopted, even though it does not correspond to the historical development. But when the machinery of both theories has been developed to a certain point, it becomes possible to apply each theory to investigate the other. The third part of the book is devoted to such applications.

In this section we describe informally the main focal points of the theory of computability. We appeal to the reader's intuitive notion of algorithms, which can be conveniently used to illuminate the structure and interrelations of the basic concepts.

When we make these concepts precise in the next section, we shall not give a description of the algorithms themselves, but rather of their results, i.e., *computable functions*. The concept of an algorithm seems to lose too much in any formalization, while the notion of algorithmic computability seems not to lose anything essential.

1.2. We now introduce several simple basic concepts. Let X and Y be two sets. A *partial function* (or mapping) from X to Y is any pair $\langle D(f), f\rangle$ consisting of a subset $D(f) \subset X$ and a mapping $f: D(f) \to Y$. Here $D(f)$ (instead of the earlier dom f) is called the domain of definition of f; f is defined at a point $x \in X$ if $x \in D(f)$; f is nowhere defined if $D(f)$ is empty; and there exists a unique nowhere defined partial function.

We let $\mathbf{Z}^+ = \{1, 2, 3, \ldots\}$ denote the set of natural numbers, *excluding zero*. (It is not necessary, only convenient, to exclude zero.) If $n \geqslant 1$, we let $(\mathbf{Z}^+)^n$ denote the n-fold direct product of $\mathbf{Z}^+$ with itself, i.e., the set of ordered n-tuples $\langle x_1, \ldots, x_n\rangle$, $x_i \in \mathbf{Z}^+$. It is convenient to let $(\mathbf{Z}^+)^0$ denote the set consisting of an arbitrary element, denoted "$\cdot$." The basic objects of our concern will be partial functions from $(\mathbf{Z}^+)^m$ to $(\mathbf{Z}^+)^n$ for various m and n. When we classify these functions according to their computability, the reader can think of the word "program" as referring to a program for a universal computer which is written without regard to time or memory limitations. Here every program for computing a function has a special "blank space" in which to insert the value of the argument.

1.3. *The basic informal definitions*. (a) A partial function f from $(\mathbf{Z}^+)^m$ to $(\mathbf{Z}^+)^n$ is called *computable* if there exists a "program" which, whenever a vector $x \in (\mathbf{Z}^+)^m$ is entered in the input, gives as output

$$\begin{array}{ll} f(x), & \text{if } x \in D(f); \\ 0, & \text{if } x \notin D(f). \end{array}$$

Here 0 merely indicates that f is not defined at x; we could allow the output in this case to be anything *not in* $(\mathbf{Z}^+)^n$.

(b) A partial function f from $(\mathbf{Z}^+)^m$ to $(\mathbf{Z}^+)^n$ is called *semi-computable* if there exists a "program" which, whenever a vector $x \in (\mathbf{Z}^+)^m$ is entered in the input, gives $f(x)$ as output if $x \in D(f)$, and either gives 0 as output or else works infinitely long without stopping if $x \notin D(f)$.

In particular, *computable functions are semi-computable*, and *everywhere defined semi-computable functions are computable*.

(c) A partial function f is called *noncomputable* if it does not satisfy condition (b) (and *a fortiori* (a)).

1.4. *Comments*

(a) The most basic of these three concepts is semi-computability, since computability reduces to this property. In fact, to determine whether a semi-computable function is computable, we proceed as follows.

Let $X \subset Y$ be two sets. By the *characteristic function* of X in Y we mean the function $\chi_X : Y \to \mathbf{Z}^+$ such that

$$\chi_X(x) = \begin{cases} 1, & \text{if } x \in X; \\ 2, & \text{if } x \notin X. \end{cases}$$

Note that χ_X is everywhere defined on Y.

Now let f be a semi-computable function from $(\mathbf{Z}^+)^m$ to $(\mathbf{Z}^+)^n$. If f were computable as well, then the characteristic function of $D(f)$ would also be computable: simply add to the program which computes f the instructions "send 0 to 2, and anything not 0 to 1, and print as output." Conversely, if $\chi_{D(f)}$ is computable, then so is f: in front of the program which semi-computes f, put the program which computes $\chi_{D(f)}$ and then the instruction to give 0 as output immediately if $\chi_{D(f)}(x) = 2$ and to continue with the program for f with x as the argument if $\chi_{D(f)}(x) = 1$. Thus, since the everywhere defined function $\chi_{D(f)}$ is computable if and only if it semi-computable, we have f is computable $\Leftrightarrow f$ is semi-computable and $\chi_{D(f)}$ is semi-computable. Later, we shall first formalize the concept of semi-computability, and then take the right side of this equivalence as the formalization of computability.

(b) *There exist noncomputable functions.* In fact, any program is a finite text in a finite alphabet, so that the set of programs is countable, while the set of all functions $\mathbf{Z}^+ \to \mathbf{Z}^+$ is uncountable. (For a critical discussion of this argument, see 1.5 below.)

An example of a noncomputable function. We consider the language of arithmetic SAr which was described in §10 of Chapter II, and number the formulas of this language as explained in §11 of Chapter II. We define a function f by stipulating that

$$f(x)\begin{cases} = 1, & \text{if the } x\text{th formula is true in the standard interpretation;} \\ \text{is not defined,} & \text{if the } x\text{th formula is false.} \end{cases}$$

The function f is noncomputable. In Chapter VII we shall see that this follows because the set $D(f)$ is not definable in arithmetic, by Tarski's theorem.

In other words, it is impossible (even in principle) to distinguish the set of all number theoretic truths by writing a single program (even a very long and complicated one) which could tell from a statement's formulation whether it is true. Of course, to prove this result requires a much deeper analysis of the concept of computability.

(c) *There exist functions which are semi-computable but not computable.* We first give a typical example of a program which semi-computes a function. We consider the following function f from $\mathbf{Z}^+$ to $\mathbf{Z}^+$, which is defined in terms of Fermat's problem:

$$f(n)\begin{cases} = 1, & \text{if there exists } x, y, z \in \mathbf{Z}^+ \text{ for which } \\ & x^{n+2} + y^{n+2} = z^{n+2}; \\ \text{is not defined}, & \text{otherwise}. \end{cases}$$

Here is a program which semi-computes f: after entering n in the input, run through all vectors $\langle x, y, z\rangle$ in a suitable order. (For example, according to increasing $x + y + z$, and for given $x + y + z$, in lexicographic order.) For each such vector verify whether or not $x^{n+2} + y^{n+2} = z^{n+2}$. If this equation holds, give 1 as output; otherwise, go on to the next $\langle x, y, z\rangle$.

Hence, f is semi-computable. But it is not known whether or not f is computable. According to Fermat's conjecture, f is nowhere defined (and hence computable!). The strongest theoretical results known concerning f —the so-called criteria of Kummer, Wieferich, Vandiver, and others—may be regarded as a sort of approximation to proving that f is *computable*, not that f is nowhere defined. That is, in order to verify the Fermat conjecture successively for various values of n, we must perform a (machine) computation (whose size grows rapidly with n) to determine $\chi_{D(f)}$ at the point n, when this determination is possible.

There is an analogous example of a semi-computable function which we actually know is not computable. In Chapter VI we prove that there exists a polynomial $P(t, x_1, \ldots, x_n)$ with integer coefficients such that the function

$$g(t)\begin{cases} = 1, & \text{if the equation } P(t, x_1, \ldots, x_n) = 0 \text{ is solvable} \\ & \text{with } x_1, \ldots, x_n \in \mathbf{Z}^q; \\ \text{is not defined}, & \text{otherwise}, \end{cases}$$

is not computable. This function is semi-computable by the same argument as in the case of the function connected with Fermat's equation.

1.5. *Critical discussion of the above proofs*. Before proceeding further, we consider from a more critical point of view, for example, the argument in 1.4(b). The first weak point that catches our attention is that we did not say precisely what a program is. But this is not essential; for any fixed definition we choose, a program must in any case be a text in a finite alphabet if it at all corresponds to our intuitive notions, and there are countably many such texts. A much stronger objection to the argument goes roughly as follows: what justification do we have for working with just *one* definition of what a program is? Could there perhaps exist an increasing hierarchy of precisely describable "methods of computation," so that, for every function from $\mathbf{Z}^+$ to $\mathbf{Z}^+$ we could choose a corresponding program which could compute this function?

A fundamental discovery in the theory of computability was that this last question has a negative answer. We now have a unique and final formal notion which corresponds to the intuitive idea of semi-computability. It can be stated as follows:

1.6. **Church's Thesis** (weakest form). *It is possible to give explicitly*:

(a) *a family of basic semi-computable functions*;
(b) *a family of elementary operations which, starting from any semi-computable functions, allow new semi-computable functions to be constructed*;

with the property that any semi-computable function can be obtained in a finite number of steps, where each step consists in applying one of the elementary operations to the functions constructed before and those in the family (a).

1.7. *Comment*. Church's thesis will be given a precise formulation in the next section: the basic functions and the elementary operations will be given explicitly. The exact mathematical theory of computability begins at that point. But it seemed important to indicate first the general significance of the discovery that such families of functions and operations exist at all and can even be given explicitly, a result that is far from obvious.

This is an experimental fact, one of the most important discovered by logic. In the next section we discuss evidence of its value and usefulness. Now we merely note that this fact is related to the finiteness of the basic logical and set theoretic principles of mathematics (implicit, for example, in L_1Set), but is not identical to this finiteness.

2 Partial recursive functions

2.1. In this section we give the precise definition and the basic properties of a class of partial functions from $(\mathbf{Z}^+)^m$ to $(\mathbf{Z}^+)^n$ which we take as an adequate formalization of the class of semi-computable functions. We give the definition in a way parallel to the statement of Church's thesis in 1.6.

2.2. *The basic functions*

$$\mathrm{suc}: \mathbf{Z}^+ \to \mathbf{Z}^+, \quad \mathrm{suc}(x) = x + 1;$$

$$1^{(n)}: (\mathbf{Z}^+)^n \to \mathbf{Z}^+, \quad 1^{(n)}(x_1, \ldots, x_n) = 1, \quad n \geqslant 0;$$

$$\mathrm{pr}_i^n: (\mathbf{Z}^+)^n \to \mathbf{Z}^+, \quad \mathrm{pr}_i^n(x_1, \ldots, x_n) = x_i, \quad n \geqslant 1.$$

2.3. *The elementary operations on partial functions*

(a) *Composition* (or *substitution*). This operation associates to every pair of partial functions f from $(\mathbf{Z}^+)^m$ to $(\mathbf{Z}^+)^n$ and g from $(\mathbf{Z}^+)^n$ to $(\mathbf{Z}^+)^p$ the function $h = g \circ f$ from $(\mathbf{Z}^+)^m$ to $(\mathbf{Z}^+)^p$ which is defined as follows:

$$D(g \circ f) = f^{-1}(D(g)) = \{x \in (\mathbf{Z}^+)^m | x \in D(f), f(x) \in D(g)\};$$
$$(g \circ f)(x) = g(f(x)).$$

(b) *Juxtaposition*. This operation associates to partial functions f_i from $(\mathbf{Z}^+)^m$ to $(\mathbf{Z}^+)^{n_i}$, $i = 1, \ldots, k$, the function $(f_1, \ldots, f_k)$ from $(\mathbf{Z}^+)^m$ to $(\mathbf{Z}^+)^{n_1} \times \cdots \times (\mathbf{Z}^+)^{n_k}$ which is defined as follows:

$$D((f_1, \ldots, f_k)) = D(f_1) \cap \cdots \cap D(f_k);$$
$$(f_1, \ldots, f_k)(x_1, \ldots, x_m) = \langle f_1(x_1, \ldots, x_m), \ldots, f_k(x_1, \ldots, x_m)\rangle.$$

(c) *Recursion*. This operation associates to a pair of partial functions f from $(\mathbf{Z}^+)^n$ to $\mathbf{Z}^+$ and g from $(\mathbf{Z}^+)^{n+2}$ to $\mathbf{Z}^+$ the partial function h from $(\mathbf{Z}^+)^{n+1}$ to $\mathbf{Z}^+$ which is defined by recursion on the last argument:

$$\begin{cases} h(x_1, \ldots, x_n, 1) = f(x_1, \ldots, x_n) & \text{(initial condition);} \\ h(x_1, \ldots, x_n, k+1) = g(x_1, \ldots, x_n, k, h(x_1, \ldots, x_n, k)), & \text{for } k \geqslant 1 \end{cases}$$

(recursive step).

The domain of definition $D(h)$ is also defined by recursion:

$$\langle x_1, \ldots, x_n, 1\rangle \in D(h) \Leftrightarrow \langle x_1, \ldots, x_n\rangle \in D(f),$$
$$\langle x_1, \ldots, x_n, k+1\rangle \in D(h) \Leftrightarrow \langle x_1, \ldots, x_n, k\rangle \in D(h) \quad \text{and}$$
$$\langle x_1, \ldots, x_n, k, h(x_1, \ldots, x_n, k)\rangle \in D(g) \quad \text{for } k \geqslant 1.$$

(d) *The μ-operator*. This operation associates to a partial function f from $(\mathbf{Z}^+)^{n+1}$ to $\mathbf{Z}^+$ the partial function h from $(\mathbf{Z}^+)^n$ to $\mathbf{Z}^+$ which is defined as follows:

$$D(h) = \{\langle x_1, \ldots, x_n\rangle | \exists x_{n+1} \geqslant 1, f(x_1, \ldots, x_n, x_{n+1}) = 1 \text{ and}$$
$$\langle x_1, \ldots, x_n, k\rangle \in D(f) \text{ for all } k \leqslant x_{n+1}\};$$
$$h(x_1, \ldots, x_n) = \min\{x_{n+1} | f(x_1, \ldots, x_n, x_{n+1}) = 1\}.$$

The general role of μ is to introduce "implicitly defined" functions, as is often done in many areas of mathematics. Three remarks about the definition of μ should be made at this point. First, we obviously chose the minimal y with $f(x_1, \ldots, x_n, y) = 1$ in order to ensure that the function h is single-valued. The second observation is that, at first glance, it might seem that the domain of definition of h is artificially narrow. If, for example, we have $f(x_1, \ldots, x_n, 2) = 1$ and $f(x_1, \ldots, x_n, 1)$ is not defined, then we have taken $h(x_1, \ldots, x_n)$ to be undefined, rather than equal to 2. This is done because we want to preserve intuitive semicomputability in going from f to h, as will be discussed in somewhat greater detail below (see 2.7(a)).

Finally, we note that all the operations before μ, if applied to everywhere defined functions, give an everywhere defined function. This is obviously not the case for μ. Thus, μ is the only one of the operations which causes partial functions to arise unavoidably.

2.4. Definition.

(a) A sequence of partial functions $f_1, \ldots, f_N$ is called a partial recursive (respectively primitive recursive) description of the function $f_N = f$ if

f_1 belongs to the family of basic functions;

f_i, $i \geqslant 2$, either belongs to the family of basic functions, or else is obtained by applying one of the elementary operations (respectively one of the elementary operations other than μ) to certain of the functions $f_1, \ldots, f_{i-1}$.

(b) A function f is called partial recursive (respectively primitive recursive) if it admits a partial recursive (respectively primitive recursive) description.

(The analogy with the definition of a deduction in a formal language immediately catches our attention, and can sometimes be of use.)

2.5. Church's Thesis (usual form)

(a) *A function f is semi-computable if and only if it is partial recursive.*

(b) *A function f is computable if and only if both f and $\chi_{D(f)}$ are partial recursive.*

Remark on terminology. Everywhere defined partial recursive functions are also called *general recursive* functions. If the domain of definition is either clear or not essential in a given context, we simply use the term "recursive." (Note that every primitive recursive function is general recursive.)

2.6. *Use of Church's thesis*. Before discussing in detail the arguments supporting Church's thesis, we indicate how it is used in practice in mathematics. Two basic applications are especially evident in the literature.

(a) *Church's thesis used for a definition of algorithmic undecidability*. Suppose we have a countable sequence of mathematical "problems" $P_1, P_2, \ldots$. Further, suppose that each problem has a "yes" or "no" answer, and that the conditions in P_n are written out "effectively" as a function of n. Such a sequence $P = (P_n)$ is called a "mass problem." We associate to such a problem a function f from $\mathbf{Z}^+$ to $\mathbf{Z}^+$:

$$D(f) = \{i \in \mathbf{Z}^+ \mid P_i \text{ has "yes" for an answer}\};$$

$$f(i) = 1, \quad \text{if } i \in D(f).$$

A mass problem P is called *algorithmically decidable* if the functions f and $\chi_{D(f)}$ are partial recursive. Otherwise P is called *algorithmically undecidable*. We also distinguish the case when only $\chi_{D(f)}$ is not partial recursive from the case when even f is not partial recursive. The second type of undecidability is worse than the first; we saw examples of this in §1. Finally, a whole hierarchy of "degrees of undecidability" can be rigorously defined and investigated.

A well-known example of a mass problem is the *problem of word identities in groups*. Let G be a finitely defined group, and let $a_1, \ldots, a_r \in G$ be elements. A "reduced word" in $a_1, \ldots, a_r$ is an expression of the form $a_{i_1}^{\varepsilon_1} \cdots a_{i_k}^{\varepsilon_k}$, where $k \geqslant 1$, $\varepsilon_j = \pm 1$, and $\varepsilon_j = \varepsilon_{j+1}$ whenever $i_j = i_{j+1}$. We number all the reduced words and ask the question P_n: "Does the nth word represent the unit element of the group G?" The "mass problem" (P_n) turns out to be algorithmically decidable for certain groups G and elements $a_1, \ldots, a_n$ and algorithmically undecidable for others (Novikov, Boone, Higman). The function f in this case is always partial recursive, but $\chi_{D(f)}$ is not always (see Chapter VIII).

For another example of an undecidable problem, this one connected with Diophantine equations, see Chapter VI.

(b) *Church's thesis as a heuristic principle.* The intuitive notion of "semi-computability" at first seems broader than the notion of "partial recursiveness," and many problems concerning partial recursive functions become much easier if we replace the conditions in the problems by informal ideas and allow such ideas to be used to solve the problems. For example, the formula $e = \lim(1 + (1/n))$ and the Euclidean algorithm make it intuitively clear that the functions $f, g : \mathbf{Z}^+ \to \mathbf{Z}^+$ given by

$$f(n) = \text{the } n\text{th digit in the decimal expansion of } e,$$

$$g(n) = \text{the } n\text{th prime number}$$

are computable, but the verification that they are recursive requires rather painstaking constructions.

Church's thesis allows us to solve such problems in two stages: (1) finding an informal solution using any intuitive algorithms we need; and (2) formalizing the solution. The second stage presupposes a certain proficiency in finding a partial recursive description for a wide variety of semi-computable functions, and Church's thesis assures us that such a description exists.

As proofs of recursiveness become more and more numerous in the literature, it becomes increasingly common to go through only the first stage of the solution; a striking example of this is Hartley Rogers' book, *Theory of Recursive Functions and Effective Computability* (McGraw-Hill, New York, 1967). We shall also take such liberties toward the end of this book. All the same, there is a certain danger in this practice. It is possible that the habit of increasingly using informal arguments delayed the discovery of such a fundamental fact as the result that recursively enumerable sets and Diophantine sets coincide.

2.7. Arguments in support of Church's thesis

(a) First of all, the basic functions clearly must be computable, no matter how we precisely define the notion of computability. Furthermore, when the elementary operations are applied to semi-computable functions, they again give a semi-computable function. A program to semi-compute the latter function can easily be put together from the programs which semi-compute the original functions. We shall only consider the case of the μ-operator in detail, leaving the simple construction of the other three programs to the reader.

In the notation of 2.3(d), let f be a semi-computable function from $(\mathbf{Z}^+)^{n+1}$ to $\mathbf{Z}^+$. In order to compute $h(x_1, \ldots, x_n)$, we go through the vectors $\langle x_1, x_2, \ldots, x_n, 1\rangle, \langle x_1, \ldots, x_n, 2\rangle, \ldots$ in the order of increasing last coordinate, and compute the values of f at these vectors. If $\langle x_1, \ldots, x_n\rangle \in D(h)$, where h is obtained from f by applying the μ-operator, then the program for f successively computes

$$f(x_1, \ldots, x_n, 1), \ldots, f(x_1, \ldots, x_n, y-1),$$

and finally $f(x_1, \ldots, x_n, y) = 1$. The least such y, if it exists, must be given as output; it will be the value of h at the point $\langle x_1, \ldots, x_n\rangle$. On the other hand, if it turns out that one of the values $f(x_1, \ldots, x_n, k)$ (before we reach $f = 1$) is not defined, then either the program which semi-computes f will work infinitely long, or else it will give an answer not in $\mathbf{Z}^+$, which must then be given as output. But then, by definition, h is not defined at the point $\langle x_1, \ldots, x_n\rangle$, and the behavior of the program for h still agrees with the definition of h being semi-computable.

From all this we conclude that *partial recursive functions are semi-computable.* However, the stronger part of Church's thesis is the converse: *semi-computable functions are partial recursive.* (The definition of computability in terms of semi-computability is simply taken from §1 without any changes.) As has been said, this result is an experimental fact. The experimental evidence for it is divided into several classes, which we consider in (b)–(d) below.

(b) In the literature we find a huge collection of recursive descriptions of various computable and semi-computable functions. See, for example, Rózsa Péter, *Recursive Functions* (Academic Press, New York, 1967). We shall give part of this list in the next section. We also find certain techniques for composing recursive descriptions which are applicable to entire classes of (semi-) computable functions. Every time an author has tried to find a partial recursive description of a (semi-) computable function, he has met with success.

(c) Turing proposed a mathematical characterization of an abstract computer, and gave strong arguments to the effect that this computer is universal, i.e., it can (semi-) compute any (semi-) computable function. His arguments came from a detailed analysis of the characteristic features of determinate computational processes. (We again recall that we have not at all concerned ourselves with formalizing computational processes, but only with the results of such processes.) It turned out that the class of functions which are semi-computable by Turing machines exactly coincides with the class of partial recursive functions.

(d) Church, Post, Markov, Kolmogorov, Uspenskiĭ, and others have proposed other determined schemes for processing information of a general (not necessarily number theoretic) character. In all cases it has turned out that if the sets of input and output are numbered in a suitable "effective" way, these methods lead to a class of maps from $\mathbf{Z}^+$ to $\mathbf{Z}^+$ which coincides with some subclass of the partial recursive functions.

For further discussion of Church's thesis, we refer the reader to the literature; see, in particular, S. Kleene, *Introduction to Metamathematics* (Van Nostrand, New York–Toronto, 1952).

3 Basic examples of recursiveness

3.1. In this section we give a short list of recursive functions and a selection of basic techniques for proving recursiveness. Both these lists will subsequently be enlarged when needed (in particular, see Chapter VII).

3.2. (a) $\mathrm{sum}_2 : (\mathbf{Z}^+)^2 \to \mathbf{Z}^+, \quad \langle x_1, x_2 \rangle \mapsto x_1 + x_2.$

Use recursion on x_2, starting from the initial condition

$$x_1 + 1 = \mathrm{sum}_2(x_1, 1) = \mathrm{suc}(x_1)$$

and applying the recursive step

$$x_1 + k + 1 = \mathrm{sum}_2(x_1, k + 1) = \mathrm{suc}(\mathrm{sum}_2(x_1, k)).$$

(b) $\mathrm{sum}_n : (\mathbf{Z}^+)^n \to \mathbf{Z}^+, \quad \langle x_1, \dots, x_n \rangle \mapsto \sum_{i=1}^n x_i, \quad n \geqslant 3.$

Suppose that we already know that sum_{n-1} is recursive. We can obtain sum_n by juxtaposition and composition as follows:

$$\mathrm{sum}_n = \mathrm{sum}_2 \circ (\mathrm{sum}_{n-1} \circ (\mathrm{pr}_1^n, \dots, \mathrm{pr}_{n-1}^n), \mathrm{pr}_n^n).$$

Another version is to use recursion on x_n, starting from the initial condition $\mathrm{suc} \circ \mathrm{sum}_{n-1}$ and applying the recursive step

$$\sum_{i=1}^{n-1} x_i + k + 1 = \mathrm{suc}(\mathrm{sum}_n(x_1, \dots, x_{n-1}, k)).$$

This choice of recursive descriptions, even of "natural" ones, will become even more numerous as the functions become more complicated.

3.3. (a) $\mathrm{prod}_2 : (\mathbf{Z}^+)^2 \to \mathbf{Z}^+, \quad \langle x_1, x_2 \rangle \mapsto x_1 x_2.$

Use recursion on x_2, starting from the initial condition x_1 and applying the recursive step

$$x_1(k + 1) = x_1 k + x_1 = \mathrm{sum}_2(x_1 k, x_1).$$

(b) $\text{prod}_n : (\mathbf{Z}^+)^n \to \mathbf{Z}^+, \quad \langle x_1, \ldots, x_n \rangle \mapsto x_1 \cdots x_n, \quad n \geqslant 3.$

$\text{prod}_n = \text{prod}_2 \circ (\text{prod}_{n-1} \circ (\text{pr}_1^n, \ldots, \text{pr}_{n-1}^n), \text{pr}_n^n).$

3.4. (a) $\mathbf{Z}^+ \to \mathbf{Z}^+, \quad x \mapsto x \dot{-} 1 = \begin{cases} x - 1, & \text{if } x \geqslant 2; \\ 1, & \text{if } x = 1. \end{cases}$

Use recursion with the functions

$$f : (\mathbf{Z}^+)^0 \to \mathbf{Z}^+, \cdot \mapsto 1;$$

$$g = \text{pr}_1^2 : (\mathbf{Z}^+)^2 \to \mathbf{Z}^+, \quad \langle x_1, x_2 \rangle \mapsto x_1.$$

(b) $(\mathbf{Z}^+)^2 \to \mathbf{Z}^+$:

$$\langle x_1, x_2 \rangle \mapsto x_1 \dot{-} x_2 = \begin{cases} x_1 - x_2, & \text{if } x_1 > x_2; \\ 1, & \text{if } x_1 \leqslant x_2. \end{cases}$$

This "truncated difference" is obtained by applying recursion to the functions

$$f(x_1) = x_1 \dot{-} 1;$$

$$g(x_1, x_2, x_3) = x_3 \dot{-} 1.$$

3.5. $F : (\mathbf{Z}^+)^n \to \mathbf{Z}^+$, *where* F *is any polynomial in* $x_1, \ldots, x_n$ *with integer coefficients which only takes values in* $\mathbf{Z}^+$.

If all the coefficients in F are nonnegative, then F is a sum of products of the functions $\text{pr}_i^n : \langle x_1, \ldots, x_n \rangle \mapsto x_i$. Otherwise, we write $F = F^+ - F^-$, where F^+ and F^- have nonnegative coefficients, and at all points of $(\mathbf{Z}^+)^n$ the nontruncated difference coincides with the truncated difference $F^+ \dot{-} F^-$ because of the assumption concerning F.

We shall often use the recursiveness of the function $(x_1 - x_2)^2 + 1$, or $h = (f - g)^2 + 1$, where f and g are recursive. This technique allows us to identify the set on which $f = g$ with the "level set of h at 1," i.e., the set on which $h = 1$.

3.6. "Step functions": for each $a, b, x_0 \in \mathbf{Z}^+$, the function defined by

$$s_{x_0}^{a,b}(x) = \begin{cases} a, & \text{for } x \leqslant x_0, \\ b, & \text{for } x > x_0. \end{cases}$$

If $x_0 = 1$, we obtain this function by recursion with initial value a and all the succeeding values b. In the general case we set

$$s_{x_0}^{a,b}(x) = s_1^{a,b}(x + 1 \dot{-} x_0).$$

3.7. $\text{rem}(x, y) =$ *the remainder in* $[1, x]$ (since we cannot use zero!) *when* y *is divided by* x.

We have:

$$\text{rem}(x, 1) = 1,$$

$$\text{rem}(x, y + 1) = \begin{cases} 1, & \text{if } \text{rem}(x, y) = x; \\ \text{suc} \circ \text{rem}(x, y), & \text{if } \text{rem}(x, y) \neq x. \end{cases}$$

We now apply a somewhat artificial technique. We consider the step function $s = s_1^{2,1}$, i.e., $s(1) = 2$ and $s(x) = 1$ if $x \geqslant 2$, and we set

$$\phi(x, y) = s\big((\mathrm{rem}(x, y) - x)^2 + 1\big).$$

Obviously,

$$\mathrm{rem}(x, y) \neq x \Leftrightarrow \phi(x, y) = 1,$$
$$\mathrm{rem}(x, y) = x \Leftrightarrow \phi(x, y) = 2,$$

so that

$$\mathrm{rem}(x, y + 1) = 2\,\mathrm{suc}(\mathrm{rem}(x, y)) \dot{-} \phi(x, y)\,\mathrm{suc}(\mathrm{rem}(x, y)).$$

This gives a recursive definition of rem.

We next describe this technique in a more general form.

3.8. Suppose h is defined by "recursion with conditions," i.e.,

$$h(x_1, \ldots, x_n, 1) = f(x_1, \ldots, x_n);\ h(x_1, \ldots, x_n, k + 1)$$
$$= g_i(x_1, \ldots, x_n, k, h(x_1, \ldots, x_n, k)),$$

if the condition $C_i(x_1, \ldots, x_n, k, h)$ holds, $i = 1, \ldots, m$, where the exhaustive and mutually exclusive conditions C_i are given in the form

$$C_i \text{ is fulfilled} \Leftrightarrow \phi_i(x_1, \ldots, x_n, k, h(x_1, \ldots, x_n, k)) = 1,$$

with ϕ_i an everywhere defined recursive function which only takes the values 1 and 2. Then we can write the recursive step as follows:

$$h(x_1, \ldots, x_n, k + 1) = 2 \sum_{i=1}^{m} g_i(x_1, \ldots, x_n, k, h(x_1, \ldots, x_n k))$$
$$\dot{-} \sum_{i=1}^{m} (g_i\phi_i)(x_1, \ldots, x_n, k, h(x_1, \ldots, x_n, k)).$$

This device allows us to show that the following functions, which will be needed later, are primitive recursive:

3.9. $$\mathrm{qt}(x, y) = \begin{cases} \text{the integral part of } y/x, & \text{if } y/x \geqslant 1; \\ 1, & \text{if } y/x < 1. \end{cases}$$

We have:

$$\mathrm{qt}(x, 1) = 1;$$

$$\mathrm{qt}(x, y + 1) = \begin{cases} \mathrm{qt}(x, y), & \text{if } \mathrm{rem}(x, y + 1) \neq x; \\ \mathrm{qt}(x, y) + 1, & \text{if } \mathrm{rem}(x, y + 1) = x \text{ and } y + 1 \neq x; \\ 1, & \text{if } y + 1 = x. \end{cases}$$

We reduce the conditions to the standard form 3.8 using the functions

$$\tilde{s}\left((\mathrm{rem}(x, y+1) - x)^2 + 1\right),$$
$$s\left((\mathrm{rem}(x, y+1) - x)^2 + 1\right) \cdot \tilde{s}\left((x - y - 1)^2 + 1\right),$$
$$s\left((x - y - 1)^2 + 1\right),$$

where $s = s_1^{1,2}$ and $\tilde{s} = s_1^{2,1}$.

3.10. $\mathrm{rad}(x) =$ the integral part of $\sqrt{x}$.
We have:

$$\mathrm{rad}(1) = 1,$$

$$\mathrm{rad}(x+1) = \begin{cases} \mathrm{rad}(x), & \text{if } \mathrm{qt}(\mathrm{rad}(x)+1, x+1) < \mathrm{rad}(x)+1; \\ \mathrm{rad}(x)+1, & \text{if } \mathrm{qt}(\mathrm{rad}(x)+1, x+1) = \mathrm{rad}(x)+1. \end{cases}$$

The reduction of these conditions to the standard form 3.8 will be left to the reader.

3.11. (a) $\min(x, y)$:

$$\min(x, 1) = 1,$$

$$\min(x, y+1) = \begin{cases} \min(x, y), & \text{if } x \leqslant y; \\ \min(x, y) + 1, & \text{if } x > y. \end{cases}$$

(b) $\max(x, y)$: analogous.

3.12. *If $f(x_1, \ldots, x_n)$ is recursive, then*

$$Sf = \sum_{k=1}^{x_n} f(x_1, \ldots, x_{n-1}, k) \quad \text{and} \quad Pf = \prod_{k=1}^{x_n} f(x_1, \ldots, x_{n-1}, k)$$

are recursive.

In fact,

$$Sf(x_1, \ldots, x_{n-1}, x_n + 1) = Sf(x_1, \ldots, x_n) + f(x_1, \ldots, x_n + 1),$$
$$Pf(x_1, \ldots, x_{n-1}, x_n + 1) = Pf(x_1, \ldots, x_n) \cdot f(x_1, \ldots, x_n + 1).$$

3.13. *If $f(x_1, \ldots, x_n)$ is recursive, then so are the functions obtained from f by:*

(a) *any permutation of the arguments;*
(b) *adding any number of "dummy" arguments;*
(c) *identifying the elements of any subset of the arguments ($f(x, x)$ instead of $f(x, y)$, and so on).*

In fact, all of these functions can be obtained from f and the various pr_i^m using composition and juxtaposition.

3.14. *A map* $f : (\mathbf{Z}^+)^m \to (\mathbf{Z}^+)^n$ *is recursive if and only if all of its components* $\mathrm{pr}_i^n \circ f$ *are recursive.*

This is obvious.

In conclusion, we note that all the specific functions described above are primitive recursive, and that all the above general operations, when applied to primitive recursive functions, yield primitive recursive functions. Starting in the next section, we shall make essential use of the μ-operator, which was defined in 2.3(d).

4 Enumerable and decidable sets

4.1. **Definition.** A set $E \subset (\mathbf{Z}^+)^n$ is called *recursively enumerable* if there exists a partial recursive function f such that $E = D(f)$ (the domain of definition of f).

The discussion in §1 and §2 showed that recursive enumerability has the following intuitive meaning: there exists a program which identifies the elements x in E but which might not identify the elements not in E. Later, in 4.12 and 4.18, we shall give another intuitive description of recursively enumerable sets which is more closely related to the etymology of the name: these are *sets all of whose elements can be obtained using a suitable "generating" program (perhaps with repetitions and with no indication of the order in which the elements occur).*

The concept of a recursively enumerable set occupies a central place in the theory of computability, alongside the concept of a partial recursive function. It will later be clear, in particular from Proposition 4.15, that either of these concepts can be reduced to the other one. However, only by using both ideas together do we obtain the flexibility necessary for efficient proofs.

We begin with the following simple fact.

Recall that the *level set* at m (or simply the m-level) of a function f from $(\mathbf{Z}^+)^n$ to $\mathbf{Z}^+$ is the set $E \subset D(f)$ such that

$$x \in E \Leftrightarrow f(x) = m.$$

4.2. **Proposition.** *The following three classes of sets coincide*:

(a) *Recursively enumerable sets.*
(b) *Level sets of partial recursive functions.*
(c) *Level sets at* 1 *of partial recursive functions.*

(a) $\subset$ (c). Suppose that E is recursively enumerable, so that $E = D(f)$, where f is partial recursive. Then $E =$ the 1-level of the function $1^{(1)} \circ f$.

(b) = (c). The m-level of f coincides with the 1-level of $(f - m)^2 + 1$. The function $(f - m)^2 + 1$ is partial recursive whenever f is, by Proposition 3.5.

(c) $\subset$ (a). Suppose that E is the 1-level of a partial recursive function $f(x_1, \ldots, x_n)$. Set

$$g(x_1, \ldots, x_n) = \min\left\{ y \mid (f(x_1, \ldots, x_n) - 1)^2 + y = 1 \right\}.$$

Obviously, g is partial recursive and $E = D(g)$. □

The following much more difficult assertion, along with its corollaries, constitutes the central result of this section.

4.3. **Theorem.** *The following two classes of sets coincide:*

(a) *Recursively enumerable sets.*
(b) *Projections of level sets of primitive recursive functions with values in* $\mathbf{Z}^+$.

4.4. First part of the proof. We first recall that, if we are given a set $E \subset (\mathbf{Z}^+)^{n+m}$, then its projection ("onto the space of the first n coordinates") is the set $F \subset (\mathbf{Z}^+)^n$ which is defined as follows:

$$\langle x_1, \ldots, x_n \rangle \in F$$
$$\Leftrightarrow \exists \langle y_1, \ldots, y_m \rangle \in (\mathbf{Z}^+)^m, \qquad \langle x_1, \ldots, x_n, y_1, \ldots, y_m \rangle \in E.$$

(From this point on, we shall not adhere to the practice in Part I of using different notation for "variable coordinates" and for particular values of the coordinates.) We similarly define the projection "onto the coordinates with indices $(i_1, \ldots, i_n) \subset (1, \ldots, n+m)$." The number m is called the *codimension* of the projection. The canonical map $E \to F$ (as well as its image) is also customarily called a projection, but this is not likely to cause any confusion.

For the time being we shall call projections of level sets of primitive recursive functions *primitive enumerable* sets. The first part of the proof consists in showing that primitive enumerable sets are recursively enumerable; the second part consists in verifying the converse implication.

Thus, let $f(x_1, \ldots, x_n, x_{n+1}, \ldots, x_{n+m})$ be a primitive recursive function, and let E be the projection of its 1-level onto the first n coordinates. (We need only consider 1-levels because of the consideration used once before: the k-level of f coincides with the 1-level of $f' = (f-k)^2 + 1$.) We explicitly construct a partial recursive function g such that $E = D(g)$.

We distinguish three cases, depending on the codimension of the projection: $m = 0$, $m = 1$, and $m \geqslant 2$.

Case (a): $m = 0$. Then $E =$ the 1-level of $f \Leftrightarrow E$ is recursively enumerable, by Proposition 4.2 (where g is constructed explicitly).

Case (b): $m = 1$. Let

$$g(x_1, \ldots, x_n) = \min\left\{ x_{n+1} \mid f(x_1, \ldots, x_n, x_{n+1}) = 1 \right\}.$$

Obviously, g is partial recursive, and $D(g) = E$. (Notice that we have used here the fact that $D(f) = (\mathbf{Z}^+)^{n+1}$.)

Case (c): $m \geqslant 2$. We reduce this case to the previous one using the following lemma, which is also important in many other situations and is of interest in its own right (as a statement that there is no notion of dimension in "recursive geometry").

4.5. **Lemma.** *For each $m \geqslant 1$ there exists a one-to-one mapping $t^{(m)} : \mathbf{Z}^+ \to (\mathbf{Z}^+)^m$ such that*

(a) *The function $t_i^{(m)} = \mathrm{pr}_i^m \circ t^{(m)}$ is primitive recursive for all $1 \leqslant i \leqslant m$.*
(b) *The inverse function $\tau^{(m)} : (\mathbf{Z}^+)^m \to \mathbf{Z}^+$ is primitive recursive.*

4.6. *How the lemma is used.* Suppose that the lemma is true. We apply it to the situation in case (c) in 4.4 as follows. For $m \geqslant 2$ we set

$$g(x_1, \ldots, x_n, y) = f(x_1, \ldots, x_n, t_1^{(m)}(y), \ldots, t_m^{(m)}(y)).$$

Obviously, g is primitive recursive if f is. It is easy to see that E coincides with the projection of the 1-level of g onto the first n coordinates. Since this is a projection of codimension 1, we have reduced this case to the previous one.

4.7. PROOF OF THE LEMMA. The case $m = 1$ is trivial. We use induction on m, starting with $m = 2$.

Construction of $t^{(2)}$. We first construct $\tau^{(2)} : (\mathbf{Z}^+)^2 \to \mathbf{Z}^+$ explicitly by setting

$$\tau^{(2)}(x_1, x_2) = \tfrac{1}{2}\big((x_1 + x_2)^2 - x_1 - 3x_2 + 2\big).$$

It is easy to see that, if we list the pairs $\langle x_1, x_2 \rangle \in (\mathbf{Z}^+)^2$ in "Cantor order," i.e., according to increasing $x_1 + x_2$ and, among those with given $x_1 + x_2$, according to increasing x_1, then $\tau^{(2)}(x_1, x_2)$ will be precisely the index of the pair $\langle x_1, x_2 \rangle$ in this list. Thus, $\tau^{(2)}$ is a one-to-one correspondence and, moreover, is primitive recursive (where we use Proposition 3.5 and then the recursiveness of qt in 3.9 to take care of the $\frac{1}{2}$).

The calculation of the pair $\langle x_1, x_2 \rangle$ as a function of its index y is an elementary problem, and results in the following formulas for the inverse function $t^{(2)}$:

$$t_1^{(2)}(y) = y - \tfrac{1}{2}\left[\sqrt{2y - \tfrac{7}{4}} - \tfrac{1}{2}\right]\left(\left[\sqrt{2y - \tfrac{7}{4}} - \tfrac{1}{2}\right] + 1\right)$$
$$t_2^{(2)}(y) = \left[\sqrt{2y - \tfrac{7}{4}} - \tfrac{1}{2}\right] - t_1^{(2)}(y) + 2.$$

Here $[z]$ denotes the integral part of z. The verification that these functions are primitive recursive using the results (and techniques) of §3 is left to the reader as an exercise.

Construction of $t^{(m)}$, $m \geqslant 3$. Suppose that $t^{(m-1)}$ and $\tau^{(m-1)}$ have already been constructed with the required properties. We first set

$$\tau^{(m)}(x_1, \ldots, x_m) = \tau^{(2)}\big(\tau^{(m-1)}(x_1, \ldots, x_{m-1}), x_m\big).$$

It is clear that $\tau^{(m)}$ is one-to-one and primitive recursive. Solving the equation $\tau^{(2)}(\tau^{(m-1)}(x_1, \ldots, x_{m-1}), x_m) = y$ in two steps, we obtain the following formulas for the inverse function $t^{(m)}$:

$$t_m^{(m)}(y) = t_2^{(2)}(y),$$
$$t_i^{(m)}(y) = t_i^{(m-1)}(t_1^{(2)}(y)), \quad 1 \leqslant i \leqslant m-1.$$

The $t_i^{(m)}$ are primitive recursive by the induction assumption. This completes the proof of the lemma, and by the same token the first part of the proof of Theorem 4.3. □

SECOND PART OF THE PROOF. We must now show that every recursively enumerable set is primitive enumerable. We begin with the following property of the class of primitive enumerable sets.

4.8. **Lemma.** *The class of primitive enumerable sets is closed with respect to the following operations: finite direct product, finite intersection, finite union, and projection.*

PROOF. Let $E, E' \subset (\mathbf{Z}^+)^n$ and $E_1 \subset (\mathbf{Z}^+)^m$ be three primitive enumerable sets which are projections of the 1-levels of the primitive recursive functions f, f', and f_1, respectively:

$$x = \langle x_1, \ldots, x_n \rangle \in E \Leftrightarrow \exists y = \langle y_1, \ldots, y_r \rangle, \qquad f(x, y) = 1,$$
$$x = \langle x_1, \ldots, x_n \rangle \in E^1 \Leftrightarrow \exists z = \langle z_1, \ldots, z_q \rangle, \qquad f'(x, z) = 1,$$
$$u = \langle u_1, \ldots, u_m \rangle \in E_1 \Leftrightarrow \exists v = \langle v_1, \ldots, v_s \rangle, \qquad f_1(u, v) = 1.$$

We then have:

$E \times E_1$ = a projection of the 1-level of the function
$$g(x, u; y, v) = f(x, y) \cdot f_1(u, v);$$

$E \cup E'$ = a projection of the 1-level of the function
$$g(x; y, z) = (f(x, y) - 1)(f'(x, z) - 1) + 1;$$

$E \cap E'$ = a projection of the 1-level of the function
$$g(x; y, z) = f(x, y) \cdot f'(x, z).$$

Closure with respect to the projection operation is clear from the definition. Lemma 4.8 is proved. □

Now let E be a recursively enumerable set. We realize E as the 1-level of a partial recursive function f from $(\mathbf{Z}^+)^n$ to $\mathbf{Z}^+$ using Proposition 4.2, and we note that, to prove that E is primitive enumerable, it suffices to show that the *graph* $\Gamma_f \subset (\mathbf{Z}^+)^n \times \mathbf{Z}^+$ of f is primitive enumerable. In fact, it is clear that E = the 1-level of f = the projection of the set $\Gamma_f \cap [(\mathbf{Z}^+)^n \times \{1\}]$ onto the first n coordinates. Here the set $\{1\} \subset \mathbf{Z}^+$ is primitive enumerable (for example, by 3.6) so that, if we prove that Γ_f is primitive enumerable, it will follow from Lemma 4.8 that the same is true for E. Thus, our problem is finally reduced to the following form: we must prove that *the graph of a partial recursive function f is primitive enumerable.* To do this we verify that first, the graphs of the simplest functions are primitive enumerable; and second, if we apply any of the elementary operations to functions having primitive enumerable graphs, then the resulting function also has a primitive enumerable graph.

Graphs of the basic functions

$$\Gamma_{\mathrm{suc}} \subset (\mathbf{Z}^+)^2 = \text{the 1-level of } (x_1 + 1 - x_2)^2 + 1,$$

$$\Gamma_{1^{(n)}} \subset (\mathbf{Z}^+)^{n+1} = \text{the 1-level of } x_{n+1},$$

$$\Gamma_{\mathrm{pr}_i^n} \subset (\mathbf{Z}^+)^{n+1} = \text{the 1-level of } (x_i - x_{n+1})^2 + 1.$$

Stability under juxtaposition. Let f and g be partial functions from $(\mathbf{Z}^+)^m$ to $(\mathbf{Z}^+)^p$ and $(\mathbf{Z}^+)^q$, respectively. Suppose that Γ_f and Γ_g are primitive enumerable. Then $\Gamma_{(f,g)} \subset (\mathbf{Z}^+)^m \times (\mathbf{Z}^+)^p \times (\mathbf{Z}^+)^q$ coincides with the intersection

$$\left(\Gamma_f \times (\mathbf{Z}^+)^q\right) \cap \operatorname{perm}\left(\Gamma_g \times (\mathbf{Z}^+)^p\right),$$

where $\operatorname{perm} : (\mathbf{Z}^+)^m \times (\mathbf{Z}^+)^q \times (\mathbf{Z}^+)^p \to (\mathbf{Z}^+)^m \times (\mathbf{Z}^+)^p \times (\mathbf{Z}^+)^q$ is the operation of permuting the last two factors:

$$\langle x^{(m)}, y^{(q)}, z^{(p)} \rangle \mapsto \langle x^{(m)}, z^{(p)}, y^{(q)} \rangle.$$

It is clear from Lemma 4.8 that $\Gamma_{(f,g)}$ is primitive enumerable.

Stability under composition. Let g be a partial function from $(\mathbf{Z}^+)^n$ to $(\mathbf{Z}^+)^m$, let f be a partial function from $(\mathbf{Z}^+)^m$ to $(\mathbf{Z}^+)^1$, and let $h = f \circ g$. Then $\Gamma_h =$ the projection of the set $\left(\Gamma_g \times (\mathbf{Z}^+)^p\right) \cap \left((\mathbf{Z}^+)^n \times \Gamma_f\right)$ onto $(\mathbf{Z}^+)^n \times (\mathbf{Z}^+)^p$. As before, if Γ_f and Γ_g are primitive enumerable, then so is Γ_h by Lemma 4.8.

The stability relative to recursion and the μ-operator is much subtler. We shall need the following elegant and useful lemma.

4.9. **Lemma.** *There exists a primitive recursive function* $\mathrm{Gd}(k, t)$ (*Gödel's function*) *with the following property: for any* $N \in \mathbf{Z}^+$ *and any finite sequence* $a_1, \ldots, a_N \in \mathbf{Z}^+$ *of length* N, *there exists* $t \in \mathbf{Z}^+$ *such that* $\mathrm{Gd}(k, t) = a_k$ *for all* $1 \leqslant k \leqslant N$. (In other words, the function Gd allows us to consider integers as encoding arbitrarily long sequences of integers: $\mathrm{Gd}(k, t)$ is the kth member of the sequence encoded by t, and the existence assertion assures that each sequence has an encoding.)

PROOF. We first set

$$\mathrm{gd}(u, k, t) = \mathrm{rem}(1 + kt, u)$$

and show that gd has the same property as Gd if we are allowed to choose $\langle u, t \rangle \in (\mathbf{Z}^+)^2$. Once we show this, we can set $\mathrm{Gd}(k, y) =$

$\mathrm{gd}(t_1^{(2)}(y), k, t_2^{(2)}(y))$, where $t^{(2)}: \mathbf{Z}^+ \to (\mathbf{Z}^+)^2$ is the isomorphism in Lemma 4.5. (It is not really essential to remove the extra parameter in $\mathrm{gd}(u, k, t)$, but working with $\mathrm{Gd}(k, t)$ will make some of the formulas shorter.)

Thus, suppose we are given $a_1, \ldots, a_N \in \mathbf{Z}^+$. We first choose $X \in \mathbf{Z}^+$ so as to satisfy $X \geqslant N$ and $1 + kX! > a_k$ for all $1 \leqslant k \leqslant N$. We then set $t = X!$. It is easy to see that, if $k_1 \neq k_2$ and $k_1, k_2 \leqslant N$, then $1 + k_1X!$ and $1 + k_2X!$ are relatively prime, since any common divisor would have to divide $(k_1 - k_2)X!$, i.e., would have to consist of primes $\leqslant X$, but no such prime divides $1 + k_1X!$.

By the Chinese remainder theorem, there exists a solution $u \in \mathbf{Z}^+$ of the system of equations

$$u \equiv a_k \bmod(1 + kX!), \qquad 1 \leqslant k \leqslant N.$$

It is then obvious that

$$\mathrm{gd}(u, k, t) = \mathrm{rem}(1 + kt, u) = a_k, \qquad 1 \leqslant k \leqslant N. \qquad \square$$

We now continue with the proof of Theorem 4.3.

4.10. *Stability relative to the μ-operator.* Let f be a partial function from $(\mathbf{Z}^+)^{n+1}$ to $\mathbf{Z}^+$, and let

$$g(x_1, \ldots, x_n) = \min\{y \mid f(x_1, \ldots, x_n, y) = 1\}.$$

Recall that the domain of definition of g consists of those $\langle x_1, \ldots, x_n \rangle$ for which such a y exists and $\langle x_1, \ldots, x_n, k \rangle \in D(f)$ for all k less than the least such y. We want to prove that, if Γ_f is primitive enumerable, then so is Γ_g.

Suppose that Γ_f is the projection onto the first $n + 1$ coordinates of the 1-level of a primitive recursive function F:

$$\phi = f(x_1, \ldots, x_{n+1}) \Leftrightarrow \exists \langle y_1, \ldots, y_m \rangle, F(x_1, \ldots, x_{n+1}, \phi, y_1, \ldots, y_m) = 1$$

(where ϕ has been used to denote the argument of F which becomes the value of f). As in 4.4, it suffices to consider the case $m = 1$, since, if $m \geqslant 2$, then we can use Lemma 4.5 to replace the vector $\langle y_1, \ldots, y_m \rangle$ by a single y, and if $m = 0$, then we can introduce a "dummy argument" y on which F does not actually depend.

Thus, let $m = 1$. We introduce a function G of the arguments $x_1, \ldots, x_n, \gamma, y, t, t_1$ by setting $s(1) = 2$, $s(x) = 1$ for $x \geqslant 2$, and

$$F_k = F(x_1, \ldots, x_n, k, \mathrm{Gd}(k, t), \mathrm{Gd}(k, t_1)), \qquad k \geqslant 1,$$

$$G = F(x_1, \ldots, x_n, \gamma, 1, y) \prod_{k=1}^{\gamma - 1} s(\mathrm{Gd}(k, t)) \cdot F_k.$$

Here $\prod_{k=1}^{0} = 1$ by definition. It is easy to see that G is primitive recursive, since it is obtained by recursion on γ from two other functions which are obviously primitive recursive. We shall show that Γ_g is the projection of the 1-level of G onto the coordinates $(x_1, \ldots, x_n, \gamma)$.

The inclusion $\mathrm{pr}(G = 1) \subset \Gamma_g$. Let $\langle x_1, \ldots, x_n, \gamma, y, t, t_1 \rangle$ be a point in the 1-level of G. We must verify that $\langle x_1, \ldots, x_n \rangle \in D(g)$ and that $\gamma = g(x_1, \ldots, x_n)$. In other words, we must show that

$$f(x_1, \ldots, x_n, \gamma) = 1;$$

$$f(x_1, \ldots, x_n, k) \quad \text{is defined and} > 1 \text{ for all } k \leqslant \gamma - 1.$$

Since $G = 1$ at the given point, it follows that all the factors in G equal 1 there. In particular, $F(x_1, \ldots, x_n, \gamma, 1, y) = 1$, which implies that $f(x_1, \ldots, x_n, \gamma) = 1$, because Γ_f is the projection of the 1-level of F. If $\gamma = 1$, there is nothing more to be proved.

Suppose $\gamma > 1$. Since the kth factor in the product $\prod_{k=1}^{\gamma-1}$ equals 1, we obtain:

$$s(\mathrm{Gd}(k, t)) = 1 \Rightarrow \mathrm{Gd}(k, t) \geqslant 2,$$

$$F_k = 1 \Rightarrow \mathrm{Gd}(k, t) = f(x_1, \ldots, x_n, k) \geqslant 2,$$

as required.

The inclusion $\Gamma_g \subset \mathrm{pr}(G = 1)$. Let $\langle x_1, \ldots, x_n, \gamma \rangle \in \Gamma_g$. We must choose values for the remaining coordinates y, t, and t_1 in such a way as to make all the factors in G equal to 1.

First of all, $\langle x_1, \ldots, x_n, \gamma, 1 \rangle \in \Gamma_f$ by the definition of g. We find the necessary value of y by lifting this point from Γ_f to the 1-level of F. If $\gamma = 1$, we may choose arbitrary values of t and t_1.

Suppose $\gamma > 1$. We then find t from the system of equations

$$\mathrm{Gd}(k, t) = f(x_1, \ldots, x_n, k), \quad \text{for all } 1 < k \leqslant \gamma - 1.$$

(Here the right side exists by the definition of $D(g)$.)

Finally, for each $k \leqslant \gamma - 1$ we lift the point

$$\langle x_1, \ldots, x_n, k, \mathrm{Gd}(k, t) \rangle \in \Gamma_f$$

to a point on $F = 1$ having additional coordinate $y^{(k)}$, and then we find t_1 from the system of equations

$$\mathrm{Gd}(k, t_1) = y^{(k)}, \qquad 1 \leqslant k \leqslant \gamma - 1.$$

This makes all the factors in $\prod_{k=1}^{\gamma-1}$ equal to 1. In fact, $s(\mathrm{Gd}(k, t)) = 1$, since $\mathrm{Gd}(k, t) = f(x_1, \ldots, x_n, k) \geqslant 2$ for $k \leqslant \gamma - 1$, and, finally, $F_k = F(x_1, \ldots, x_n, k, \mathrm{Gd}(k, t), \mathrm{Gd}(k, t_1)) = 1$ by the definition of t and t_1. □

4.11. *Stability relative to recursion*. We now carry out the last step in the proof of Theorem 4.3.

Let f and g be partial functions of n and $n + 2$ variables, respectively, and let h be the function of $n + 1$ variables which is obtained from f and g

using recursion:

$$h(x_1, \dots, x_n, 1) = f(x_1, \dots, x_n),$$
$$h(x_1, \dots, x_n, k+1) = g(x_1, \dots, x_n, k, h(x_1, \dots, x_n, k)).$$

We must show that, if Γ_f and Γ_g are primitive enumerable, then so is Γ_h.

Let F and G be primitive recursive functions whose 1-levels project onto Γ_f and Γ_g, respectively:

$$\phi = f(x_1, \dots, x_n) \Leftrightarrow \exists y,\ F(x_1, \dots, x_n, \phi, y) = 1,$$
$$\gamma = g(x_1, \dots, x_{n+2}) \Leftrightarrow \exists z,\ G(x_1, \dots, x_{n+2}, \gamma, z) = 1,$$

where, as in 4.10, it suffices to consider the case when the projection codimension is 1.

We shall explicitly construct a function H whose 1-level projects onto Γ_h. H will be a function of the arguments $x_1, \dots, x_{n+1}, \eta, y, t, t_1$ (where η is the argument which becomes the value of h). We set:

$$\bar{s}(1) = 1, \qquad \bar{s}(x) = 2, \quad \text{for } x \geqslant 2;$$
$$G_k = G(x_1, \dots, x_n, k-1, \mathrm{Gd}(k-1, t), \mathrm{Gd}(k, t), \mathrm{Gd}(k, t_1));$$
$$H = F(x_1, \dots, x_n, \mathrm{Gd}(1, t), y) \cdot \bar{s}\left[(\eta - \mathrm{Gd}(x_{n+1}, t))^2 + 1\right] \prod_{k=2}^{x_{n+1}} G_k.$$

(We take $\prod_{k=2}^{x_{n+1}} = 1$ if $x_{n+1} = 1$.) As in 4.10, we easily verify that H is primitive recursive.

The inclusion $\mathrm{pr}(H = 1) \subset \Gamma_h$. Let $\langle x_1, \dots, x_{n+1}, \eta, y, t, t_1 \rangle$ be a point on $H = 1$. We must show that $h(x_1, \dots, x_{n+1}) = \eta$. Since the second factor in H equals 1, we first obtain $\eta = \mathrm{Gd}(x_{n+1}, t)$. If we also have $x_{n+1} = 1$, then setting the first factor in H equal to 1 gives:

$$\eta = \mathrm{Gd}(1, t) = f(x_1, \dots, x_n) = h(x_1, \dots, x_n, 1).$$

Now suppose $x_{n+1} > 1$. In this case, using the equation $G_k = 1$ we find that for all $2 \leqslant k \leqslant x_{n+1}$

$$\mathrm{Gd}(k, t) = g(x_1, \dots, x_n, k-1, \mathrm{Gd}(k-1, t)),$$

and using the equation $F = 1$ and the definition of h we find that

$$\mathrm{Gd}(1, t) = f(x_1, \dots, x_n) = h(x_1, \dots, x_n, 1).$$

If we increase k from $k = 1$ to $k = x_{n+1}$ and use the recursive definition of h, we see by induction on k that $\mathrm{Gd}(k, t) = h(x_1, \dots, x_n, k)$ and, in particular,

$$\eta = \mathrm{Gd}(x_{n+1}, t) = h(x_1, \dots, x_n, x_{n+1}).$$

The inclusion $\Gamma_h \subset \mathrm{pr}(H = 1)$. We are given a point

$$\langle x_1, \dots, x_{n+1}, h(x_1, \dots, x_{n+1}) \rangle \in \Gamma_h.$$

We let $\eta = h(x_1, \dots, x_{n+1})$. We must also choose values of y, t, and t_1 so as to make H equal to 1.

If $x_{n+1} = 1$, we choose t so that $\mathrm{Gd}(1, t) = h(x_1, \ldots, x_n, 1) = f(x_1, \ldots, x_n)$. We then lift the point $\langle x_1, \ldots, x_n, \mathrm{Gd}(1, t)\rangle \in \Gamma_f$ to a point on $F = 1$. This gives us the value of y; t_1 may be chosen arbitrarily.

Now let $x_{n+1} > 1$. We first find t from the system of equations

$$\mathrm{Gd}(1, t) = f(x_1, \ldots, x_n) = h(x_1, \ldots, x_n, 1);$$
$$\mathrm{Gd}(k, t) = h(x_1, \ldots, x_n, k) = g(x_1, \ldots, x_n, k-1, \mathrm{Gd}(k-1)),$$
$$\times 2 \leqslant k \leqslant x_{n+1}.$$

We then find y by lifting the point $\langle x_1, \ldots, x_n, \mathrm{Gd}(1, t)\rangle \in \Gamma_f$ to the 1-level of F. This makes the first two factors in H equal to 1.

We next lift the points

$$\langle x_1, \ldots, x_n, k-1, \mathrm{Gd}(k-1, t), \mathrm{Gd}(k, t)\rangle \in \Gamma_g, \qquad 2 \leqslant k \leqslant x_{n+1}$$

to the 1-level of G by adding coordinates $z^{(k)}$, and then solve the following system of equations for t_1:

$$\mathrm{Gd}(k, t_1) = z^{(k)}, \qquad 2 \leqslant k \leqslant x_{n+1}.$$

This makes the G_k factors in H equal to 1.

The proof of Theorem 4.3 is complete. □

4.12. *Explanation of the term "recursively enumerable set."* Theorem 4.3 shows that, if E is recursively enumerable, then there exists a program which "generates" E (see 4.1). In fact, suppose E is the projection onto the first n coordinates of the 1-level of the primitive recursive function $f(x_1, \ldots, x_n, y)$. The program that generates E must run through the vectors $\langle x_1, \ldots, x_n, y\rangle$, say in Cantor order, compute f at each vector, and give $\langle x_1, \ldots, x_n\rangle$ as output if and only if f equals 1 (compare with Corollary 4.18 below). Unlike programs of the type described in §1, which can become stuck forever on an element not in E, a generating program sooner or later gives us any given element of E, and nothing other than such elements. However, if E is empty, we might never find this out.

We conclude this section by discussing the properties of the so-called decidable sets. Intuitively, $E \subset (Z^+)^n$ is decidable if there exists a program which for every element of $(\mathbf{Z}^+)^n$ tells whether or not it belongs to E.

4.13. **Definition.** A set $E \subset (\mathbf{Z}^+)^n$ is called *decidable* if both it and its complement are recursively enumerable.

In §5 and in the next chapter we show that there exist sets which are recursively enumerable but not decidable. This result is closely connected with Gödel's incompleteness theorem, which is the subject of Chapter VII.

4.14. **Theorem.** *The following three classes of sets coincide*:

(a) *sets whose characteristic function is recursive*;

(b) *level sets of general recursive (i.e., everywhere defined partial recursive) functions;*
(c) *decidable sets.*

PROOF. The relations (a) = (b) and (b) ⊂ (c) are obvious from what has already been proved. It thus remains to show that (c) ⊂ (a).

Let $E \subset (\mathbf{Z}^+)^n$ be a decidable set, and let E' be its complement. By definition, $E = D(f)$ and $E' = D(f')$ for certain partial recursive functions f and f'. We may even assume that $f \equiv 1$ and $f' \equiv 2$ (where they are defined). We consider $\Gamma_f \cup \Gamma_{f'} \subset (\mathbf{Z}^+)^n \times \mathbf{Z}^+$. This union is obviously the graph Γ_g of the characteristic function g of the set E. It is clear from the proof of Lemma 4.8 that Γ_g is recursively enumerable whenever Γ_f and $\Gamma_{f'}$ are. Hence, the partial recursiveness of g is implied by the following result, which is also of independent interest.

4.15. **Proposition.** *In order for a partial function g from $(\mathbf{Z}^+)^n$ to $\mathbf{Z}^+$ to be partial recursive, it is necessary and sufficient that its graph Γ_g be recursively enumerable.*

PROOF. Necessity has already been proved.

We verify sufficiency. Since Γ_g is recursively enumerable, there exists a primitive recursive function $G(x_1, \ldots, x_n, y, z)$ (see 4.10) such that $\Gamma_g =$ the projection of the 1-level of G onto $(x_1, \ldots, x_n, y)$. We set

$$H(x_1, \ldots, x_n, u) = G(x_1, \ldots, x_n, t_1^{(2)}(u), t_2^{(2)}(u)),$$

where $u \mapsto \langle t_1^{(2)}(u), t_2^{(2)}(u)\rangle$ is the primitive recursive isomorphism $\mathbf{Z}^+ \to (\mathbf{Z}^+)^2$ described in 4.5 and 4.7. H is obviously primitive recursive. Finally, we set

$$h(x_1, \ldots, x_n) = \min\{u \mid H(x_1, \ldots, x_n, u) = 1\}.$$

This is a partial recursive function whose domain of definition coincides with $D(g)$ and which easily allows us to compute g:

$$g(x_1, \ldots, x_n) = t_1^{(2)}(h(x_1, \ldots, x_n)).$$

Thus, g is partial recursive, and the proof of Proposition 4.15 and Theorem 4.14 is complete. □

4.16. **Corollary.** *Every partial recursive function g has a description in which the μ-operator is only applied once.*

4.17. **Corollary.** *Every partial recursive function g which is everywhere defined has a description $g_1, \ldots, g_N = g$ in which all the functions g_i are everywhere defined.*

In fact, the description whose last part (starting with G) was constructed in 4.15 has this property.

4.18. **Corollary.** *The class of nonempty recursively enumerable sets coincides with the class of sets of values of primitive recursive functions.*

In fact, the set of values of a function f is a projection of the graph of f. Conversely, let $E \subset (\mathbf{Z}^+)^n$ be a nonempty enumerable set that is the projection onto the $(x_1, \ldots, x_n)$-space of the 1-level of a primitive recursive function $f(x_1, \ldots, x_n, y)$. Let $\langle e_1, \ldots, e_n \rangle$ be an arbitrary member of E. Then E coincides with the set of values of the primitive recursive function

$$g(z) = \begin{cases} \langle t_1^{(n+1)}(z), \ldots, t_n^{(n+1)}(z) \rangle, & \text{if } f\big(t_1^{(n+1)}(z), \ldots, t_n^{(n+1)}(z), t_{n+1}^{(n+1)}(z)\big) = 1; \\ \langle e_1, \ldots, e_n \rangle, & \text{if not.} \end{cases}$$

□

4.19. **Corollary.**

(a) *Finite sets and their complements in $(\mathbf{Z}^+)^n$ are decidable.*
(b) *Every partial function from $(\mathbf{Z}^+)^m$ to $(\mathbf{Z}^+)^n$ with a finite domain of definition is recursive and computable.*

In fact, the one-point set $\{a\} \subset \mathbf{Z}^+$ is a level for a suitable sum of two step functions, and its complement is a level for another such sum. Decidability is preserved under finite union and intersection, so we have (a) for $n = 1$. Then the isomorphism $\tau^{(n)}$ allows us to infer this result for all n.

This also implies (b), since the graphs of the mappings in (b) are finite, and therefore enumerable.

5 Elements of recursive geometry

5.1. Let $E \subset (\mathbf{Z}^+)^m$ be an enumerable set. We consider the structure on E given by the following data:

(a) $\mathcal{E} = \{E' | E' \subset E,\ E' \text{ is enumerable}\}$.
(b) For every $E' \in \mathcal{E}$, $\mathbf{R}(E') = \{f | D(f) = E', f : E' \to \mathbf{Z}^+ \text{ is recursive}\}$.

We let $\mathfrak{R}$ = the set of pairs $\langle E', \mathbf{R}(E') \rangle$, $E' \in \mathcal{E}$.

We shall show that the structure $\{\mathcal{E}, \mathfrak{R}\}$ has much in common with the structure "a topological space together with a sheaf." This allows us to find natural interpretations for certain well-known results about enumerable sets, and to ask new questions suggested by analogies with other geometrical theories.

We begin with some simple observations.

5.2. *$\mathfrak{E}$ is a lattice, i.e., it is closed with respect to finite unions and intersections.*

Since $\mathfrak{E}$ is not closed with respect to arbitrary infinite unions, we cannot consider $\mathfrak{E}$ as the system of open subsets of E in some topology. Nevertheless, in subsection 5.9 below we show that $\mathfrak{E}$ is stable with respect to an important class of infinite unions. We shall say that $\mathfrak{E}$ determines a *quasitopology* on E (which has properties similar to those of Grothendieck topologies, but does not satisfy all the axioms of the latter).

5.3. *Let $E', E'' \in \mathfrak{E}$ and $E' \subset E''$. Then the restriction of functions to E' gives a mapping $\mathbf{R}(E'') \to \mathbf{R}(E')$: $f \mapsto f|_{E'}$.*

In fact, let $c_{E'} \in \mathbf{R}(E')$ and $c_{E'} = 1$ on E'. Then $f|_{E'} = fc_{E'}$ is recursive whenever f and $c_{E'}$ are.

5.4. *Let $E' = \bigcup_{k=1}^{n} E_k$, where $E', E_k \in \mathfrak{E}$. Suppose that the $f_k \in \mathbf{R}(E_k)$ and are compatible on intersections:*

$$\forall i, j \leqslant n, \qquad f_i|_{E_i \cap E_j} = f_j|_{E_i \cap E_j}.$$

Then there exists an (obviously unique) function $f \in \mathbf{R}(E')$ such that $\forall k \leqslant n$, $f|_{E_k} = f_k$.

We need only verify that $f \in \mathbf{R}(E')$, since there obviously exists a function $f: E' \to \mathbf{Z}^+$ which is "glued together" from the f_k. But the graph of f is the union of the finitely many enumerable graphs $\Gamma_{f_i} \subset E' \times \mathbf{Z}^+$, and so is itself enumerable. We then use Proposition 4.15.

The results 5.3 and 5.4 allow us to consider $\mathfrak{R}$ as a *sheaf* on the quasitopology $\mathfrak{E}$.

5.5. *Let E_1 and E_2 be enumerable sets, and let $f: E_1 \to E_2$ be a recursive function. Then f induces a morphism of the corresponding quasitopologies with sheaves in the following sense:*

(a) *If $E' \subset E_2$ is enumerable, then $f^{-1}(E') \subset E_1$ is enumerable.*
(b) *For every $E' \subset E_2$, composition with f determines a mapping*

$$f_{E'}^* : \mathbf{R}(E') \to \mathbf{R}(f^{-1}(E')).$$

The first part follows because $c_{f^{-1}(E')} = c_{E'} \circ f$ is recursive whenever $c_{E'}$ and f are; the second part is obvious.

One might get the impression that the pair $\langle \mathfrak{E}, \mathfrak{R} \rangle$ completely characterizes E independently of the imbedding $E \subset (\mathbf{Z}^+)^m$. However, this is not the case.

5.6. **Proposition.** *Let E_1 and E_2 be enumerable infinite sets. Then there exists a bijection $f: E_1 \xrightarrow{\sim} E_2$ such that f and f^{-1} are (partial) recursive. f induces an isomorphism $\langle \mathfrak{E}_1, \mathfrak{R}_1 \rangle \xrightarrow{\sim} \langle \mathfrak{E}_2, \mathfrak{R}_2 \rangle$.*

PROOF. We establish the following more precise facts:

(a) If $E \subset \mathbf{Z}^+$ is infinite and decidable, then there exists a general recursive bijection $f : \mathbf{Z}^+ \tilde{\to} E$ for which f^{-1} is (partial) recursive and is an increasing function. The converse is also true.

(b) If $E \subset \mathbf{Z}^+$ is infinite and enumerable, then there exists a general recursive bijection $f : \mathbf{Z}^+ \tilde{\to} E$ with f^{-1} (partial) recursive.

First suppose E is decidable, and let $g(x) = 2$ for $x \in E$, $g(x) = 1$ for $x \notin E$, and $h = c_{E'}$. We set

$$f(z) = \min\left\{ y \middle| \left(\sum_{x=1}^{y} g(x) - y - z \right)^2 + 1 = 1 \right\} = \text{the } z\text{th element of } E.$$

It is easy to see that

$$f^{-1}(x) = \left(\sum_{y=1}^{x} g(y) - x \right) h(x) \begin{cases} \text{is equal to the index of } x \\ \text{as an element of } E, \text{ if } x \in E; \\ \text{is not defined, otherwise.} \end{cases}$$

Now suppose E is enumerable. By Corollary 4.18, there exists a primitive recursive function $g : \mathbf{Z}^+ \to E$ whose image coincides with E. We shall adjust g so that it becomes bijective. We set

$$F = \{ k \in \mathbf{Z}^+ | \forall i < k, g(i) \neq g(k) \}.$$

This set is decidable, since it is the 1-level of the following primitive recursive function h:

$$h(1) = 1; \qquad h(k) = \prod_{i=1}^{k-1} s\left((g(i) - g(k))^2 + 1 \right), \quad \text{for } k \geqslant 2;$$

$$s(x) = \begin{cases} 1, & \text{for } x \geqslant 2, \\ 2, & \text{for } x = 1. \end{cases}$$

By the previous result, there exists a recursive bijection $g' : \mathbf{Z}^+ \tilde{\to} F$. Let $f = g \circ g'$. Since $g|_F : F \tilde{\to} E$ is a bijection, it follows that $f : \mathbf{Z}^+ \tilde{\to} E$ is also a bijection. The inverse function is partial recursive because

$$f^{-1}(x) = \min\{ y | (f(y) - x)^2 + 1 = 1 \}.$$

The proposition is proved. □

Because of this result we usually consider the imbedding $E \subset (\mathbf{Z}^+)^m$ to be an essential element of the structure on E. In particular, we call E_1 and E_2 *isomorphic* if there exists a bijection between them which is induced by a recursive bijection of the ambient spaces.

The complete classification of enumerable sets up to isomorphism is not known, but many subtle results have been obtained in the theory of "reducibilities." We shall only go so far as to show, using a theorem which will be proved in the next chapter, that not all enumerable sets are decidable.

5.7. *Families*. Suppose that $m \geqslant 0$ and B is a set. By a *family of m-sets* (or an *m-family*) *over the base B* we mean any mapping $B \to \mathcal{P}((\mathbf{Z}^+)^m)$. If $E_k \subset (\mathbf{Z}^+)^m$ is the image of $k \in B$ under this mapping, we also denote this family by $\{E_k\}$. We call the set $E = \{\langle x, k\rangle | x \in E_k\} \subset (\mathbf{Z}^+)^m \times B$ the *total space* of the family.

Similarly, we call a mapping $B \to \{$partial functions from $(\mathbf{Z}^+)^m$ to $\mathbf{Z}^+\}$ a *family of m-functions over the base B*. We call the function $f : \langle x, k\rangle \mapsto f_k(x)$ for $x \in D(f_k)$ the *total function* of the family.

A family of m-sets (resp. m-functions) is said to be *enumerable*, if $B \subset (\mathbf{Z}^+)^n$ for some n, and if the total space is enumerable in $(\mathbf{Z}^+)^m \times (\mathbf{Z}^+)^n$ (resp. the total function is partial recursive on $(\mathbf{Z}^+)^m \times (\mathbf{Z}^+)^n$).

If $\{E_k\}$ is enumerable, then the set $\{k \in B | E_k$ is nonempty$\}$ is enumerable, since it is a projection of the total space E. Each of the E_k is enumerable, since it is the intersection $E \cap (\mathbf{Z}^+)^m \times \{k\}$.

Similarly, if $\{f_k\}$ is enumerable, then the set $\{k \in B | f_k$ is not the nowhere defined function$\}$ is enumerable, since it is a projection of the domain of definition of the total function f. Each of the f_k is partial recursive, since it is the restriction of f to the enumerable set $D(f) \cap (\mathbf{Z}^+)^m \times \{k\}$.

If $\{f_k\}$ is an enumerable family of m-functions, then $\{D(f_k)\}$ is an enumerable family of m-sets (with total space $D(f)$), and $\{\Gamma_{f_k}\}$ is an enumerable family of $(m+1)$-sets (with total space Γ_f, or more precisely, Γ_f after a permutation of its factors).

An enumerable family $\{E_k\}$ (respectively $\{f_k\}$) is said to be *versal* if every enumerable m-set (resp. any partial recursive m-function) is among the elements of the family. [(The word "versal" is borrowed from algebraic geometry, after removing the prefix "uni" which would indicate that each term in the family could only occur once.)] In §8 of the next chapter we show that versal families *exist* for each m. This is one of the central results of the theory, since total spaces and total functions of versal families are the starting point for practically all investigations of undecidability. Here we limit ourselves to the simplest and most fundamental application:

5.8. **Theorem.** *Let $\{E_k\}$ be a versal family of 1-sets over the base $B \subset \mathbf{Z}^+$. Then the set*

$$F = \{k | k \in E_k\}$$

is enumerable, but is not decidable.

Proof. Let $E \subset \mathbf{Z}^+ \times \mathbf{Z}^+$ be the total space of the family. Then $F =$ the projection of $E \cap$ (diagonal in $\mathbf{Z}^+ \times \mathbf{Z}^+$) onto the first factor, and therefore is enumerable.

On the other hand, for every $k \in B$ we have $\overline{F} = \mathbf{Z}^+ \setminus F \neq E_k$, since k belongs to either $\overline{F}$ or E_k, but not to both. Since $\{E_k\}$ is a versal family, $\overline{F}$ cannot be enumerable. The theorem is proved. □

We now show how to use enumerable families to strengthen the results in 5.2 and 5.4. We return to the notation at the beginning of the section.

5.9. *$\mathcal{E}$ is closed with respect to taking the union of the elements of any enumerable family of subsets of E.*

In fact, suppose that $\{E'_k\}$ is such a family and E' is its total space, where $E' \subset (\mathbf{Z}^+)^m \times (\mathbf{Z}^+)^n$. Then

$$\bigcup_{k \in B} E'_k = \text{the projection of } E' \text{ on } (\mathbf{Z}^+)^m.$$

5.10. *Suppose that $\{f_k\}$ is an enumerable family of partial functions on E, $E'_k = D(f_k)$, $E' = \bigcup_{k \in B} E'_k$, and*

$$\forall i, j \in B, \qquad f_i|_{E'_i \cap E'_j} = f_j|_{E'_i \cap E'_j}.$$

Then there exists a unique function $f \in \mathbf{R}(E')$ which is glued together from the f_k.

In fact, the graph Γ_f is enumerable, since it is the union of the enumerable family of enumerable sets Γ_{f_k}.

5.11. After these remarks it is natural to consider the following system of ideas by way of analogy with the theory of spaces with sheaves.

(a) Let $E = \bigcup_{k \in B} E_k$ be a covering of an enumerable set by an enumerable family. Then for any $n \geqslant 1$ the family $\{E_{k_1} \cap \cdots \cap E_{k_n} | \langle k_1, \ldots, k_n \rangle \in B^n\}$ is also an enumerable covering of E. In fact, let E' = the total space of $\{E_k\} \subset E \times B$, and let

$$\begin{aligned} E'^n &= \{\langle x_1, \ldots, x_n, k_1, \ldots, k_n \rangle | x_i \in E_{k_i}, i = 1, \ldots, n\} \\ &\approx E' \times \cdots \times E' \ (n \text{ times}). \end{aligned}$$

Then the total space of the family $\{E_{k_1} \cap \cdots \cap E_{k_n}\}$ is isomorphic to (diagonal in E^n) $\times B^n \cap E'^n$.

(b) Using the same notation, we define the "recursive product" $\mathbf{R}\Pi_n \subset \Pi_{\langle k_1, \ldots, k_n \rangle \in B^n}(E_{k_1} \cap \cdots \cap E_{k_n})$ as follows: $\mathbf{R}\Pi_0 = \mathbf{R}(E)$, $\mathbf{R}\Pi_n$ = the set of enumerable families $\{f_{\langle k_1, \ldots, k_n \rangle}\}$ over B^n such that

$$f_{\langle k_1, \ldots, k_n \rangle} \in \mathbf{R}(E_{k_1} \cap \cdots \cap E_{k_n}), \quad \text{for } n \geqslant 1.$$

(c) For every $n \geqslant 0$ we have the following boundary mappings:

$$\begin{aligned} &\partial_i^n : \mathbf{R}\prod\nolimits_n \to \mathbf{R}\prod\nolimits_{n+1}, \qquad i = 1, \ldots, n+1 : \\ &(\partial_i^n(\cdots f_{\langle k_1, \ldots, k_n \rangle} \cdots))_{\langle l_1, \ldots, l_{n+1} \rangle} \\ &= f_{\langle l_1, \ldots, l_{i-1}, l_{i+1}, \ldots, l_{n+1} \rangle}|_{E_{l_1} \cap \cdots \cap E_{l_{n+1}}} \end{aligned}$$

(Note that we really do not have $\partial_i^n(\mathbf{R}\Pi_n) \subset \mathbf{R}\Pi_{n+1}$.)

It is possible to associate various types of "recursive Čech cohomology groups" of the covering $\bigcup_{k\in B} E_k$ to the object

$$\mathbf{R}\prod_0 = \mathbf{R}(E) \xrightarrow{\partial_0^0} \mathbf{R}\prod_1 \underset{\partial_2^1}{\overset{\partial_1^1}{\rightrightarrows}} \mathbf{R}\prod_2 \begin{smallmatrix}\to\\ \to\\ \to\end{smallmatrix} \cdots .$$

It would be interesting to study such cohomology groups. The result 5.10 shows that this complex is "exact at the first term."

The reader should not find it hard to imagine what other geometrical concepts would look like in this context. In particular, it would be worthwhile to study the quotients of enumerable sets by enumerable equivalence relations. Higman's theorem (see Chapter VIII) gives a characterization of groups in the category of such objects.

We conclude by giving several results on the structure of $\mathfrak{S}$. Because of Proposition 5.6, we need only consider subsets of $\mathbf{Z}^+$; that is, we take $\mathfrak{S} = \{E' | E' \subset \mathbf{Z}^+, E' \text{ enumerable}\}$.

5.12. **Proposition.** *There exist enumerable subsets $F \subset \mathbf{Z}^+$ having an infinite complement, such that for any infinite $E \in \mathfrak{S}$ we have $F \cap E \neq \varnothing$, so that $F \cap E$ is infinite.*

Such F are called *simple*. From a topological point of view they resemble dense open sets.

Proof. Let $\{E_k\}$ be a versal family of 1-sets over $\mathbf{Z}^+$ with total space $E \subset \mathbf{Z}^+ \times \mathbf{Z}^+$. We set $E' = E \cap \{\langle x, k\rangle | x > 2k\}$. Since E' is enumerable, there exists a primitive recursive function with image E':

$$g = (g_1, g_2) : \mathbf{Z}^+ \to E'.$$

Let $h(k) = \min\{z | g_2(z) = k\}$, let $f(k) = g_1(h(k))$, and let F denote the set of values of f. F has an infinite complement, since $f(k) > 2k$. The intersection of F with an infinite E_k is nonempty, since any value of $g_1(z)$ when $g_2(z) = k$ lies in $E_k \cap E' \neq \varnothing$. The proposition is proved. □

5.13. **Proposition.**

(a) *The quotient lattice* $\mathfrak{S}$/(finite sets) *has nontrivial maximal elements.*

(b) *Every nonsimple enumerable set with an infinite complement is contained in such a maximal element.*

(c) *There exist simple enumerable sets with an infinite complement which are not contained in any nontrivial maximal set.*

We refer the reader to Rogers' book for the proof of these and many other results.

CHAPTER VI

Diophantine sets and algorithmic undecidability

1 The basic result

1.1. In §4 of Chapter V we showed that enumerable sets are the same thing as projections of level sets of primitive recursive functions. The projections of the level sets of a special kind of primitive recursive function—polynomials with coefficients in $\mathbf{Z}^+$—are called *Diophantine sets*. We note that this class does not become any larger if we allow the coefficients in the polynomial to lie in $\mathbf{Z}$. The basic purpose of this chapter is to prove the following deep result:

1.2. **Theorem** (M. Davis, H. Putnam, J. Robinson, Ju. Matijacevič, G. Čudnovskiĭ). *All enumerable sets are Diophantine.*

The plan of proof is described in §2. §§3–7 contain the intricate yet completely elementary constructions which make up the proof itself; these sections are not essential for understanding the subsequent material, and may be omitted if the reader so desires.

In §8 we use Theorem 1.2 to prove the existence of versal families of enumerable sets and functions. Recall that in §5 of Chapter V this result was shown to imply that enumerable sets exist which are undecidable, a fact we shall use in subsection 1.3 below.

In §7, which stands somewhat apart from the rest of the chapter, we define the Kolmogorov complexity of recursive functions, establish the basic properties of this concept, and prove that the problem of computing the complexity is algorithmically undecidable.

In Chapter VII the following corollary of Theorem 1.2 will be used in an essential way: *enumerable sets are definable in* $\mathrm{L}_1\mathrm{Ar}$. In fact, by their very definition, Diophantine sets are defined by formulas of the form $\exists x_1 \cdots \exists x_n(p)$, where p is an atomic formula.

In the remainder of this section we describe the principal applications of Theorem 1.2: settling Hilbert's tenth problem, constructing polynomials which take only and all prime number values in $\mathbf{Z}^+$, and so on.

1.3. *Hilbert's tenth problem.* Hilbert stated it as follows:

> "Suppose we are given a Diophantine equation with an arbitrary number of unknowns and with rational integer coefficients. Give a way in which it is possible to determine after a finite number of operations whether or not this equation is solvable in rational integers."

We show that the combination of Theorem 1.2, Theorem 5.8 of Chapter V (which follows from Theorem 1.2), and Church's thesis implies that this problem is undecidable.

First of all, any natural number is the sum of four integer squares (Lagrange). Hence $f(x_1, \ldots, x_n) = 0$ is solvable in $(\mathbf{Z}^+)^n$ if and only if the equation $f(1 + \Sigma_{i=1}^4 y_{i1}^2, \ldots, 1 + \Sigma_{i=1}^4 y_{in}^2) = 0$ is solvable in $(\mathbf{Z})^{4n}$. Consequently, it is sufficient to show that the mass problem "determining whether or not there are solutions in $(\mathbf{Z}^+)$" (see subsection 2.6 of Chapter V) is algorithmically undecidable.

Let $E \subset \mathbf{Z}^+$ be an enumerable set which is not decidable. We represent E as the projection onto the t-coordinate of the 0-level of the polynomial $f_t = f(t; x_1, \ldots, x_n)$, where $f \in Z[t, x_1, \ldots, x_n]$. The equation $f_{t_0} = 0$, $t_0 \in \mathbf{Z}^+$, has a solution if and only if $t_0 \in E$. By the discussion in §2 of Chapter V, the corresponding mass problem for the family $\{f_t\}$ is algorithmically decidable if and only if the characteristic function of E is computable. But, by our choice of E, this characteristic function is only semi-computable.

Thus, solvability in integers cannot even be determined algorithmically for a suitable one-parameter family of equations. The number of unknowns in the equation, and, in general, the codimension of the projection in Theorem 1.2, can be reduced to 13 (Matijacevič, Robinson). The precise minimum is not known, although it is an interesting problem.

Finally, it should be noted that the construction of a Diophantine representation for any enumerable set E is completely effective in the sense that, given a recursive description of f with $D(f) = E$ or of g with $g(\mathbf{Z}^+) = E$, we can write out the corresponding polynomial explicitly. The same holds for the construction of versal families, of an enumerable undecidable set, and so on. These are all constructive assertions, and not simple existence theorems.

1.4. *Polynomials which represent the prime numbers.* The search for "explicit formulas" for prime numbers was a traditional occupation of dedicated number theory enthusiasts for many centuries. Euler found the polynomial $x^2 + x + 41$, which takes a long series of only prime values. But it has long been known that the set of values at integer points of a polynomial f in $\mathbf{Z}[x_1, \ldots, x_n]$ cannot consist entirely of prime numbers: for example, if p and q are two sufficiently large primes, then the congruence $f \equiv 0 \bmod pq$ can be solved (in infinitely many ways). On the other hand, the problem becomes solvable in the class of primitive recursive functions: the function $\{i \mapsto$ the ith prime$\}$ is itself primitive recursive (see §1 of Chapter VII), but for trivial reasons.

The nontrivial statement of the problem and the problem's solution involve Theorem 1.2: the set of prime numbers is the set of all *positive values at points in* $(\mathbf{Z}^+)^n$ *of a certain polynomial in* $\mathbf{Z}[x_1, \ldots, x_n]$ (or, if we prefer, n may be replaced by $4n$; see the reduction step in 1.3). Matijacevič showed that there is a suitable polynomial of degree 37 in 24 variables.

This is actually a general result which has nothing to do with the specific properties of prime numbers:

1.5. **Proposition.** *Let $E \subset \mathbf{Z}^+$ be a Diophantine set. Then there exists a polynomial $g \in \mathbf{Z}[x_0, \ldots, x_n]$ such that E coincides with the set of positive values of g at points in $(\mathbf{Z}^+)^{n+1}$.*

PROOF. Let E be the projection of the 0-level of the polynomial $f(x_0, x_1, \ldots, x_n)$ onto the x_0-coordinate. We set

$$g = x_0\left[1 - f^2(x_0, x_1, \ldots, x_n)\right].$$

Clearly, the positive values of g are precisely the elements of E. □

It remains only to use the fact that the set of prime numbers is decidable, and hence Diophantine by Theorem 1.2.

The following sets are also sets of positive integer values of polynomials:

1.6. *The sequences* $\{1, 10, 100, \ldots, 10^k, \ldots\}$ and $\{1, 2^2, 3^{3^3}, \ldots, n^{n^{\cdot^{\cdot^{n}}}}$ (n times), $\ldots\}$.

It is amazing that the values of the corresponding polynomials can drop to zero and below in neighborhoods of points where these values are so large.

1.7. *The Fermat set* $\{n \mid n > 2$ *and* $x^n + y^n + z^n = 0$ *is solvable in* $\mathbf{Z}\}$. Thus, the variable n can be moved from the exponent to the coefficients of a Diophantine equation.

1.8. The set $\{10\varepsilon_1, 10^2\varepsilon_2, \ldots, 10^n\varepsilon_n, \ldots\}$, where ε_i is the ith digit after the decimal point in the decimal expansion of e (or π or $\sqrt[3]{2}$, or any other "computable" irrational number).

1.9. *The set of all partial fractions in the continued fraction expansion of e, or π, or $\sqrt[3]{2}$.*

We recall that in the case of $\sqrt[3]{2}$ it is not known whether this set is finite or infinite.

These examples show that many number theoretic questions reduce to problems of the solvability of Diophantine equations. In Chapter VII we shall see that, in a certain sense, "almost all of mathematics" reduces to such problems.

2 Plan of proof

2.1. In this section we introduce some auxiliary notions and give the plan of proof for Theorem 1.2.

We shall temporarily introduce a class of sets which are intermediate between enumerable and Diophantine sets. In order to define this class, we consider the map which to every subset $E \subset (\mathbf{Z}^+)^n$ corresponds the set $F \subset (\mathbf{Z}^+)^n$ which is given by the following rule:

$$\langle x_1, \ldots, x_n \rangle \in F \Leftrightarrow \forall k \in [1, x_n], \langle x_1, \ldots, x_{n-1}, k \rangle \in E.$$

We shall say that F is obtained from E by applying the bounded universal quantifier to the nth coordinate. We define similarly the operation of applying the bounded universal quantifier to any coordinate.

2.2. **Definition-Lemma.** *Consider the following three classes of subsets of* $(\mathbf{Z}^+)^n$ *for each* n.

(I) *Projections of level sets of primitive recursive functions.*
(II) *The least class of sets which contains the level sets of polynomials with integer coefficients and which is closed with respect to taking finite direct products, finite unions, finite interesections, projections, and applying the bounded universal quantifier.*
(III) *Projections of level sets of polynomials with integer coefficients.*

The following assertions hold for these classes:

(a) *The class* (I) *coincides with the class of enumerable sets, and the class (III) coincides with the class of Diophantine sets. We shall call sets in the class (II) D-sets.*
(b) (I) $\supset$ (II) $\supset$ (III).

PROOF.

(a) In Theorem 4.3 of Chapter V we showed that the class of primitive enumerable sets coincides with the class of enumerable sets. The rest of (a) merely consists of definitions.

(b) Only the inclusion (II) $\subset$ (I) is not completely obvious. First of all, the m-level set of a polynomial f is the same as the 1-level set of the primitive recursive function $(f-m)^2+1$. Hence, to verify (II) $\subset$ (I) it suffices to show that the class (I) is closed with respect to (finite) direct product, union, intersection, and the bounded universal quantifier. All except for the last of these were established in Lemma 4.8 of Chapter V.

Finally, suppose F is the image of a primitive enumerable set E under the bounded universal quantifier:

$$\langle x_1, \ldots, x_{n-1}, x_n \rangle \in F \Leftrightarrow \forall k \leqslant x_n, \quad \langle x_1, \ldots, x_{n-1}, k \rangle \in E.$$

Starting with the function $f(x_1, \ldots, x_{n-1}, x_n; y_1, \ldots, y_m)$ whose 1-level projects onto E, we want to construct a function g whose 1-level projects onto F. A natural idea is to consider as an approximation to g the product

$$\prod_{k=1}^{x_n} f(x_1, \ldots, x_{n-1}, k; y_{1k}, \ldots, y_{mk}),$$

where the y_{ik} are "independent variables." The only problem is that the number of arguments of this "function" increases with x_n. To deal with this, we apply the Gödel function $\mathrm{Gd}(k, t)$, which was defined in subsection 4.9 of Chapter V. The function g will now depend on $x_1, \ldots, x_n$ and on m additional arguments $t_1, \ldots, t_m$:

$$g(x_1, \ldots, x_n; t_1, \ldots, t_m)$$
$$= \prod_{k=1}^{x_n} f(x_1, \ldots, x_{n-1}, k; \mathrm{Gd}(k, t_1), \ldots, \mathrm{Gd}(k, t_m)).$$

This function is primitive recursive, because the kth factor is obtained from f and Gd by substitution and identifying arguments, and then g is constructed from such factors by recursion.

We now verify that the set F is the projection of the 1-level of g onto the $\langle x_1, \ldots, x_n \rangle$-coordinates. In fact, if $g(x_1, \ldots, t_m) = 1$, then for all $1 \leqslant k \leqslant x_n$ we have $f(x_1, \ldots, k, \mathrm{Gd}(k, t_1), \ldots, \mathrm{Gd}(k, t_m)) = 1$, i.e., for all $1 \leqslant k \leqslant x_n$ the point $\langle x_1, \ldots, x_{n-1}, k \rangle$ belongs to E. This means that $\langle x_1, \ldots, x_n \rangle \in F$.

Conversely, if $\langle x_1, \ldots, x_n \rangle \in F$, then for $1 \leqslant k \leqslant x_n$ we can lift the point $\langle x_1, \ldots, x_{n-1}, k \rangle$ to the 1-level of f. Let the y-coordinates of the resulting point be $y_{1,k}, \ldots, y_{m,k}$. We solve the following system of equations for the t_i:

$$\mathrm{Gd}(k, t_i) = y_{i,k}, \quad \text{for all } 1 \leqslant k \leqslant x_n.$$

This is possible by the fundamental property of Gd. The resulting values for the t_i, along with $x_1, \ldots, x_n$, make g equal to one. This completes the proof of Lemma 2.2. □

2.3. The plan for the rest of the proof of Theorem 1.2 is as follows. In §3 we show that the classes (I) and (II) coincide, and in §§4–7 we show that (II) and (III) coincide.

2.4. *Remark*. In the course of proving Lemma 2.2, we obtained the following facts, which should always be kept in mind in what follows:

(a) In the definitions of the classes (I)–(III) we may always replace "level sets" by "1-level sets" (by going from f to $(f-m)^2+1$).
(b) All of the classes (I)–(III) are closed with respect to (finite) products, intersections, unions, and also projections. (The proof of this for the class (I) in Lemma 4.8 of Chapter V is also applicable to the class (III).)

We encounter much greater difficulty in treating the bounded universal quantifier. Indeed, the most technical part of the proof in §§4–7 is concerned with showing that the class of Diophantine sets is closed with respect to the bounded universal quantifier.

3 Enumerable sets are D-sets

Let $f : (\mathbf{Z}^+)^n \to \mathbf{Z}^+$ be a primitive recursive function. Its 1-level can be represented as the projection onto the first n coordinates of the set $\Gamma_f \cap \left[(\mathbf{Z}^+)^n \times \{1\} \right]$, where Γ_f is the graph of f. Thus, an enumerable set can be obtained as a projection of the intersection of the graphs of two primitive recursive functions. Since, by definition, the class of D-sets is closed with respect to projections and intersections, the assertion in the title of this section follows from the following fact:

3.1. **Proposition.** *The graphs of primitive recursive functions are D-sets.*

Proof. The graphs of the basic functions are Diophantine. The stability of the property of graphs "being D-sets" relative to the composition and juxtaposition of functions is verified by the same arguments as in the proof of Lemma 4.8 of Chapter V. It remains to prove the stability under recursion. We shall first of all need information about the graph of Gödel's function. Here it is more convenient to use gd instead of Gd.

3.2. **Lemma.** *The graph of the Gödel function* $\mathrm{gd}(u, k, t) = \mathrm{rem}(1 + kt, u)$ *is Diophantine, and a fortiori, a D-set.*

Proof. The set

$$\Gamma_{\mathrm{gd}} = \{ \langle u, k, t, \gamma \rangle | \gamma \text{ is the remainder when } u \text{ is divided by } 1 + kt \}$$

is the intersection of the following two sets in $(\mathbf{Z}^+)^4$:

$$E_1 : \gamma \leqslant 1 + kt;$$
$$E_2 : u - \gamma \geqslant 0 \quad \text{and is divisible by } 1 + kt.$$

Both E_1 and E_2 are Diophantine. In fact, E_1 is a projection of the 0-level of the polynomial $2 + kt - \gamma - y_1$, and E_2 is a projection of the 0-level of the polynomial $u - \gamma - (1 + kt)(y_2 - 1)$. The lemma is proved. □

3.3. **Corollary.** *Let f and g be functions of n and $n+2$ arguments, respectively, whose graphs are D-sets. Then the following equations determine D-sets in the $(x_1, \ldots, x_{n+1}, u, t, \ldots)$-coordinate space (where any additional coordinates may follow the t):*

$$E : \mathrm{gd}(u, 1, t) = f(x_1, \ldots, x_n);$$

$$F : \mathrm{gd}(u, x_{n+1} + 1, t) = g(x_1, \ldots, x_{n+1}, \mathrm{gd}(u, x_{n+1}, t)).$$

PROOF. Introducing extra coordinates after the t amounts to taking the direct product with $(\mathbf{Z}^+)^p$, and this, of course, takes D-sets to D-sets.

E can be represented as a projection of the intersection of the sets $\mathrm{gd}(u, k, t) = w$, $f(x_1, \ldots, x_n) = w$, and $k = 1$ (where k and w are auxiliary coordinates). Since Γ_{gd} and Γ_f are D-sets, the same is true for E.

Similarly, F can be represented as a projection of the intersection of the sets

$$\mathrm{gd}(u, x_{n+1} + 1, t) = w_1,$$
$$\mathrm{gd}(u, x_{n+1}, t) = w_2,$$
$$g(x_1, \ldots, x_{n+1}, w_2) = w_1.$$

These are D-sets, because Γ_g and Γ_{gd} are D-sets. □

3.4. PROOF OF PROPOSITION 3.1. Recall that it remains to verify the following assertion: Let h be the function defined recursively from functions f and g by the equations

$$h(x_1, \ldots, x_n, 1) = f(x_1, \ldots, x_n),$$
$$h(x_1, \ldots, x_n, k+1) = g(x_1, \ldots, x_n, k, h(x_1, \ldots, x_n, k));$$

then the graph Γ_h

$$\langle x_1, \ldots, x_{n+1}, \eta \rangle \in \Gamma_h \Leftrightarrow \eta = h(x_1, \ldots, x_{n+1})$$

is a D-set whenever the graphs Γ_f and Γ_g are D-sets.

First step. We set $\Gamma_h = \Gamma^1 \cup \Gamma^2$, where $x_{n+1} = 1$ on Γ^1 and $x_{n+1} \geqslant 2$ on Γ^2. Since

$$\langle x_1, \ldots, x_{n+1}, \eta \rangle \in \Gamma^1 \Leftrightarrow x_{n+1} = 1 \quad \text{and} \quad \eta = f(x_1, \ldots, x_n),$$

it follows that Γ^1 is the intersection of $\Gamma_f \times \mathbf{Z}^+$ and a D-set, and therefore is a D-set. It remains to verify that Γ^2 is also a D-set.

Second step. In the $(x_1, \ldots, x_{n+1}, \eta, u, t)$-coordinate space we consider the sets

$$E_1 : \eta = \mathrm{gd}(u, x_{n+1}, t),$$
$$E_2 : \mathrm{gd}(u, 1, t) = f(x_1, \ldots, x_n),$$
$$E_3 : x_{n+1} > 1, \quad \mathrm{gd}(u, k, t) = g(x_1, \ldots, x_n, k-1, \mathrm{gd}(u, k-1, t))$$
for all $2 \leqslant k \leqslant x_{n+1}$.

It is easy to see that $\Gamma^2 = \text{pr} \bigcap_{i=1}^{3} E_i$. In fact, as in §4 of Chapter V, we obtain inclusion in one direction by comparing E_2 and E_3 with the inductive definition of h, and in the other direction by suitably choosing the parameters u and t in Gödel's function. Thus, it remains to show that the E_i are D-sets.

Third step. E_1 is the graph of gd with some additional coordinates. E_2 was shown to be a D-set in the proof of Corollary 3.3.

Finally, E_3 is "almost" obtained from the set F in Corollary 3.3 by applying the bounded universal quantifier to the x_{n+1}-coordinate. More precisely (for brevity, we ignore the η-coordinate):

$$\langle x_1, \ldots, x_{n+1}, u, t\rangle \in E_3 \Leftrightarrow \forall k \in [2, x_{n+1}], \langle x_1, \ldots, x_n, k-1, u, t\rangle \in F$$

$$\Leftrightarrow \forall k \in [1, x_{n+1}-1], \langle x_1, \ldots, x_n, k, u, t\rangle \in F.$$

Consequently, if we apply to F the bounded universal quantifier in the x_{n+1}-coordinate, we obtain a D-set which is the same as E_3 with the x_{n+1}-coordinates of all its points decreased by 1. So it remains to see that the operation of shifting back by 1 preserves the property of "being a D-set," and this follows easily from the definitions. The proof is complete. □

4 The reduction

4.1. The next three sections are devoted to proving that the class of D-sets coincides with the class of Diophantine sets. As noted at the end of §2, it suffices to show that the class of Diophantine sets is closed with respect to the bounded universal quantifier.

Let $f(x_1, \ldots, x_n, k, y_1, \ldots, y_m)$ be any nonconstant polynomial with integer coefficients. f will be fixed for the duration of this section. Let d be the degree of f, and let c be the sum of the absolute values of its coefficients.

We define the set E by the condition

$$\langle x_1, \ldots, x_n, y\rangle \in E \Leftrightarrow \forall k \leqslant y \; \exists \langle y_1, \ldots, y_m\rangle,$$
$$f(x_1, \ldots, x_n, k, y_1, \ldots, y_m) = 0.$$

We want to show that E is Diophantine. In this section we prove the following reduction step, which is due to Davis, Putnam, and Robinson.

4.2. **Proposition.** *E is Diophantine if the following three sets are Diophantine:*

$$x_1 = x_2^{x_3};$$
$$x_1 = x_2!;$$
$$\frac{x_1}{x_2} = \binom{x_3/x_4}{x_5}, \qquad x_3 \geqslant x_4 x_5,$$

where $\binom{n}{k} = n(n-1)\cdots(n-k+1)/k!$ is the "binomial coefficient."

The proof of this and all subsequent propositions of this type follows a standard pattern. To show that E is Diophantine, we introduce auxiliary sets E_i with the following properties:

$$\text{(a) } E = \bigcap_{i=1}^{N} E_i;$$

$$\text{(b) the } E_i \text{ are Diophantine.}$$

But usually we are not able to establish directly that all the E_i are Diophantine, so we apply the same procedure to certain of the E_i. Thus, the proof that E is Diophantine has a tree-like pattern.

The exposition of each step will consist of the following stages: the introduction of auxiliary variables, which disappear when we project; explicit construction of the sets E_i; the proof of the inclusion $E \subset \text{pr} \bigcap_{i=1}^{N} E_i$; and the proof of the inclusion $E \supset \text{pr} \bigcap_{i=1}^{N} E_i$.

4.3. Proof of Proposition 4.2. We denote the auxiliary variables by the symbols $Y, N, K, Y_1, \ldots, Y_m$. We introduce the sets E_i in the $\langle x_1, \ldots, x_n, y, Y, N, K, Y_1, \ldots, Y_m \rangle$-space by the following relations:

$$E_1: \quad N \geqslant c \cdot (x_1 \cdots x_n y Y)^d, \qquad Y < Y_1, \ldots, Y < Y_m$$

(intuitively speaking, the right side of the first inequality gives a rough estimate for the value of the polynomial f at the point $\langle x_1, \ldots, x_n, y, y_1, \ldots, y_m \rangle$ if all $y_i \leqslant Y$).

$$E_2: \quad 1 + KN! = \prod_{k=1}^{y} (1 + kN!)$$

(this is a "large modulus"; $f = 0$ will be replaced by divisibility by this modulus).

$$E_3: \quad f(x_1, \ldots, x_n, K, Y_1, \ldots, Y_m) \equiv 0 \bmod (1 + KN!);$$

$$E_{3+i}: \quad \prod_{j \leqslant Y} (Y_i - j) \equiv 0 \bmod (1 + KN!), \quad i = 1, \ldots, m.$$

We define the set E' as $\bigcap_{i=1}^{m+3} E_i$.

Proof of the inclusion $E \subset \text{pr } E'$. Given a point $\langle x_1, \ldots, x_n, y \rangle \in E$, we must choose values for the other coordinates so that the relations $E_1, \ldots, E_{m+3}$ are fulfilled.

By the definition of E, each point $\langle x_1, \ldots, x_n, k \rangle$, $k \leqslant y$, can be lifted to the 0-level of f:

$$f(x_1, \ldots, x_n, k, y_{1k}, \ldots, y_{mk}) = 0.$$

For Y we take the maximum of y and the y_{ik}. Then, as before, we find the Y_i and N by solving the system of Gödel equations

$$\text{gd}(Y_i, k, N!) = y_{ik}, \quad \text{for all } 1 \leqslant k \leqslant y.$$

The proof of Gödel's lemma shows that the Y_i and N may be taken arbitrarily large: in particular, so as to satisfy E_1. The number K is uniquely determined by E_2.

All the choices have now been made. The relation E_{3+i} holds, because, by the definition of Y_i and gd, we can find a number $Y_i - j$ with $j \leqslant Y$, namely $j = y_{ik}$, such that $Y_i - j \equiv 0 \bmod(1 + kN!)$, for every $k \leqslant y$. Hence, the product on the left in E_{3+i} is divisible by all the $1 + kN!$, $1 \leqslant k \leqslant y$, which are pairwise relatively prime, since $N \geqslant y$ by E_1. Therefore, this product is divisible by $1 + KN!$.

Finally, to verify E_3 we note that E_2 implies the congruence $K \equiv k \bmod(1 + kN!)$, $1 \leqslant k \leqslant y$, because $(1 + KN!) - (1 + kN!) \equiv 0 \bmod(1 + kN!)$. But then, since $y_{ik} \equiv Y_i \bmod(1 + kN!)$ by our choice of Y_i, we find that

$$f(x_1, \ldots, x_n, K, Y_1, \ldots, Y_m) \equiv f(x_1, \ldots, x_n, k, y_{ik}, \ldots, y_{mk}) \equiv 0 \bmod(1 + kN!).$$

Since the moduli $1 + kN!$ are pairwise relatively prime, this congruence implies E_3.

Proof of the inclusion pr $E' \subset E$. Given a point

$$\langle x_1, \ldots, x_n, y, Y, N, K, Y_1, \ldots, Y_m \rangle$$

whose coordinates satisfy the relations $E_1, \ldots, E_{m+3}$, we must find a vector $\langle y_{1k}, \ldots, y_{mk} \rangle$ for each $k \leqslant y$ such that

$$f(x_1, \ldots, x_n, k, y_{1k}, \ldots, y_{mk}) = 0.$$

To do this we let p_k denote any prime divisor of $1 + kN!$, and we set

$$y_{ik} = \textit{the remainder when } Y_i \textit{ is divided by } p_k.$$

We claim that these y_{ik} give us the required equality. In fact, E_3 implies that $f(x_1, \ldots, x_n, k, y_{1k}, \ldots, y_{mk}) \equiv 0 \bmod p_k$. It suffices to show that the number on the left is less than p_k. We have

$$\begin{aligned} p_k \text{ divides } \prod_{j \leqslant Y} (Y_i - j) \quad & \text{by } E_{3+i} \\ \Rightarrow p_k \text{ divides } Y_i - j \quad & \text{for some } j \leqslant Y \\ \Rightarrow y_{ik} = \text{the remainder when } & Y_i \text{ is divided by } p_k \leqslant Y \\ \Rightarrow f(x_1, \ldots, x_n, k, y_{1k}, \ldots, y_{mk}) & \leqslant c(x_1 \cdots x_n yY)^d \leqslant N < p_k, \end{aligned}$$

where the second inequality in the last line follows from E_1, and the third inequality follows because p_k divides $1 + kN!$.

Conclusion of the proof. It remains to show that the sets $E_1, \ldots, E_{m+3}$ are Diophantine if the sets in Proposition 6.1 are Diophantine. In fact, if we trivially introduce new variables and make substitutions, we can first reduce the verification that all the E_i are Diophantine to showing that the following sets are Diophantine:

$$x_1 = x_2!;$$

$$x_1 = \prod_{k \leq x_2} (1 + kx_3);$$

$$x_1 = \prod_{j \leq x_3} (x_2 - j), \qquad x_2 > x_3.$$

It then remains to notice that the second of these relations can be written in the form

$$x_1 = x_3^{x_2} \begin{bmatrix} \frac{1}{x_3} + x_2 \\ x_2 \end{bmatrix},$$

and the third relation can be written as

$$x_1 = x_3! \binom{x_2 - 1}{x_3}, \qquad x_2 > x_3.$$

This completes the proof of Proposition 4.2. □

5 Construction of a special Diophantine set

5.1. In this section we begin the proof that the three sets in Proposition 4.2 are Diophantine. In order that the reader may better appreciate this stage in the proof, we mention that the most troublesome obstacle here is the rapid growth of one of the coordinates in comparison to the others (for example, $x_1 = x_2!$). J. Robinson had the following key idea. She proved that, if we know that *any* specific set in $(\mathbf{Z}^+)^2$ is Diophantine and has one coordinate which grows faster than any power of the other but slower than, say, x^x (for example, exponentially), we may then conclude that *all* enumerable sets are Diophantine. After this, Matijacevič and Čudnovskiĭ were able to show that a certain set of that type (connected with Fibonacci numbers) is Diophantine. For a history of the question, see Matijacevič's article "Diophantine Sets" in *Uspehi Mat. Nauk*, vol. XXVII, No. 5 (1972) (translated in *Russian Math. Surveys*).

In this section we give a construction which is an improved version of the original construction. Its idea is based on the following observation. Let $x^2 - dy^2 = 1$ be Pell's equation (where $d \in \mathbf{Z}^+$ is not a perfect square). Its solutions $\langle x, y \rangle \in (\mathbf{Z}^+)^2$ form a semigroup with composition law

$$(x_1 + y_1\sqrt{d})(x_2 + y_2\sqrt{d}) = x_3 + y_3\sqrt{d}.$$

This is a cyclic semigroup. That is, let $\langle x_1, y_1\rangle$ be the solution with the least first coordinate. Then any other solution has the form $\langle x_n, y_n\rangle$, where $n \in \mathbf{Z}^+$, and

$$x_n + y_n\sqrt{d} = \left(x_1 + y_1\sqrt{d}\right)^n.$$

We call n the *number* of the solution $\langle x_n, y_n\rangle$.

The coordinates x_n and y_n grow exponentially with n, so that the set of solutions of Pell's equation, and also the projections of this set on the x- and y-axes, are Diophantine sets having logarithmic density. This is not yet enough: we still have the problem of including the solution number n among the coordinates of a Diophantine set. Only then can we apply Robinson's technique. This is what will be done below.

5.2. *Notation*. We consider Pell's equation with variable d. Its first solution generally varies as a function of d in an uncontrollable fashion, so that it is convenient to choose only those d whose first solutions have the simple special form $\langle a, 1\rangle$, $a \in \mathbf{Z}^+$. Obviously, then $d = a^2 - 1$.

We shall call the equation $x^2 - (a^2 - 1)y^2 = 1$ the *a-equation*. We define the two sequences $x_n(a)$ and $y_n(a)$ as the coordinates of its nth solution:

$$x_n(a) + y_n(a)\sqrt{a^2 - 1} = \left(a + \sqrt{a^2 - 1}\,\right)^n.$$

For each n, a formal definition of $x_n(a)$ and $y_n(a)$ as polynomials in a can easily be given by induction on n. Then the expressions $x_n(a)$ and $y_n(a)$ will make sense for all $n \in \mathbf{Z}$ and $a \in \mathbf{C}$. In particular,

$$x_n(1) = 1, \qquad y_n(1) = n;$$

and all the formulas given below remain true. The basic result of this section is

5.3. **Proposition.** *The set*

$$E: \quad y = y_n(a), \qquad a > 1$$

in the $\langle y, n, a\rangle$-space is Diophantine.

The proof uses the elementary number theoretic properties of the sequences $x_n(a)$ and $y_n(a)$, most of which will be verified at the end of the section (see 5.8). The idea for determining n in a Diophantine way from $\langle y, a\rangle$ is to observe that $y_n(a) \equiv n \bmod (a - 1)$ (Lemma 5.4). This uniquely determines n as long as $n < a - 1$. To pass to the general case, we introduce an auxiliary A-equation with A large, and find formulas for its nth solution (using y) in which n only appears in a Diophantine context.

Formally, the proof that E is Diophantine follows the pattern described in 4.2. In addition to the basic variables y, n, a, we introduce six auxiliary variables: x, x_1, y_1, A, x_2, y_2. We set:

$$E_1: \quad y \geqslant n, \qquad a > 1;$$

$$E_2: \quad x^2 - (a^2 - 1)y^2 = 1;$$

$$E_3: \quad y_1 \equiv 0 \bmod 2x^2y^2;$$

$$E_4: \quad x_1^2 - (a^2 - 1)y_1^2 = 1;$$

$$E_5: \quad A = a + x_1^2(x_1^2 - a);$$

$$E_6: \quad x_2^2 - (A^2 - 1)y_2^2 = 1;$$

$$E_7: \quad y_2 - y \equiv 0 \bmod x_1^2;$$

$$E_8: \quad y_2 \equiv n \bmod 2y.$$

Let $E' = \bigcap_{i=1}^{8} E_i$. We show that pr $E' = E$.

The inclusion $E \subset \text{pr } E'$. Given $\langle y, n, a\rangle \in E$, we must find values for the other variables so that $E_1, \ldots, E_8$ hold. As before, we shall not introduce any new symbols for these values; after we choose, say, a value for x, the letter x will become the name for this value.

E_1 is automatically satisfied: $y_n(a) \geqslant n$ for all $a \geqslant 1$, $n \geqslant 1$ (induction on n). We find x uniquely from $E_2 : x = x_n(a)$. We take $\langle x_1, y_1/2x^2y^2\rangle$ to be any solution of the Pell equation $X^2 - (a^2 - 1)(2x^2y^2)^2Y^2 = 1$; this gives E_4. A is found uniquely from E_5. We take $\langle x_2, y_2\rangle$ to be the nth solution of the A-equation. Now all choices have been made. To verify E_7 and E_8 we need two lemmas.

5.4. **Lemma.** $y_k(a) \equiv k \bmod (a - 1)$.

5.5. **Lemma.** *If* $a \equiv b \bmod c$, *then* $y_n(a) \equiv y_n(b) \bmod c$.

These lemmas will be proved in 5.8.

We use these lemmas as follows. From E_5 we find:

$$A = a + \left(1 + (a^2 - 1)y_1^2\right)\left(1 + (a^2 - 1)y_1^2 - a\right) \equiv 1 \bmod 2y,$$

because of E_3. Lemma 5.4 then gives $y_2 = y_n(A) \equiv n \bmod 2y$; this is E_8. Lemma 5.5 gives $y_n(A) \equiv y_n(a) \bmod x_1^2$ (because of E_5); this is E_7.

The inclusion pr $E' \subset E$. From the relations $E_1, \ldots, E_8$ we have only to prove that n is the number of the solution $\langle x, y\rangle$. Note that n only occurs in E_8.

For the time being we let N, N_1, and N_2 denote the numbers of the solutions $\langle x, y\rangle$, $\langle x_1, y_1\rangle$, and $\langle x_2, y_2\rangle$, respectively. We shall prove that

$$n \equiv N \quad \text{or} \quad n \equiv -N \bmod 2y.$$

Since we also have $y \geqslant n$ (by E_1) and $y \geqslant N$ (by the definition of N), it follows that $n = N$, as required. The number N_2 will be the "stepping stone" to get from n to N.

First of all, as before, it follows from E_5 that $A \equiv 1 \bmod 2y$, and then it follows from the definition of N_2 and Lemma 5.4 that $y_2 \equiv N_2 \bmod 2y$. But by E_8 we have $y_2 \equiv n \bmod 2y$; hence,

$$N_2 \equiv n \bmod 2y.$$

Next, $A \equiv a \bmod x_1^2$ by E_5, and then $y_2 = y_{N_2}(A) \equiv y_{N_2}(a) \bmod x_1^2$ by Lemma 5.5. Using E_7, we have $y = y_N(a) \equiv y_2 \bmod x_1^2$. Hence,

$$y_N(a) \equiv y_{N_2}(a) \bmod x_1^2.$$

We now need two more lemmas, which will be proved in 5.8.

5.6. **Lemma.** *If $y_i(a) \equiv y_j(a) \bmod x_n(a)$, where $a > 1$, then either $i \equiv j$ or $i \equiv -j \bmod 2n$.*

5.7. **Lemma.** *If $y_i(a)^2$ divides $y_j(a)$, then $y_i(a)$ divides j.*

If we apply Lemma 5.6 with N, N_2, and N_1 in place of i, j, and n, and use the last congruence proved, we obtain:

$$N \equiv \pm N_2 \bmod 2N_1.$$

If we apply Lemma 5.7 with N and N_1 in place of i and j, and use E_3, we obtain $y | N_1$. Hence,

$$N \equiv \pm N_2 \bmod 2y,$$

and, since we have already shown that $N_2 \equiv n \bmod 2y$, this completes the proof. □

5.8. Proof of the lemmas. We shall write x_n and y_n instead of $x_n(a)$ and $y_n(a)$. Using the formula

$$x_{nk} + y_{nk}\sqrt{a^2-1} = \left(x_n + y_n\sqrt{a^2-1}\right)^k,$$

we find that

$$y_{nk} = \sum_{\substack{j \leq k \\ j \equiv 1 (\mathrm{mod}\ 2)}} \binom{k}{j} x_n^{k-j} y_n^j (a^2-1)^{(j-1)/2}.$$

In particular,

$$y_{nk} \equiv k x_n^{k-1} y_n \bmod (a^2-1),$$

which gives Lemma 5.4 if we set $n = 1$. In addition, we have

$$y_{nk} \equiv k x_n^{k-1} y_n \bmod y_n^3.$$

If we replace nk, k, and n by n, n/k, and k, respectively, we obtain

$$y_n \equiv \frac{n}{k} x_k^{n/k-1} y_k \bmod y_k^3.$$

Since x_k and y_k are relatively prime, we have

$$y_n \equiv 0 \bmod y_k^2 \Rightarrow \frac{n}{k} \equiv 0 \bmod y_k \Rightarrow n \equiv 0 \bmod y_k,$$

which gives Lemma 5.7.

If we write $y_n(a)$ as a polynomial in a with integer coefficients whose degree and coefficients only depend on n, we immediately obtain Lemma 5.5. It remains to prove Lemma 5.6.

First of all, the equation

$$x_{n \pm m} + \sqrt{a^2-1}\; y_{n\pm m} = \left(x_n + \sqrt{a^2-1}\; y_n\right)\left(x_m \pm \sqrt{a^2-1}\; y_m\right)$$

gives us

$$x_{n\pm m} = x_n x_m \pm (a^2-1) y_n y_m,$$
$$y_{n\pm m} = \pm x_n y_m + x_m y_n.$$

Hence,

$$y_{2n\pm m} = y_{n+(n\pm m)} \equiv x_{n\pm m} y_n \bmod x_n \equiv \pm (a^2-1) y_n^2 y_m \bmod x_n$$
$$\equiv \mp y_m \bmod x_n,$$

and, similarly,

$$y_{4n\pm m} = y_{2n+(2n\pm m)} \equiv -y_{2n\pm m} \bmod x_n \equiv y_{\pm m} \bmod x_n.$$

This means that the class $y_k \bmod x_n$ has period $4n$ as a function of k, and within $[1, 4n]$ its behavior is determined by its values on the first quarter-period $[1, n]$:

$$y_{2n\pm m} \equiv \mp y_m, \qquad y_{\pm m} \equiv \pm y_m, \quad \text{for } 1 \leqslant m \leqslant n.$$

If $a \geqslant 3$, it is clear that Lemma 5.6 follows from these facts and from the inequality $y_m < \frac{1}{2} x_n$ for $1 \leqslant m \leqslant n$, which, in turn, follows because

$$4y_m^2 < (a^2-1) y_n^2 + 1 = x_n^2.$$

If $a = 2$, then we only have $y_m < \frac{1}{2} x_n$ for $m \leqslant n-1$, but this is still enough to complete the proof of the lemma in this case. □

6 The graph of the exponential is Diophantine

6.1. **Proposition.** *The set*

$$E: \quad m = a^n$$

in the $\langle m, a, n\rangle$-space is Diophantine.

Proof. It suffices to show that $E_0 = E \cap \{a | a > 1\}$ is Diophantine. If $a > 1$, we easily obtain by induction on n

$$(2a-1)^n \leqslant y_{n+1}(a) \leqslant (2a)^n,$$

in the notation of §5. Hence, for any $N \geqslant 1$ we have

$$a^n\left(1 - \frac{1}{2Na}\right)^n = \frac{(2Na-1)^n}{(2N)^n} \leqslant \frac{y_{n+1}(Na)}{y_{n+1}(N)} \leqslant \frac{(2Na)^n}{(2N-1)^n}$$
$$= a^n\left(1 - \frac{1}{2N}\right)^{-n}.$$

Thus, if we choose N large enough so that both

$$\left(1 - \frac{1}{2N}\right)^{-n} - 1 < \frac{1}{a^n} \quad \text{and} \quad 1 - \left(1 - \frac{1}{2Na}\right)^n < \frac{1}{a^n},$$

then we obtain: $a^n = [y_{n+1}(Na)/y_{n+1}(N)]$ (where the brackets here and below denote the integral part of a number). E_0 is therefore a projection of the set E_1:

$$a > 1,$$
$$0 \leqslant y_{n+1}(Na) - y_{n+1}(N)m < y_{n+1}(N),$$
$$N > ?,$$

where a suitable lower bound for N must be inserted in place of ?, in such a way as to keep the last relation Diophantine. An elementary calculation shows that it suffices to set $N > 4n(y+1)$. The results in §5 then imply that E_1 is Diophantine if we trivially introduce the auxiliary relations

$$y' = y_{n+1}(N) \quad \text{and} \quad y'' = y_{n+1}(Na). \qquad \square$$

7 The factorial and binomial coefficient graphs are Diophantine

In this section we carry out the last series of arguments.

7.1. **Proposition.** *The set*

$$E: \quad r = \binom{n}{k}, \quad n \geqslant k,$$

in the $\langle r, k, n\rangle$-space is Diophantine.

Here, by definition, $\binom{n}{k} = n(n-1)\cdots(n-k+1)/k!$. We shall need the following

7.2. **Lemma.** *If $u > n^k$, then $\binom{n}{k}$ = the remainder when $[(u+1)^n/u^k]$ is divided by u.*

PROOF. We have

$$(u+1)^n/u^k = \sum_{i=k+1}^{n} \binom{n}{i} u^{i-k} + \binom{n}{k} + \sum_{i=0}^{k-1} \binom{n}{i} u^{i-k}.$$

The first sum is divisible by u, and the last sum is less than 1 if $u > n^k$. $\square$

7.3. Proof of Proposition 7.1. We introduce the auxiliary variables u and v, and take the relations

$$E_1: \quad u > n^k;$$

$$E_2: \quad v = \left[(u+1)^n / u^k\right];$$

$$E_3: \quad r \equiv v \bmod u;$$

$$E_4: \quad r < u;$$

$$E_5: \quad n \geqslant k.$$

Lemma 7.2 immediately implies that $E = \mathrm{pr} \bigcap_{i=1}^{5} E_i$. E_1 is Diophantine because of Proposition 6.1; E_3, E_4, and E_5 are obviously Diophantine. It also becomes obvious that E_2 is Diophantine if we write E_2 in the form

$$(u+1)^n \leqslant u^k v < (u+1)^n + u^k$$

and again use Proposition 6.1. This completes the proof. □

7.4. **Proposition.** *The set* $E: m = k!$ *is Diophantine.*

7.5. **Lemma.** *If* $k > 0$ *and* $n > (2k)^{k+1}$, *then* $k! = \left[n^k / \binom{n}{k}\right]$. (*This is proved by some simple estimates.*)

Proof of Proposition 7.4. We take the auxiliary variable n and the relations

$$E_1: \quad n > (2k)^{k+1};$$

$$E_2: \quad m = \left[n^k / \binom{n}{k}\right].$$

The rest is obvious (using Propositions 6.1 and 7.1). □

7.6. **Proposition.** *The set*

$$E: \quad \frac{x}{k} = \binom{p/q}{k}, \quad p > qk,$$

in the $\langle x, y, p, q, k\rangle$*-space is Diophantine.*

The proof which follows is a slightly more complicated version of the argument in 7.2 and 7.3.

7.7. **Lemma.** *Let* $a > 0$ *be an integer such that* $a \equiv 0 \bmod q^k k!$ *and* $a > 2^{p-1} p^{k+1}$. *Then*

$$\binom{p/q}{k} = a^{-1}\left[a^{2k+1}(1+a^{-2})^{p/q}\right] - a\left[a^{2k-1}(1+a^{-2})^{p/q}\right].$$

This is proved using the binomial Taylor series for $(1+a^{-2})^{p/q}$. The inequality $a > 2^{p-1} p^{k+1}$ allows us to throw away all the terms in the first sum starting with the $(k+1)$th and all the terms in the second sum starting with the kth when we take the integral part. The congruence $a \equiv$ mod $q^k k!$ ensures that the partial sums are integers. □

7.8. Proof of Proposition 7.6. We use the auxiliary variables a, u_1, u_2, and v, and the following relations:

$$E_1: \quad a \equiv 0 \bmod q^k k!;$$

$$E_2: \quad a > 2^{p-1} p^{k+1};$$

$$E_3: \quad u_1/u_2 = a^{-1}\left[a^{2k+1}(1+a^{-2})^{p/q}\right];$$

$$E_4: \quad v = a\left[a^{2k-1}(1+a^{-2})^{p/q}\right];$$

$$E_5: \quad xu_2 = y(u_1 - vu_2).$$

It follows from Lemma 7.7 that $E = \mathrm{pr} \bigcap_{i=1}^{5} E_i$. E_1 and E_2 are immediately seen to be Diophantine from Propositions 6.1 and 7.1. E_3 and E_4 are shown to be Diophantine just as at the end of 7.3, except that this time we must raise the inequalities to the qth power after clearing denominators. E_5 is obviously Diophantine.

This concludes the proof of Theorem 1.2, that enumerable sets coincide with Diophantine sets. □

8 Versal families

Versal families were defined and first used in subsection 5.7 of Chapter V. The purpose of this section is to prove their existence, using the result that enumerable sets are Diophantine (Theorem 1.2).

8.1. **Theorem.** *For any $m \geqslant 0$, versal enumerable families of m-sets and m-functions over the base $\mathbf{Z}^+$ exist and can be effectively constructed.*

Proof. We divide the proof into several steps. Recall that $\tau^{(2)} : (\mathbf{Z}^+)^2 \tilde{\Rightarrow} \mathbf{Z}^+$ is the primitive recursive isomorphism constructed in §4 of Chapter V, and $\langle t_1^{(2)}, t_2^{(2)} \rangle$ is its inverse. We shall write t_1 and t_2 for brevity.

(a) *A versal family of polynomials in* $\mathbf{Z}^+[x_1, x_2, x_3, \ldots]$. We define polynomials $f[l] \in \mathbf{Z}^+[x_1, x_2, x_3, \ldots]$ by recursion on $l \in \mathbf{Z}^+$, $l \geqslant 4$:

$$f[1] = f[2] = f[3] = 1;$$

$$f[4k] = k;$$

$$f[4k+1] = x_k;$$

$$f[4k+2] = f[t_1(k)] + f[t_2(k)];$$

$$f[4k+3] = f[t_1(k)]f[t_2(k)].$$

The definition is correct, since $t_1(k), t_2(k) < 4k+2$. The image of the map

$k \mapsto f[k]$ coincides with all of $\mathbf{Z}^+[x_1, x_2, x_3, \ldots]$, since it contains $\mathbf{Z}^+$ (in the $4k$-places) and all the x_k (in the $4k+1$-places), and, whenever it contains two polynomials $f[k_1]$ and $f[k_2]$, it contains their sum (in the $4\tau^{(2)}(k_1, k_2)+2$-place) and their product (in the $4\tau^{(2)}(k_1, k_2)+3$-place). (Compare with the numbering of constructible sets by ordinals in Chapter V.)

(b) *Construction of a versal* 1-*family over* $\mathbf{Z}^+$. Let E_k be the projection onto the x_1-coordinate of the 0-level of the polynomial $f[t_1(k)] - f[t_2(k)]$. Since all the elements of $\mathbf{Z}[x_1, x_2, x_3, \ldots]$ can be represented as such a difference, it is clear that the family $\{E_k\}$ contains all enumerable sets.

(c) $\{E_k\}$ *is enumerable*. We must show that the total space $E = \{\langle i,j\rangle | i \in E_j\} \subset \mathbf{Z}^+ \times \mathbf{Z}^+$ is enumerable. We write the condition $i \in E_j$ in the form of an $\mathfrak{L}_1$ type formula, in which all the quantified variables take values in $\mathbf{Z}^+$. We use the fact that $f[t_1(j)] - f[t_2(j)] \in \mathbf{Z}[x_1, \ldots, x_j]$. We have:

$$\langle i,j\rangle \in E \Leftrightarrow i \in E_j \Leftrightarrow \exists x_1 \ldots \exists x_j \big(x_1 = i \wedge f[t_1(j)] = f[t_2(j)]\big)$$

$$\Leftrightarrow \exists t\Big(\big(\exists x_1 \ldots \exists x_j \; \forall k \leqslant j \; (f[k] = \mathrm{Gd}(k, t))\big)$$

$$\wedge \mathrm{Gd}(5, t) = i \wedge \mathrm{Gd}(t_1(j), t) = \mathrm{Gd}(t_2(j), t)\Big),$$

where $\mathrm{Gd}(k, t)$ is Gödel's function (see §4 of Chapter V). Furthermore, by the definition of $f[k]$:

$$\exists x_1 \cdots \exists x_j \; \forall k \leqslant j \; (f[k] = \mathrm{Gd}(k, t))$$

$$\Leftrightarrow \forall k \leqslant j \; \Big((k \leqslant 3 \wedge \mathrm{Gd}(k, t) = 1) \vee \exists l \Big((k = 4l \wedge \mathrm{Gd}(k, t) = l)$$

$$\vee \big(k = 4l + 2 \wedge \mathrm{Gd}(k, t) = \mathrm{Gd}(t_1(l), t) + \mathrm{Gd}(t_2(l), t)\big)$$

$$\vee \big(k = 4l + 3 \wedge \mathrm{Gd}(k, t) = \mathrm{Gd}(t_1(l), t)\, \mathrm{Gd}(t_2(l), t)\big)\Big)\Big).$$

Here the part of the formula after $\exists l$ defines a decidable set in $\langle k, t, l\rangle$-space. The quantifier $\exists l$ projects this set onto the $\langle k, t\rangle$-coordinates, thereby taking it to an enumerable set, and the bounded quantifier $\forall k \leqslant j$ preserves enumerability (see §2). Returning to the formula which defines E, we find that the set we have constructed so far must be intersected with two other decidable sets and then projected along the t-axis, so that the result is again enumerable.

(d) *Construction of a versal m-family over* $\mathbf{Z}^+$. The case $m = 0$ is trivial, and the case $m = 1$ has already been discussed. The case $m \geqslant 2$ reduces to the case $m = 1$ using the isomorphism $\tau^{(m)} : (\mathbf{Z}^+)^m \tilde{\Rightarrow} \mathbf{Z}^+$. In fact, let $E_k = E_k^{(1)}$ be a versal 1-family, and set $E_k^{(m)} = (\tau^{(m)})^{-1}(E_k^{(1)})$. The family

$\{E_k^{(m)}\}$ is enumerable because

$$E^{(m)} = \{\langle x, k\rangle | x \in E_k^{(m)}\} = \{\langle (\tau^{(m)})^{-1}(x), k\rangle | x \in E_k^{(1)}\}$$
$$= (\tau^{(m)}, \mathrm{pr}_1^1)^{-1} E^{(1)}.$$

(e) *Construction of a versal family of* 1-*functions*. We take a versal 2-family $\{E_k^{(2)}\}$ with total space

$$E^{(2)} = \{\langle x, y, k\rangle | \langle x, y\rangle \in E_k^{(2)}\} \subset (\mathbf{Z}^+)^3.$$

Let $g(x, y, k, z)$ be a primitive recursive function such that the projection of its 1-level onto the $\langle x, y, k\rangle$-coordinates coincides with $E^{(2)}$. We set

$$f(x, k) = t_1^{(2)}\left(\min\{u | g(x, t_1^{(2)}(u), k, t_2^{(2)}(u)) = 1\}\right).$$

We claim that $\{f_k | f_k(x) = f(x, k)\}$ is a versal family of 1-functions. The total function is obviously partial recursive. We need only verify that every partial recursive 1-function f occurs in the family.

Let Γ_f be the graph of f, and let $\Gamma_f = E_{k_0}^{(2)}$, where $k_0 \in \mathbf{Z}^+$. We show that $f = f_{k_0}$. In fact,

$$\langle x, f(x)\rangle \in \Gamma_f = E_{k_0}^{(2)} \Leftrightarrow \langle x, f(x), k_0\rangle \in E^{(2)}$$
$$\Leftrightarrow \exists z \in \mathbf{Z}^+,$$
$$g(x, f(x), k_0, z) = 1.$$

Among the $z \in \mathbf{Z}^+$ which make $g(x, f(x), k_0, z) = 1$, we choose the z for which the number u given by $\langle f(x), z\rangle = \langle t_1^{(2)}(u), t_2^{(2)}(u)\rangle$ is minimal. For this u we have $f_{k_0}(x) = t_1^{(2)}(u) = f(x)$, which proves the claim.

(f) *Construction of a versal family of m-functions*. The case $m = 0$ is trivial. If $\{f_k^{(1)}\}$ is a versal family of 1-functions, then for $m \geqslant 2$ we set

$$f_k^{(m)}(x_1, \ldots, x_m) = f_k^{(1)}(\tau^{(m)}(x_1, \ldots, x_m)),$$

thereby obtaining a versal family of m-functions.

The theorem is proved. □

8.2. The choice of versal families is far from unique. If $m > 1$, there does not exist a versal family which contains each function or each set exactly once (i.e., a universal family). Nevertheless, there are important methods of extracting invariant information from data about the position of a function or set in a versal family. The next section is devoted to this question.

9 Kolmogorov complexity

9.1. Let $u = \{u_k\}$ be an enumerable family of m-functions over $\mathbf{Z}^+$, and let f be a partial recursive m-function. We define the *complexity of f relative to the family u* as

$$C_u(f) = \begin{cases} \min\{k | u_k = f\}, & \text{if such a } k \text{ exists;} \\ \infty, & \text{otherwise.} \end{cases}$$

We call the enumerable family u (asymptotically) *optimal* if, for any other enumerable family v, there exists a constant $c_{u,v} > 0$ such that for every partial recursive m-function f we have

$$C_u(f) \leqslant c_{u,v} C_v(f).$$

If we take v to be any versal family, we see that an optimal family must be versal, i.e., $C_u(f)$ never takes the value ∞.

9.2. **Theorem** (Kolmogorov)

(a) *For any $m \geqslant 0$, optimal families exist and can be effectively constructed.*

(b) *If u and v are optimal families of m-functions, then for any m-function f*

$$c_{v,u}^{-1} \leqslant C_u(f)/C_v(f) \leqslant c_{u,v}.$$

9.3. *Remarks*

(a) The measure of complexity $C_u(f)$ involves the following intuitive ideas. In order to define any enumerable family u, it is only necessary to give a finite amount of information, for example, a program which semi-computes the total function of u. Therefore, in order to define a specific function f which occurs in the family u, it suffices to give no more than

$$\log_2 C_u(f) + \text{const}$$

bits of information, namely, the program for u and the number of f in u.

(b) A family being optimal means that it can be used to compute *any* m-function, and that the loss in using it rather than any other family to compute a function is bounded by a constant which does not depend on the function.

(c) Finally, the inequality 9.2(b), which follows trivially from the definition of an optimal family, shows that, to within an additive term which is bounded in absolute value, the logarithmic measure of complexity

$$K_u(f) = [\log_2 C_u(f)] + 1 \qquad (\text{where "[ ]"="integral part"})$$

does not depend on the choice of the optimal family u, and so is an asymptotic invariant of f.

9.4. Proof of Theorem 9.2. We first choose a recursive imbedding $\theta : \mathbf{Z}^+ \times \mathbf{Z}^+ \to \mathbf{Z}^+$ which has a recursive inverse function and which satisfies the following linear growth condition in one of its arguments:

$$\theta(k,j) \leqslant k \cdot \phi(j), \quad \text{for all } k, j \in \mathbf{Z}^+ \text{ and some suitable } \phi : \mathbf{Z}^+ \to \mathbf{Z}^+.$$

For example, we could let $\theta_1(k,j) = (2k-1)2^j$ with $\phi_1(j) = 2^{j+1}$, or, following Kolmogorov, we could let

$$\theta_2\left(\overline{k_1 k_2 \cdots k_r}, \overline{j_1 j_2 \cdots j_s}\right) = \overline{j_1 j_1 \cdots j_s j_s 01 k_1 \cdots k_r},$$

where $k_\alpha, j_\beta \in \{0, 1\}$ and the bar denotes the binary expansion of a number. Here $\phi_2(j) < \text{const} \cdot j^2$, so that this function grows more slowly. (See also subsection 9.8 below.)

Now let U be any versal family of $(m+1)$-functions. We define a family u of m-functions by setting

$$u(x_1, \ldots, x_m, k) = U(x_1, \ldots, x_m, \theta^{-1}(k)).$$

We show that the family u is optimal, with the following bound for the constants

$$c_{u,v} \leqslant \phi(C_U(v)).$$

In fact, let f be a recursive m-function. It suffices to consider the case when f occurs in the family v. Then

$$\begin{aligned} f(x_1, \ldots, x_m) &= v(x_1, \ldots, x_m; C_v(f)) \\ &= U(x_1, \ldots, x_m, C_v(f); C_U(v)) \\ &= u(x_1, \ldots, x_m, \theta(C_v(f), C_U(v))), \end{aligned}$$

so that

$$C_u(f) \leqslant \theta(C_v(f), C_U(v)) \leqslant C_v(f)\phi(C_U(v)).$$

The theorem is proved. □

9.5. Example. A 0-function f can be identified with the single value it takes, i.e., with a positive integer n. In this case, Theorem 9.2 gives us an almost invariant complexity $C_u(n)$ for the integers. We have:

(a) $C_u(n) \leqslant \text{const} \cdot n$ for all n, since the function "n" appears in the nth place in the simplest versal family $u_n(\cdot) = n$.

(b) $C(n) \sim \min\{2^{j-1}(2k-1) | n$ is the kth value of the jth function in some versal family of 1-functions$\}$. (We write $f \sim g$ if f and g have the same domain of definition, and $f \leqslant \text{const} \cdot g$ and $g \leqslant \text{const} \cdot f$ for suitable constants. In relations of the type $C_u(f_k) \sim g(k)$, we often omit the designation of the optimal family u, which we take to be arbitrary, but fixed.)

It is clear from (b) that the complexity of the numbers p_n (the nth prime), n^2, or

$$n^{n^{\cdot^{\cdot^{n}}}} \qquad (n \text{ times})$$

as $n \to \infty$ is asymptotically no greater than $\text{const} \cdot n$, since each of these is the nth value of a fixed recursive function. In 9.7(b) below, we shall lower this estimate to $\text{const} \cdot C(n)$.

Instead of integers, Kolmogorov and his collaborators considered finite binary sequences and constructed a theory which showed that the most complex binary sequences are those which approach random behavior. See the survey article by A. K. Zvonkin and L. A. Levin in *Uspehi Matem. Nauk*, vol. XXV, No. 6 (1970) (translated in *Russian Mathematical Surveys*), which contains a large bibliography.

9.6. **Proposition.**

(a) *Let*

$$F = f_0(f_1(x_1, \ldots, x_m), \ldots, f_n(x_1, \ldots, x_m), x_{n+1}, \ldots, x_p),$$

where the f_i are recursive functions. Then

$$C(F) \leqslant \mathrm{const} \cdot \prod_{i=1}^{n} C(f_i)\left(\log \prod_{i=1}^{n} C(f_i)\right)^{n-1},$$

if f_0 is fixed and f_i runs through all possible m-functions. Here const *depends on f_0 and on the families used to compute the complexity, but does not depend on $f_1, \ldots, f_n$.*

(b) *If f_0 is also allowed to vary, then $\prod_{i=1}^{n}$ must be replaced by $\prod_{i=0}^{n}$ and* $\log^{n-1}$ *must be replaced by* $\log^n$ *on the right.*

9.7. *Special cases*

(a) If, for example, we set $f_0 = \mathrm{sum}_2$ or prod_2, then we have:

$$C(f_1 + f_2), C(f_1 f_2) \leqslant \mathrm{const}\, C(f_1)C(f_2) \log(C(f_1)C(f_2)).$$

(b) If we set $n = 1$ and $m = 0$, we find that for any enumerable family $\{f_k\}$:

$$C(f(k, x_1, \ldots, x_p)) \leqslant \mathrm{const}\, C(k).$$

9.8. Proof of Proposition 9.6. First of all, for every $n \geqslant 1$ we define the following recursive bijection with a recursive inverse:

$$\theta^{(n)}(k_1, \ldots, k_n) = \begin{cases} \text{the index of the } n\text{-tuple } \langle k_1, \ldots, k_n \rangle \\ \quad \text{if we order } n\text{-tuples according to increasing } \prod_{i=1}^{n} k_i, \\ \quad \text{and in alphabetical order for fixed } \prod_{i=1}^{n} k_i. \end{cases}$$

It is easy to see (by induction on n) that

$$\theta^{(n)}(k_1, \ldots, k_n) \leqslant \mathrm{const} \prod_{i=1}^{n} k_i \left(\log \prod_{i=1}^{n} k_i\right)^{n-1}.$$

We define the function $\Theta : (\mathbf{Z}^+)^{n+1} \to \mathbf{Z}^+$ as follows:

$$\Theta(l_1, \ldots, l_{n+1}) = \theta(\theta^{(n)}(l_1, \ldots, l_n), l_{n+1}),$$

where θ is as described in 9.4.

We now consider two optimal families $v(x_1, \ldots, x_p, l)$ and $u(x_1, \ldots, x_m, k)$ of p-functions and m-functions, respectively. We use these two families to construct the families

$$W(x_1, \ldots, x_p; k_1, \ldots, k_n, l)$$
$$= v(u(x_1, \ldots, x_m, k_1), \ldots, u(x_1, \ldots, x_m, k_n), x_{n+1}, \ldots, x_p, l),$$
$$w(x_1, \ldots, x_p, k) = W(x_1, \ldots, x_p, \Theta^{-1}(k)).$$

The function F occurs in the

$$\theta(\theta^{(n)}(C_u(f_1), \ldots, C_u(f_n)), C_v(f_0))$$

place in the family w. Then the estimate $\theta(k, j) \leqslant k \cdot \phi(j)$, along with the estimate for $\theta^{(n)}$, give assertion (a).

We similarly obtain (b), if we replace Θ by $\theta^{(n+1)}$ in the definition of w. □

Remark. The function $\theta^{(n)}$ gives us the most economical estimate for $C(F)$ which is symmetrical in the $C(f_1), \ldots, C(f_n)$. In specific situations it might make sense to improve the estimate in certain of the $C(f_i)$ at the expense of worsening the estimate with respect to the others; this is done by suitably changing θ. For example, Kolmogorov's θ gives

$$C(f_1 + f_2) \leqslant \text{const}\, C(f_1)C(f_2)^2,$$

which is better than

$$\text{const}\, C(f_1)C(f_2)\log(C(f_1)C(f_2))$$

if $C(f_2)$ grows much more slowly than $C(f_1)$.

9.9. **Theorem.** *The function $C(f)$ is not computable. More precisely, let $g(k)$ be any unbounded partial recursive function, and let $\{f_k\}$ be any enumerable family. Then it is false that $C(f_k)|_{D(g)} \sim g(k)$.*

Thus, $C(f_k)$ can only be computable (even up to $\sim$) on a set of indices k such that there are only finitely many different functions among the functions f_k; otherwise, $C(f_k)$ is not bounded on this set.

Proof. Suppose that $C(f_k)|_{D(g)} \sim g(k)$. We show that there exists a general recursive function $h : \mathbf{Z}^+ \to \mathbf{Z}^+$ whose image is contained in $D(g)$ and such that $g \circ h$ is monotonically increasing. We then obtain a contradiction as follows. By 9.7(b), for all k we have

$$C(f_{h(k)}) \leqslant \text{const}\, C(k),$$

and, by our assumption and by the fact that $g \circ h$ is increasing,

$$C(f_{h(k)}) \geqslant \text{const}\, g(h(k)) \geqslant \text{const} \cdot k.$$

But these two inequalities are incompatible, because $\liminf C(k)/k = 0$ (for example, $C(k^2)/k^2 \leqslant \text{const}/k$).

It remains to construct h. We choose a general recursive bijection $h_1 : \mathbf{Z}^+ \stackrel{\sim}{\Rightarrow} D(g)$, using Proposition 5.6 of Chapter V, and we set

$$E = \{k | \forall i < k, g(h_1(i)) < g(h_1(k))\}.$$

This set is decidable and infinite, and $g \circ h_1$ is an increasing function on E.

Let $h_2 : \mathbf{Z}^+ \to E$ be an increasing general recursive bijection (again using Proposition 5.6 of Chapter V). Then $h = h_1 \circ h_2$ has the necessary properties. The theorem is proved. □

9.10. *Remarks*

(a) Theorem 9.9 shows that computing complexity is a problem demanding creativity: even if we find the number of a place where f occurs in an optimal family $\{u_k\}$, there is no algorithm which could tell us whether or not this function occurs even sooner.

(b) Since $C(k) \neq C(l) \Rightarrow k \neq l$, it follows that for all x and B

$$\text{card}\{y | y \leqslant x, C(y) \leqslant x/B\} \leqslant x/B,$$

i.e., most numbers have a large complexity.

Nevertheless, it is not possible to give effectively a sequence of numbers which asymptotically have maximal complexity. More precisely, let $\{k_i\}$ be any increasing sequence with $C(k_i) \geqslant k_i/B$ for some constant B. Then the set $\{k_i\}$ does not contain a single infinite enumerable set E. Otherwise, we would be able to find an increasing general recursive function $h : \mathbf{Z}^+ \to E$, and would obtain a contradiction, as in Theorem 9.9.

(c) Let $u = \{u_k\}$ be any optimal family of m-functions. The "moments of first appearance" $\{k | \forall i < k, u_i \neq u_k\}$ actually form a sequence of asymptotically maximal complexity, since, by the definition and by 9.7(b), they satisfy

$$k = C_u(u_k) \leqslant \text{const} \cdot C(k).$$

Thus, we might say that in an optimal family the functions first appear "at random moments."

The problem of computing $C(u_k)$ is complicated by the fact that, at least in the specific families in the proof of Theorem 9.2, any function appears infinitely often, so that if we are not lucky we might first notice the function arbitrarily far out from the place where it first appeared.

(d) Finally, we mention that at least one essential aspect of the complexity of computations has not been touched upon in our discussion of C_u. Namely, $\log_2 C(k)$ measures the *length of a program* that could compute k, but says nothing about the *time* it takes for such a program to work, let alone the possibilities for shortening the time by performing parallel computations, lengthening the program, and so on.

The concept of complexity is rather far removed from practical uses. But it seems to be such a fundamental idea that its role in theoretical mathematics is likely to grow.

III

PROVABILITY AND COMPUTABILITY

CHAPTER VII

Gödel's incompleteness theorem

1 Arithmetic of syntax

1.1. In this section we show how the syntax of formal languages reduces in principle to arithmetic. We do this by identifying the symbols, expressions, and texts in a finite or countable alphabet A with certain natural numbers (i.e., by *numbering* them) in such a way that the syntactic operations (juxtaposition, substitution, etc.) are represented by recursive functions, and the syntactic relations (occurrence in an expression, "being a formula," etc.) are represented by decidable or enumerable sets.

In Chapter II we described how this technique works for Smullyan's language of arithmetic, but now we shall investigate it more systematically. Our first task is to show that the computability of syntactic operations and the decidability (enumerability) of syntactic relations on the sets of expressions and texts do not depend on how we number them, as long as we adhere to certain weak natural restrictions.

This independence of the method of numbering allows us to consider this numbering not only as a technical device, but also as a reflection of a deep equivalence between arithmetic and the combinatorial properties of formal texts. In modern computers, where a single store-location may serve consecutively as a number, a name (code), and a command, this equivalence between syntax and arithmetic is realized "in the flesh" and is accepted as a basic principle. This was not the case, however, in 1931, when Gödel first introduced the concept of numbering.

1.2. *Numbering*. Let S be a finite or countable set. By a *numbering* of S we mean any *injective* map $N : S \to \mathbf{Z}^+$ whose image is *decidable*. We call $N(s)$ the *N-number* of an element $s \in S$. We call two numberings N and M of a set S *equivalent* if the partial functions $N \circ M^{-1}$ and $M \circ N^{-1}$ from $\mathbf{Z}^+$ to

$\mathbf{Z}^+$ are *partial recursive*. These functions are automatically *computable* (not only semi-computable), since their domains of definition are decidable (see §1–2 of Chapter V).

The intuitive meaning of these definitions is clear: requiring the set of $N(s)$ to be decidable ensures that it is possible to determine whether or not a natural number has the property of "being the number of an element of S," and two numberings are equivalent when each of them can be effectively recovered from the other for any $s \in S$.

1.3. **Lemma.**

(a) *The relation of equivalence between numberings is reflexive, symmetric, and transitive.*

(b) *Any injective map from a finite set S to $\mathbf{Z}^+$ is a numbering, and any two numberings of a finite S are equivalent.*

(c) *Any numbering of an infinite set is equivalent to a numbering whose image is all of $\mathbf{Z}^+$.*

All this either is obvious or has already been proved. In particular, (c) follows from Proposition 5.2 in Chapter V.

1.4. Let S_1 and S_2 be two sets, and let $N_i : S_i \to \mathbf{Z}^+$, $i = 1, 2$, be numberings of them. We call a partial function $f : S_1 \to S_2$ *partial recursive relative to* $\langle N_1, N_2 \rangle$ if the map $N_2 \circ f \circ N_1^{-1}$ is partial recursive. A tautological example: any numbering function $N : S \to \mathbf{Z}^+$ is partial recursive relative to $\langle N$, identity$\rangle$.

A subset $T \subset S$ is said to be decidable, enumerable, arithmetical (i.e., definable in $\mathrm{L_1Ar}$, see Chapter II, §2) relative to the numbering N_1 if the set $N_1(T)$ has the corresponding property.

1.5. **Lemma.** *If $\langle N_1, N_2 \rangle$ is replaced by a pair of equivalent numberings $\langle N_1', N_2' \rangle$ in 1.4, then the classes of recursive functions $f : S_1 \to S_2$ and of decidable, enumerable, and arithmetical subsets of S_1 do not change.*

Proof. The composition of computable recursive functions is recursive and computable. The inverse image of a decidable (respectively enumerable) set with respect to a computable function is decidable (respectively enumerable). Finally, suppose that $f : \mathbf{Z}^+ \to \mathbf{Z}^+$ is a partial recursive function, and that $E \subset \mathbf{Z}^+$ is an arithmetical set. Then $f^{-1}(E) = \mathrm{pr}_1((\mathbf{Z}^+ \times E) \cap \Gamma_f)$ (in $\mathbf{Z}^+ \times \mathbf{Z}^+$). Since $\mathbf{Z}^+ \times E$ is arithmetical and Γ_f is also arithmetical (even Diophantine), it follows that $f^{-1}(E)$ is arithmetical. □

1.6. Let S_i be sets with numberings N_i, $i = 1, \ldots, r$. A numbering $N : S_1 \times \cdots \times S_r \to \mathbf{Z}^+$ is said to be *compatible* with $\langle N_1, \ldots, N_r \rangle$ if the projection $\mathrm{pr}_i : S_1 \times \cdots \times S_r \to S_i$ is recursive relative to $\langle N, N_i \rangle$ for all

$i = 1, \ldots, r$, and if the partial function

$$(\mathbf{Z}^+)^r \overset{(N_1^{-1}, \ldots, N_r^{-1})}{\Longrightarrow} S_1 \times \cdots \times S_r \overset{N}{\Longrightarrow} \mathbf{Z}^+$$

is recursive. In other words, the N_i-numbers of the coordinates are computed from the N-number of the vector, and conversely.

1.7. **Lemma.**

(a) *In the notation of* 1.5, *for any* $\langle N_1, \ldots, N_r \rangle$ *there exists a numbering* N *which is compatible with them. For example, for* $s_i \in S_i$, $i = 1, \ldots, r$, *we may set*

$$N(s_1, \ldots, s_r) = \tau^{(r)}(N_1(s_1), \ldots, N_r(s_r))$$

(for the definition of $\tau^{(r)}$, see subsection 4.5 in Chapter V).

(b) *If* N *is compatible with* $\langle N_1, \ldots, N_r \rangle$, N *is equivalent to* M, *and* N_i *is equivalent to* M_i *for* $i = 1, \ldots, r$, *then* M *is compatible with* $\langle M_1, \ldots, M_r \rangle$.

(c) *If* N *is compatible with* $\langle N_1, \ldots, N_r \rangle$ *and* M *is compatible with* $\langle N_1, \ldots, N_r \rangle$, *then* N *and* M *are equivalent. If* N *is compatible with* $\langle N_1, \ldots, N_r \rangle$ *and also with* $\langle M_1, \ldots, M_r \rangle$, *then* N_i *and* M_i *are equivalent for all* $i = 1, \ldots, r$.

What all this says is that the relationship of compatibility gives a one-to-one correspondence between families consisting of r equivalence classes of numberings of the sets $S_1, \ldots, S_r$ and certain equivalence classes of numberings of $S_1 \times \cdots \times S_r$. This lemma is proved by mechanically checking the definitions.

1.8. Let $A^l = A \times \cdots \times A$ (l times), and let $S(A) = A^1 \cup A^2 \cup \cdots \cup A^l \cup \cdots$. If A is an alphabet, then $S(A)$ is the set of expressions in the alphabet. Here $A^0 = \{\wedge\}$ consists of the empty expression. The function $S(A) \to \mathbf{Z}^+$ which takes the value p on each element of A^p is called the *length* of the expression. The "ith coordinate" partial function from $\mathbf{Z}^+ \times S(A)$ to A^1 given by $\langle i, \langle a_1, \ldots, a_p \rangle \rangle \mapsto a_i$ is defined on the subset $\bigcup_{i=1}^{\infty} \{i\} \times (A^i \cup A^{i+1} \cup \cdots)$. The "juxtaposition" function from $S(A) \times S(A)$ to $S(A)$ takes

$$\langle \langle a_1, \ldots, a_p \rangle, \langle b_1, \ldots, b_q \rangle \rangle \text{ to } \langle a_1, \ldots, a_p, b_1, \ldots, b_q \rangle.$$

A numbering N of $S(A)$ is called *admissible* if the length function, the ith coordinate function, and the juxtaposition function are partial recursive relative to $\langle N, \mathrm{id} \rangle$, $\langle \langle \mathrm{id}, N \rangle, N \rangle$, and $\langle \langle N, N \rangle, N \rangle$, respectively. A numbering N of $S(A)$ is said to be *compatible* with a numbering N_0 of A if it is admissible and if the restriction of N to A^1 is equivalent to N_0 on A (where we identify A^1 with A).

Here is the basic result of this section:

1.9. **Proposition.**

(a) *If N is admissible, then any numbering equivalent to N is also admissible.*
(b) *If N is compatible with N_0, N' is equivalent to N, and N_0' is equivalent to N_0, then N' is compatible with N_0'.*
(c) *If N and N' are both compatible with N_0 then they are equivalent.*
(d) *For any numbering N_0 of A, there exists a compatible numbering N of $S(A)$, whose equivalence class is uniquely determined by the class of N_0 because of* (c).

PROOF. We obtain (a) and (b) formally from Lemma 1.6. To prove (c), we find the N-number of an expression from its N'-number as follows. Let $m * n = N(N^{-1}(m)N^{-1}(n))$ (where the argument of N is the *juxtaposition* of the two expressions $N^{-1}(m)$ and $N^{-1}(n)$). The partial function from $\mathbf{Z}^+ \times \mathbf{Z}^+$ to $\mathbf{Z}^+$ defined by $\langle m, n\rangle \mapsto m * n$ is recursive and associative, since N is admissible. Further, let $(k)_i = N$ (the ith coordinate of $N^{-1}(k)$). The partial function $\mathbf{Z}^+ \times \mathbf{Z}^+ \to \mathbf{Z}^+ : \langle k, i\rangle \mapsto (k)_i$ is recursive for the same reason. We similarly define $(k)_i'$ in terms of N'. Finally, let $l' : \mathbf{Z}^+ \to \mathbf{Z}^+$ be the partial function "the length of $N'^{-1}(k)$." It is also recursive.

Then we have:

$$N \circ N'^{-1}(k) = N \circ N'^{-1}((k)_1') * \cdots * N \circ N'^{-1}((k)_{l'(k)}').$$

But the N'-numbers of the one-letter expressions $\{(k)_i'\}$ form a decidable subset of $\mathbf{Z}^+$ (namely, the 1-level of the computable function l'). The restriction of $N \circ N'^{-1}$ to this subset is a recursive function, since the restrictions of N and N' to A^1 are equivalent. We obtain (c) from this and from the recursiveness of $*$, $(k)_i'$, and l' (by applying induction on x to $*_{i=1}^{x} N \circ N'^{-1}((k)_i')$ and then substituting $x = l'(k)$).

We prove (d) using an explicit construction of Gödel (the idea of which, incidentally, goes back to Leibniz).

(d_1) *Construction of N compatible with N_0*:

$$N(a_1, \ldots, a_m) = p_1^{N_0(a_1)} \cdots p_m^{N_0(a_m)},$$

where $p_1 = 2, p_2 = 3, \ldots$ are the prime numbers. Here $N(\wedge) = 1$. We verify that N has the required properties.

(d_2) *N is a numbering.* First of all, $N : S(A) \to \mathbf{Z}^+$ is an imbedding because $N_0 : A \to \mathbf{Z}^+$ is injective, and we have unique factorization in $\mathbf{Z}^+$.

We show that the image of N is decidable. In the first place, the set of prime numbers in $\mathbf{Z}^+$ is decidable, since it is the 2-level of the everywhere defined recursive function

$$n \mapsto \text{the number of divisors of } n = \sum_{k=1}^{n} d(k, n) \dot{-} n,$$

where (see §3 of Chapter V)

$$d(k, n) = s\big((\text{rem}(k, n) - k)^2 + 1\big) = \begin{cases} 2, & \text{if } k|n, \\ 1, & \text{otherwise,} \end{cases}$$

$$s(1) = 2, \qquad s(\geqslant 2) = 1.$$

Thus, the function $i \mapsto p_i$ is recursive (see the proof of Proposition 5.2 in Chapter V).

We now set

$$f(n, i, y) = s\big((\text{rem}(p_i^y, n) - p_i^y)^2 + 1\big).$$

This function is recursive, and hence so is the function of (n, i)

$$v_i(n) = \min\{y | f(n, i, y) = 1\} = (\text{the power of } p_i \text{ which divides } n) + 1.$$

This implies that the "length" function is recursive:

$$l(n) = \text{the number of prime divisors of } n = \sum_{i=1}^{n} s(v_i(n)) \dot{-} n$$

(automatically $p_m \nmid n$ when $m > n$, since $p_m > m$).

Now let E be the image of N_0 in $\mathbf{Z}^+$. Then

$$\text{image of } N = \{n | \forall i \leqslant l(n), v_i(n) \in E + 1\}.$$

But the set $F = \{\langle i, n\rangle | v_i(n) \in E + 1\}$ is decidable, since it is the preimage of $E + 1$ under v, and applying the bounded universal quantifier preserves the decidability. In fact, let $\chi_F(i, n) = 1$ if $\langle i, n\rangle \in F$ and $\chi_F(i, n) = 2$ if $\langle i, n\rangle \notin F$. Then the image of N is the 2-level of the following function of n: $s(\prod_{i=1}^{l(n)} \chi_F(i, n))$.

(d$_3$) *N is admissible*. We have already shown that the length function is recursive. The ith coordinate function is represented by $[p_i^{v_i(n)}/p_i]$ (the integral part). Finally, juxtaposition is represented by the function

$$m * n = m \prod_{j=1}^{l(n)} p_{l(m)+j}^{v_j(n)-1},$$

which is recursive by what has already been proved.

We note that our number-theoretic functions are defined on all of $\mathbf{Z}^+$, not only on the Gödel numbers of any specific numbering. In what follows we shall only point out when such an extension of the domains of definition is possible if there is a special reason for mentioning this possibility.

(d$_4$) *N is compatible with N_0*. The functions $x \mapsto 2^x$ and $y \mapsto \log_2(y)$ ($y \in 2^{\mathbf{Z}^+}$) tell us how to go from one numbering to the other on one-letter expressions. These functions are obviously recursive.

This completes the proof of Proposition 1.9. □

1.10. *Concluding remarks*. Proposition 1.9 shows that, if we are given an equivalence class of numberings of an alphabet A of a formal language, then this uniquely determines an equivalence class of numberings of the set of expressions $S(A)$, of the set of texts $S(S(A))$, and so on, all of which are compatible with the numberings of A in the given class. Hence, the set of recursive operations and the set of decidable or enumerable relations are invariantly defined on the expressions and texts. The only nonuniqueness that remains is the choice of the equivalence class of the numbering of A.

In all cases of which the author is aware, this choice is also determined canonically in the following way. Namely, *A is realized as a decidable subset of the expressions in some finite "protoalphabet"* A_0, where decidability is understood in the sense of any numbering of $S(A_0)$ which is compatible with any numbering of A_0. It follows from Lemmas 1.3 and 1.5 and Proposition 1.9 applied to A_0 that the resulting class of numberings of A will not depend on either the imbedding of A in $S(A_0)$, the numbering of A_0, or even the choice of A_0 (where we recall that, if $A_0 \subset A_1$ are finite, then $S(A_0) \subset S(A_1)$ is decidable).

From this point of view, it is natural to consider the nine-letter alphabet of SAr which was described in §10 of Chapter II to be a protoalphabet. Then $x, x', x'', x''', \ldots$ are elements of the "real alphabet." Smullyan's particular numbering system is very convenient for proving Tarski's theorem, but the "undefinability of truth" in SAr does not depend on the special form of this numbering, as should by now be completely clear.

More generally, any complete printed description of any alphabet A realizes A in the protoalphabet of available typographical symbols, which is of course finite, and thereby determines a canonical equivalence class of numberings of A.

2 Incompleteness principles

2.1. Gödel's theorem on the incompleteness of formal theories can be given many precise formulations, none of which entirely exhausts its content. In this section, using the results obtained in §1, we shall try to separate the conceptual aspects of the theorem from the technical details needed to prove it for various languages.

2.2. Let A be a finite or countable alphabet with its canonical equivalence class of numberings, and let $S(A)$ be the set of expressions in A. We suppose that the following two subsets of $S(A)$ have somehow been defined:

(a) $T \subset S(A)$, the set of "true" expressions. For example, we might have been given a language with A as its alphabet, some sort of semantics for the language, and a truth function.

(b) $D \subset S(A)$, the set of "provable" or "deducible" expressions. This set might be described by giving "axioms" and "rules of deduction," or in some other way. We shall always assume that $D \subset T$, as the terminology suggests (it is only possible to prove what is true).

There is every reason to expect that, if D and T have been constructed "in a natural way," in the process of formalizing some fragment of modern mathematics, then the following principles hold true.

2.3. *The set D is enumerable.* The intuitive arguments to support this assertion are as follows. Suppose that the "provable" expressions are those for which "proofs" exist. Here "proofs" are certain texts which, perhaps, are written in another alphabet B, i.e., they are elements of $S(S(B))$. (For example, theorems in L_1Ar may be proved in L_1Set.) One minimal requirement for formal mathematical proofs is that it must be possible mechanically to determine that they are proofs, i.e., they must form a decidable subset of $S(S(B))$. (Here it would actually be sufficient to require that the set of "proofs" be enumerable.) Another unavoidable requirement is that from every proof we must be able to obtain mechanically the "expression proved" in $S(A)$. In other words, the partial function from $S(S(B))$ to $S(A)$ given by "proof" $\mapsto$ "expression proved" must be (semi-) computable. But then the image of this function is enumerable. In §5 we show that the set of deducible formulas in $\mathfrak{L}_1$ is enumerable, in accordance with these informal considerations.

We note that a time aspect has implicitly entered into the discussion. A "proof" is understood to mean a "proof using the means accepted at the present time and (semi-) identifiable as being accepted." If, for example, we introduce a new axiom of set theory and it becomes widely accepted, then the concept of a proof becomes broader, as happened with the axiom of choice (or, rather, the principle of transfinite induction, Zorn's lemma, . . .). See the discussion in §7.

2.4. *The set T is not enumerable if the semantics of truth is rich enough to include elementary arithmetic.* We clearly have in mind some version of Tarski's theorem, which, in fact, even tells us that T is not an arithmetical set. In the next section we give several precise formulations of this principle. (See also subsections 7.3–7.4 below.)

2.5. **Gödel's incompleteness theorem** (General form). *All formal theories of mathematics satisfy the principles* 2.3 *and* 2.4. *Therefore, if a theory is sufficiently rich, it always contains true expressions which are not provable.*

3 Nonenumerability of true formulas

The following criteria are all variations on a single theme, even if this is not obvious at first, namely, "self-reference, or the diagonal process."

3.1. *The language* SAr. We refer the reader to §10 of Chapter II for the description of this language and its standard interpretation. In §11 of

Chapter II we showed that the set of numbers of true formulas in Smullyan's numbering system is nonarithmetical. This set is *a fortiori* nonenumerable, since enumerable sets are even Diophantine.

3.2. *The language* $\mathrm{L_1Ar}$. Here we give two versions of the argument, one of which gives the stronger result and the other of which gives the more concrete result. A third version, which is closer to Gödel's original proof, will be described in §7.

(a) *Tarski's theorem for* $\mathrm{L_1Ar}$. The proof that the set of true formulas in $\mathrm{L_1Ar}$ is nonarithmetical can be reduced to Tarski's theorem for SAr in the following way. In the first place, the sets of formulas in $\mathrm{L_1Ar}$ and SAr are decidable in the set of all expressions (this will be shown for $\mathrm{L_1Ar}$ in §4). In the second place, the translation map $\{\text{formulas in SAr}\} \overset{\mathrm{tr}}{\Longrightarrow} \{\text{formulas in } \mathrm{L_1Ar}\}$, which was described in §10 of Chapter II, is recursive (as is easily shown using the arguments in the next section). Since the map tr preserves the truth function, we have $T_S = \mathrm{tr}^{-1}(T_{L_1})$ in the obvious notation. But then, if T_{L_1} were arithmetical, it would follow that T_S is also arithmetical (see the proof of Lemma 1.5), which contradicts Tarski's theorem for SAr. It would be a useful exercise for the reader, after first reading §4, to carry out this proof in complete detail.

The following argument is simpler and more precise, but it only shows that T_{L_1} is nonenumerable, and not that it is nonarithmetical.

(b) Let $E \subset \mathbf{Z}^+$ be an enumerable but undecidable set (which exists by §5 of Chapter V). Let E be defined by the formula $P(x)$ in $\mathrm{L_1Ar}$ which has one free variable x. For $n \geqslant 2$ we set $\bar{n} = +\Big(+\big(\bar{1} + (\bar{1}, \bar{1})\big) \cdots \Big)$, which is the term-name for the integer n in the obvious canonical $\mathfrak{L}_1$ type notation. We consider the family of closed formulas $\{\neg(P(\bar{n}))|n \in \mathbf{Z}^+\}$ in $\mathrm{L_1Ar}$.

3.3. **Proposition.**

(a) *The function* $\mathbf{Z}^+ \to \{$*formulas in* $\mathrm{L_1Ar}\}$ *given by* $n \mapsto \neg P(\bar{n})$ *is recursive.*

(b) *The set* $\{n|\neg(P(\bar{n}))$ *is true*$\}$ *is nonenumerable.*

Corollary. T_{L_1} *is nonenumerable; more precisely, the set of true formulas in the family* $\{\neg(P(\bar{n}))\}$ *is nonenumerable.*

(If T_{L_1} were enumerable, its preimage in $\mathbf{Z}^+$ would also be enumerable.)

PROOF.

(a) Let the formula $\neg(P(x))$ have the form $R_1 \times R_2 \times \cdots \times R_{S+1}$, where x does not occur in the expressions R_i. Using the same notation as in the proof of Proposition 1.9, for a fixed numbering N of the set of expressions with juxtaposition function $*$ we have:

$$N(\neg(P(\bar{n}))) = N(R_1) * N(\bar{n}) * N(R_2) * \cdots * N(R_{S+1}).$$

Hence, it suffices to show that the function $n \mapsto N(\bar{n})$ is recursive. But, since $\overline{n+1} = +(\bar{1}, \bar{n})$, it follows that for $n \geqslant 1$

$$N(\overline{n+1}) = N(+) * N("(") * N(\bar{1}) * N(\bar{n}) * N(")"),$$

which expresses $N(\overline{n+1})$ recursively in terms of $N(\bar{n})$.

(b) $\{n | \neg(P(\bar{n})) \in T_{L_1}\} = \mathbf{Z}^+ \setminus E$ by the definition of the formula $P(x)$ defining E. But the complement of E is nonenumerable, since E is undecidable.

The proposition and the corollary are proved. □

3.4. *Languages at least as rich as* L_1Ar. Let L be an arbitrary language with a (finite or countable) alphabet A, in which we are given a set T of "true" expressions. We suppose that L is no poorer than a language of arithmetic in the following sense: *There exists a translation map*

$$\text{tr} : \{\textit{formulas in } \mathrm{L_1Ar}\} \Rightarrow \{\textit{expressions in } A\}$$

which takes T_{L_1} *to* T, *takes the complement of* T_{L_1} *to the complement of* T, *and is recursive.*

Then T *is nonenumerable.*

Such a translation map can be constructed for L_1Set, for example. Proposition 3.3 shows that, actually, we need only know how to translate into L the formulas in the family $\neg(P(\bar{n}))$; this allows us to use a very modest language of arithmetic.

3.5. *Remarks*

(a) The series of Diophantine problems "Is $P(\bar{n})$ true?," i.e., "Does the Diophantine equation $F(n; x_1, \ldots, x_r) = 0$ have a solution in $\mathbf{Z}^+$?" (where F is a suitable polynomial with integer coefficients, see Chapter VI) has the property that no finitely describable collection of means of proof is adequate to answer this series of questions completely. One might say that even the theory of Diophantine equations is infinitely complicated.

(b) In some sense any problem in mathematics reduces to a Diophantine problem. In fact, after translating the problem into a suitable formal language, we may just ask, "Is the formula P or the formula $\neg P$ provable?" But this is precisely the same as asking whether the number of P (the number of $\neg P$) belongs to the enumerable set D of provable formulas, i.e., whether the Diophantine equation corresponding to D in the given series is decidable.

This gives somewhat unexpected support for Gauss's opinion regarding the queenly status of arithmetic. There even exists a "queen of the Diophantine equations" whose graph projects onto the set of numbers of formulas in L_1Set which are deducible from the Zermelo–Fraenkel axioms.

But, of course, we normally ask "Is P true?" and not "Is P provable?." From this point of view, the most creative activity in mathematics is the discovery of new principles of proof which do not reduce to the "legacy of the past" and which again must be taken on faith. Set theory as a whole was the most recent such principle in the modern development of mathematics. The dramatic history of its creation and of the disputes surrounding its acceptance is worthy of a discovery of this magnitude.

It is amazing that within formal mathematics it is possible to say something about such informal things. See also §7 below.

4 Syntactic analysis

4.1. This section contains the preliminary technical material which will be needed in §5, when we prove that the set of deducible formulas in a language of $\mathfrak{L}_1$ is enumerable.

Let L be a fixed language in $\mathfrak{L}_1$ having a finite or countable alphabet A. In order to shorten the technical work somewhat, we assume that we are working with a dialect which contains only the connectives $\neg$ and $\rightarrow$ and the quantifier $\forall$. This is not in any sense essential. As in §1, we have a canonical equivalence class of numberings of A, which determines numberings of $S(A)$, $S(S(A))$, and so on. The terms "recursive," "decidable" etc. will be understood to refer to this class. Thus, we may omit explicit mention of the numbering in the statements of the basic results. But in the proofs it will be more convenient to work directly with a numbering. We therefore fix one of the numberings $N: S(A) \rightarrow \mathbf{Z}^+$ with juxtaposition function $*$, length function l, and ith coordinate function $(k)_i$, as in the proof of Proposition 1.9. We shall assume that $m * n > \max(m, n)$, i.e., the number of any part of an expression is strictly less than the number of the whole expression. Such an N is called a *Gödel numbering*.

In addition to the conditions given in §1, we require that N satisfy the following conditions regarding recognition of the syntactic characteristics of the symbols of the alphabet:

(a) *The sets of variables, of constants, of operations, and of relations in A are decidable.*
(b) *The "degree" function on the set of operations and relations is recursive.*

We are now ready to begin. But before reading further the reader is advised to review §1 of Chapter II.

4.2. *The partial function from $S(A) \times \mathbf{Z}^+$ to $\mathbf{Z}^+$ given by*

$$\langle \text{an expression } P, i\rangle \mapsto \begin{cases} \text{the place in } P \text{ containing the right parenthesis which} \\ \text{corresponds to the left parenthesis in the } i\text{th place in } P \end{cases}$$

is compatable, i.e., it is recursive and has a decidable domain of definition.

Proof. It will be convenient to use the following notation: if Q is a statement about integers in $\mathbf{Z}^+$, then

$$\|Q\| = \begin{cases} 1, & \text{if } Q \text{ is true,} \\ 2, & \text{if } Q \text{ is false.} \end{cases}$$

This is a truth function which has been adjusted so as to take values in $\mathbf{Z}^+$, which does not contain zero.

We construct a function $\text{Par}(k, i) : \mathbf{Z}^+ \times \mathbf{Z}^+ \to \mathbf{Z}^+$ as follows: if $(k)_i$ is not defined, or if $(k)_i \neq N(\text{“(”})$, or if $(k)_i = N(\text{“(”})$ but $\forall j \in [i, l(k)]$, $\Sigma^j_{m=i}\|(k)_m = N(\text{“(”})\| \neq \Sigma^j_{m=i}\|(k)_m = N(\text{“)”})\|$, let $\text{Par}(k, i) = 1$; otherwise, let $\text{Par}(k, i) = \min\{j | j \leqslant l(k) \text{ and } \Sigma^j_{m=i}\|(k)_m = N(\text{“(”})\| = \Sigma^j_{m=i}\|(k)_m = N(\text{“)”})\|\}$. Obviously, when restricted to $N^{-1}(S(A)) \times \mathbf{Z}^+$, the function $\text{Par}(k, i)$ gives the place in the expression $N^{-1}(k)$ containing the “)” which corresponds to the “(” in the ith place if this is possible, and gives 1 when this is not possible. (Compare with Lemma 1.2 in §1 of Chapter II.) Hence, it suffices to show that $\text{Par}(k, i)$ is recursive. But $\text{Par}(k, i)$ has been defined by gluing together a finite number (four) of recursive functions having decidable domains of definition (by the properties of N). Thus, $\text{Par}(k, i)$ is recursive. □

4.3. *The partial function $S(A) \to \mathbf{Z}^+$ given by (an expression P)$\mapsto$ (the number of terms in L which are juxtaposed to get P) is computable.*

We recall that this number is uniquely defined (§1 of Chapter II).

Proof. We first construct a formula which defines the function

$$LT(k) = \begin{cases} l(k) + 1, & \text{if } N^{-1}(k) \text{ is not a} \\ \text{the number of terms whose} & \text{juxtaposition of terms;} \\ \quad \text{juxtaposition is } N^{-1}(k), & \text{otherwise} \end{cases}$$

from $\mathbf{Z}^+$ to $\mathbf{Z}^+$ recursively in terms of its values on smaller values of the argument. The way to carry out this syntactic analysis of $N^{-1}(k)$ can be described verbally as follows: first see whether or not $N^{-1}((k)_1)$ is a variable or constant and, if it is, whether or not $N^{-1}((k)_2 * \cdots (k)_{l(k)})$ is a juxtaposition of terms; if $N^{-1}((k)_1)$ is not a variable or constant, check whether it is an operation, and, if it is, whether it is followed by “(,” whether there is a corresponding “),” whether a juxtaposition of the required number of terms lies between the “(” and the “),” and whether “)” is followed by a juxtaposition of terms.

To describe this procedure systematically, we set

$$f_1(k) = \begin{cases} (k)_2 * \cdots * (k)_{l(k)}, & \text{if } l(k) \geqslant 2; \\ 1, & \text{otherwise;} \end{cases}$$

$$f_2(k) = \begin{cases} (k)_3 * \cdots * (k)_{\text{Par}(k, 2)-1}, & \text{if } 4 \leqslant \text{Par}(k, 2); \\ 1, & \text{otherwise;} \end{cases}$$

$$f_3(k) = \begin{cases} (k)_{\text{Par}(k, 2)+1} * \cdots * (k)_{l(k)}, & \text{if } 1 < \text{Par}(k, 2) < l(k); \\ 1, & \text{otherwise.} \end{cases}$$

All of these functions are recursive.

We now write the following recipe for computing $LT(k)$ recursively:

$$l(k)=1 \quad \text{and} \quad \begin{cases} N^{-1}(k) \text{ is a variable} \Rightarrow LT(k)=1, \\ N^{-1}(k) \text{ is a constant} \Rightarrow LT(k)=1, \\ N^{-1}(k) \text{ is neither a variable nor a constant} \Rightarrow LT(k)=2; \end{cases}$$

$l(k)>1$ and $N^{-1}((k)_1)$ is a variable $\Rightarrow LT(k)=1+LT(f_1(k))$;

$l(k)>1$ and $N^{-1}((k)_1)$ is a constant $\Rightarrow LT(k)=1+LT(f_1(k))$;

$l(k)>1$, $N^{-1}((k)_1)$ is an operation, $(k)_2=N(\text{"("})$,
$4 \leqslant \text{Par}(k, 2)=l(k)$,
degree $N^{-1}((k)_1)=LT(f_2(k)) \leqslant l(f_2(k)) \Rightarrow LT(k)=1$;

$l(k)>1$, $N^{-1}((k)_1)$ is an operation, $(k)_2=N(\text{"("})$,
$4 \leqslant \text{Par}(k, 2)<l(k)$,
degree $N^{-1}((k)_1)=LT(f_2(k)) \leqslant l(f_2(k))$,
$LT(f_3(k)) \leqslant l(f_3(k)) \Rightarrow LT(k)=1+LT(f_3(k))$;

$l(k)>1$ and none of the previous additional conditions hold
$\Rightarrow LT(k)=1+l(k)$.

To show that LT is recursive, we first note that, for each of the above eight alternatives, we can easily construct a recursive function $h_i(k, x, y, z)$ with the following property:

$$\|k \text{ satisfies the } i\text{th alternative}\| = h_i(k, LT(f_1(k)), LT(f_2(k)), LT(f_3(k))),$$

and we can also construct a recursive function $v_i(k, x, y, z)$ with the property that k satisfies the ith alternative $\Rightarrow$

$$LT(k)=v_i(k, LT(f_1(k)), LT(f_2(k)), LT(f_3(k))).$$

We therefore have the equation:

$$LT(k)=2\sum_{i=1}^{8} v_i(k, LT(f_1(k)), LT(f_2(k)), LT(f_3(k))) \dot{-} \sum_{i=1}^{8} (h_i v_i)(k, LT(f_1(k)), LT(f_2(k)), LT(f_3(k))).$$

Since $f_i(k)<k$ for $k>1$, this formula allows us successively to compute the values of $LT(k)$, starting with $LT(1)$. But the recursion here computes the value at k not in terms of the value at $k-1$, but in terms of several earlier values. It is this which presents the basic difficulty in showing that the syntactic functions are recursive. We now describe the device for overcoming this difficulty here and in all future cases.

In general, let $\phi_1(k), \ldots, \phi_s(k)$ be recursive functions having the property that $\phi_i(k) < k$ for all $i \leqslant s$ and $k \geqslant 2$. Further, let $h(x_1, \ldots, x_m, k, y_1, \ldots, y_s)$ be a recursive function, and let $g(x_1, \ldots, x_n, k)$ be defined by the relations

$$g(x_1, \ldots, x_n, 1) = \text{some known recursive function,}$$

$$g(x_1, \ldots, x_n, k+1) = h\big(x_1, \ldots, x_n, k, g(x_1, \ldots, x_n, \phi_1(k)), \ldots, g(x_1, \ldots, x_n, \phi_s(k))\big).$$

Using the juxtaposition function $*$, we let

$$G(x_1, \ldots, x_n, k) = \mathop{*}_{i=1}^{k} g(x_1, \ldots, x_n, i).$$

Since

$$g(x_1, \ldots, x_n, i) = \big(G(x_1, \ldots, x_n, k)\big)_i$$

for all $i \leqslant l(G(x_1, \ldots, x_n, k)) = k$, and in particular for the greatest such i, it follows that, to verify that g is recursive, it suffices to show that G is recursive. But for $k \geqslant 2$ we have

$$\begin{aligned} G(x_1, \ldots, x_n, k+1) &= G(x_1, \ldots, x_n, k) * g(x_1, \ldots, x_n, k+1) \\ &= G(x_1, \ldots, x_n, k) \end{aligned}$$

$$* \, h\big(x_1, \ldots, x_n, k, (G(x_1, \ldots, x_n, k))_{\phi_1(k)}, \ldots, (G(x_1, \ldots, x_n, k))_{\phi_s(k)}\big),$$

which is in the standard form for a recursive equation.

If we apply this device to LT, setting $n = 0$, $s = 3$, and $\phi_i(k) = f_i(k+1)$, we obtain the recursiveness of LT. □

Corollary. *The set of terms is decidable.*

In fact, this set is the 1-level of the computable function LT.

4.4. *The set of atomic formulas is decidable.*

In fact,

$$N^{-1}(k) \text{ is an atomic formula} \Leftrightarrow (k)_1 \text{ is a relation, } (k)_2 = N(\text{“(”}),$$

$$\text{Par}(k, 2) = l(k) \geqslant 4,$$

$$\text{and degree } N^{-1}((k)_1) = LT(f_2(k)) \leqslant l(f_2(k)),$$

where $f_2(k)$ was defined in 4.3.

4.5. *The set of formulas is decidable.*

In fact, in our dialect, which has been simplified to include only $\neg$, $\rightarrow$, and $\forall$, we have:

$N^{-1}(k)$ is a formula

$\Leftrightarrow N^{-1}(k)$ is an atomic formula, or is of the form $\neg(P)$, $(P) \Rightarrow (Q)$, or $\forall x(P)$, where P and Q are formulas and x is a variable.

Using the procedure in 4.3, we define the recursive functions

$$f_4(k) = \begin{cases} (k)_3 * \cdots * (k)_{l(k)-1}, & \text{if } l(k) \geqslant 4; \\ 1, & \text{otherwise;} \end{cases}$$

$$f_5(k) = \begin{cases} (k)_2 * \cdots * (k)_{\mathrm{Par}(k,\,1)-1}, & \text{if } \mathrm{Par}(k, 1) \geqslant 3; \\ 1, & \text{otherwise;} \end{cases}$$

$$f_6(k) = \begin{cases} (k)_{\mathrm{Par}(k,\,1)-3} * \cdots * (k)_{l(k)-1}, & \text{if } 3 \leqslant \mathrm{Par}(k, 1) \leqslant l(1) - 1; \\ 1, & \text{otherwise;} \end{cases}$$

$$f_7(k) = \begin{cases} (k)_4 * \cdots * (k)_{l(k)-1}, & \text{if } l(k) \geqslant 5; \\ 1, & \text{otherwise;} \end{cases}$$

$$\mathrm{At}(k) = \begin{cases} 1, & \text{if } N^{-1}(k) \text{ is an atomic formula;} \\ 2, & \text{otherwise.} \end{cases}$$

The function

$$\mathrm{Fm}(k) = \begin{cases} 1, & \text{if } N^{-1}(k) \text{ is a formula,} \\ 2, & \text{otherwise} \end{cases}$$

is computed using the following recursive relation (where $s(1) = 1$ and $s(k) = 2$ for $k \geqslant 2$):

$$\begin{aligned} \mathrm{Fm}(k) = s \circ \min\{ & \mathrm{At}(k);\ \|(k)_1 = N(\text{``}\neg\text{''})\| \cdot \|(k)_2 = N(\text{``(''})\| \cdot \|l(k) \geqslant 4\| \\ & \cdot \mathrm{Fm}(f_4(k)); \\ & \|(k)_1 = N(\text{``(''})\| \cdot \|\mathrm{Par}(k, 1) \geqslant 3\| \cdot \mathrm{Fm}(f_5(k)) \\ & \cdot \|(k)_{\mathrm{Par}(k,\,1)+1} = N(\text{``}\rightarrow\text{''})\| \cdot \|(k)_{\mathrm{Par}(k,\,1)+2} = N(\text{``(''})\| \\ & \cdot \|\mathrm{Par}(k, \mathrm{Par}(k, 1) + 2) = l(k)\| \cdot \mathrm{Fm}(f_6(k)); \\ & \|(k)_1 = N(\text{``}\forall\text{''})\| \cdot \|(k)_2 = N(\text{a variable})\| \cdot \|(k)_3 = N(\text{``(''})\| \\ & \cdot \|\mathrm{Par}(k, 3) = l(k) \geqslant 5\| \cdot \mathrm{Fm}(f_7(k))\}. \end{aligned}$$

$\mathrm{Fm}(k)$ is now shown to be recursive using the device described in 4.3.

Corollary. *The sets of formulas of the form* $\neg(P)$, $(P)\rightarrow(Q)$, *and* $\forall x(P)$ *are decidable.*

4.3. *The following function from* $S(A) \times \mathbf{Z}^+ \times S(A)$ *to* $S(A)$ *is computable*: $\langle P, i, Q\rangle \mapsto$ *the result of substituting* P *for the* ith *symbol in* Q.

We set

$$\mathrm{Sub}(k, i, m) = \begin{cases} (m)_1 * \cdots * (m)_{i-1} * k * (m)_i * \cdots * (m)_{l(m)}, \\ \text{if } i \leqslant l(m); \\ 1, \quad \text{otherwise.} \end{cases}$$

This function is clearly recursive, and coincides with the required map on the set of $\langle k, i, m\rangle$ with $k, m \in N^{-1}(S(A))$. □

4.7. *The following relation in* $\mathbf{Z}^+ \times S(A) \times S(A)$ *is decidable*: "*the one-letter expression* x *is a free variable in the* ith *place in the formula* P."

If fact, we set

$$\mathrm{Fr}(i, k, l) = \begin{cases} 1, & \text{if the condition in 4.7 holds for } P = N^{-1}(k) \\ & \text{and } \langle x\rangle = N^{-1}(l); \\ 2, & \text{otherwise.} \end{cases}$$

Then we have:

$$N^{-1}(k) \text{ is not a formula}, \quad \text{or} \quad N^{-1}(l) \text{ is not a variable}, \quad \text{or} \quad i > l(k) \Rightarrow \mathrm{Fr}(i, k, l) = 2.$$

Now suppose that $N^{-1}(k)$ is a formula, $N^{-1}(l)$ is a variable, and $i \leqslant l(k)$. Then the following alternatives remain:

$$l \neq (k)_i \Rightarrow \mathrm{Fr}(i, k, l) = 2;$$

$$l = (k)_i, \quad \mathrm{At}(k) = 1 \Rightarrow \mathrm{Fr}(i, k, l) = 1;$$

$$l = (k)_i, \quad N^{-1}(k) \text{ has the form } \neg(P) \Rightarrow \mathrm{Fr}(i, k, l) = \mathrm{Fr}(i, f_5(k), l);$$

$$l = (k)_i, \quad N^{-1}(k) \text{ has the form } (P)\rightarrow(Q),\ i < \mathrm{Par}(k, 1) \Rightarrow \mathrm{Fr}(i, k, l) = \mathrm{Fr}(i, f_5(k), l)$$

$$l = (k)_i, \quad N^{-1}(k) \text{ has the form } (P)\rightarrow(Q),\ i > \mathrm{Par}(k, 1) + 2 \Rightarrow \mathrm{Fr}(i, k, l) = \mathrm{Fr}(i, f_6(k), l);$$

$$l = (k)_i, \quad N^{-1}(k) \text{ has the form } \forall x(P),\ (k)_2 = l \Rightarrow \mathrm{Fr}(i, k, l) = 2;$$

$$l = (k)_i, \quad N^{-1}(k) \text{ has the form } \forall x(P),\ (k)_2 \neq l \Rightarrow \mathrm{Fr}(i, k, l) = \mathrm{Fr}(i, f_7(k), l).$$

Here the functions f_5, f_6, and f_7 were defined in 4.5. The rest of the proof that Fr is recursive follows the same procedure as in 4.4 and 4.5. □

4.8. *The set* $\{\langle x, P, t\rangle | x$ *is a variable*, P *is a formula*, t *is a term, and* x *does not bind* t *in* $P\}$ *is decidable*.

In terms of the numbers $\langle i, k, m\rangle$, this condition means that:

$$\forall j \leqslant l(k)\{\text{either } (k)_j \neq i, \text{ or else } (k)_j = i \wedge \mathrm{Fr}(j, k, i) = 2,$$

$$\text{or else } (k)_j = i \wedge \mathrm{Fr}(j, k, i) = 1$$

$$\wedge \forall n \in [1, l(m)](\mathrm{Fr}(j + n \mathbin{\dot{-}} 1, \mathrm{Sub}(m, j, k), \mathrm{Sub}(m, j, k)_{j+n \dot{-} 1})$$

$$= \|(m)_n \text{ is a variable}\|)\}.$$

That is, if t is substituted in place of any free occurrence of x in P, all the variables in t remain free. □

4.9. *The following partial function is computable*: $\langle x, P, t\rangle \mapsto$ *the result of substituting t in place of all free occurrences of x in P.*

Let $\langle i, k, m\rangle$ be the numbers of x, P, and t. We set

$$f(j, k, i, m) = \begin{cases} (k)_j, & \text{if } \mathrm{Fr}(j, k, i) = 2; \\ m, & \text{if } \mathrm{Fr}(j, k, i) = 1. \end{cases}$$

This is a recursive function. We further set

$$\mathrm{Sub}\, t(i, k, m) = \mathop{*}_{j=1}^{l(k)} f(j, k, i, m).$$

This is the number of the expression obtained by substituting t in place of all free occurrences of x in P. □

5 Enumerability of deducible formulas

5.1. *General setup*. Let L be any language with a numbered countable alphabet A. We suppose that the following data is fixed:

(a) An enumerable set of "axioms" $\mathrm{Ax} \subset S(A)$.
(b) A partial recursive function $\mathrm{Inf} : \mathbf{Z}^+ \times S(S(A)) \to S(A)$, i.e., an enumerable family of "rules of deduction."

We shall say that an expression $P \in S(A)$ *is a direct consequence of the expressions* $P_1, \ldots, P_r$ *by the* ith *rule of deduction, if* $\langle i, \langle P_1, \ldots, P_r\rangle\rangle \in D(\mathrm{Inf})$ and $\mathrm{Inf}(i, \langle P_1, \ldots, P_r\rangle) = P$. We shall call an expression P *deducible* (from the "axioms") if there exists a finite sequence of expressions $P_1, \ldots, P_n = P$ such that for each $j \leqslant n$ either $P_j \in \mathrm{Ax}$ or there exist $i \in \mathbf{Z}^+$ and $\{P_{k_1}, \ldots, P_{k_r}\} \subset \{P_1, \ldots, P_{j-1}\}$ such that P_j is a direct consequence of $P_{k_1}, \ldots, P_{k_r}$ by the ith rule of deduction. We let D denote the set of all deducible expressions.

5.2. **Proposition.** *D is enumerable.*

Proof. Let $a : \mathbf{Z}^+ \to S(A)$ be a recursive function whose image coincides with Ax, and let $\mathrm{inf} : \mathbf{Z}^+ \to S(A)$ be the partial recursive function given by $\mathrm{inf}(n) = \mathrm{Inf}(t_1^{(2)}(n), N_1^{-1}(t_2^{(2)}(n)))$, where $N_1 : S(S(A)) \to \mathbf{Z}^+$ is any numbering of the texts which is compatible with the given numbering of the expressions.

We construct a recursive function $d : \mathbf{Z}^+ \to S(A)$ as follows:

$$d(2n-1) = a(n),$$
$$d(2n) = \mathrm{inf}(n), \qquad n \geqslant 1.$$

We claim that its image is D. In fact, it suffices to verify that (a)

$\mathrm{Ax} \subset$ image of d; and, (b) if $P_1, \ldots, P_r \in$ image of d and P is a direct consequence of $P_1, \ldots, P_r$ by the ith rule of deduction, then $P \in$ image of d.

But (a) is obvious, since all the axioms are written out in the odd numbered places. To verify (b), we choose n so that

$$t_1^{(2)}(n) = i, \qquad t_2^{(2)}(n) = N_1(\langle P_1, \ldots, P_r \rangle).$$

Then $d(2n) = P$. The proposition is proved. □

We now verify that the general setup in 5.1 can always be realized in languages of $\mathfrak{L}_1$.

5.3. *The rules of deduction* Gen *and* MP. We define the map $\mathrm{Inf} : \mathbf{Z}^+ \times S(S(A)) \to S(A)$ as follows:

$$D(\mathrm{Inf}) = \{\langle 1, \langle P, (P) \to (Q) \rangle \rangle | P \text{ and } Q \text{ are formulas}\}$$
$$\cup \{\langle i, \langle P \rangle \rangle | P \text{ is a formula}, i \geqslant 2\},$$
$$\mathrm{Inf}\langle 1, \langle P, (P) \to (Q) \rangle \rangle = Q,$$
$$\mathrm{Inf}\langle i, \langle P \rangle \rangle = \forall x_{i-1}(Q),$$

where x_j is the jth variable in L in any fixed numbering of the variables which has image $\mathbf{Z}^+$ and is compatible with the numbering of A. It is clear that Inf is recursive and exhausts the rules of deduction Gen and MP.

5.4. *The axioms*. We verify that the following sets are enumerable in any language in $\mathfrak{L}_1$:

(a) The tautologies.
(b) The logical quantifier axioms.
(c) The axioms of equality.

Two other sets we show to be enumerable are:

(d) The special axioms of $\mathrm{L}_1\mathrm{Ar}$.
(e) The special axioms of $\mathrm{L}_1\mathrm{Set}$.

Actually, using the methods of §4 it is not hard to prove that all of these sets are even decidable. But the proof of enumerability is somewhat shorter, and will suffice for our purposes.

5.5. *The tautologies*. In §5 of Chapter II we constructed a finite list of basis tautologies and showed that all the other tautologies can be deduced from them using MP. Thus, by Proposition 5.2, it is sufficient to verify that the basis tautologies are enumerable.

Each of the basis tautologies determines a set of formulas of the form

$$Q_1 P_{i_1} Q_2 P_{i_2} \cdots P_{i_r} Q_{r+1},$$

where the Q_i are fixed expressions which are nonempty (with the possible exception of Q_1 and Q_{r+1}); $i_1, \ldots, i_r \in \{1, \ldots, m\}$ and the $\langle P_1, \ldots, P_m\rangle$ varies over all ordered m-tuples of formulas in L. Since the set of such m-tuples is decidable by 4.5 above, and since the operation of juxtaposition is recursive, it is clear that we obtain an enumerable set of formulas.

5.6. *The logical quantifier axioms.* In case our dialect of $\mathfrak{L}_1$ does not have $\exists$, these axioms can be expressed as the following two axiom schemes:

(a) $(\forall x(P(x)))\to(P(t))$, if x does not bind the term t in the formula P.
(b) $(\forall x((P)\to(Q)))\to((P)\to(\forall x(Q)))$, if x does not occur freely in P.

By 4.8, the set of triples $\{\langle x, P, t\rangle | x$ does not bind t in $P\}$ is decidable, and, by 4.9, the map $\langle x, P, t\rangle \mapsto P(t)$ is recursive. Since juxtaposition is also recursive, the set of axioms (a) is the image of a decidable set under a recursive function, and so is enumerable.

We may similarly conclude that (b) is enumerable if we verify that the condition "x does not occur freely in P" is decidable. But this is equivalent to the following condition: "the formula obtained from P by substituting either of the variables x_1 or x_2 in place of all free occurrences of x in P coincides with P," where $\langle x_1, x_2\rangle$ is any fixed pair of distinct variables. This condition is decidable by 4.9.

5.7. *The axioms of equality.* By the definition in 4.6 of Chapter II, it suffices to show that the set of formulas of the form

$$(x=y)\Rightarrow(P(x,x)\Rightarrow P(x,y))$$

is enumerable, where P runs through the atomic formulas in the language, x and y are variables, and $P(x,y)$ is obtained from P by replacing x by y in any subset of the occurrences of x in P. This set of formulas can be obtained, for example, as the image of the following function, which is partial recursive by the results in 4.4 and 4.6;

$S(A)\times A^1\times A^1\times S(\mathbf{Z}^+)\to S(A)$;

$\langle P, \langle x\rangle, \langle y\rangle, \langle i_1, \ldots, i_r\rangle\rangle \mapsto$ the expression obtained by substituting y in the $i_1, \ldots, i_r$ places in the atomic formula P if x occurs in those places.

5.8. *The special axioms of* $\mathrm{L_1Ar}$ *and* $\mathrm{L_1Set}$. Most of these axioms only contain variables of the language, and not "metalanguage" variables for formulas. This is true of all the axioms of arithmetic except for induction and all the axioms of set theory except for replacement. Each set of axioms not containing variable formulas is decidable because it can be described by a condition such as "the set of formulas of length 40 in which "(" is in the first place, "$\forall$" is in the second place, a variable is in the third place, "(" is in the fourth place, . . . , ")" is in the 39th place, and ")" is in the 40th place; in which the variables in the 3rd, 8th, and 16th places are the same, in the 9th and 36th places are the same, and in the 17th and 37th places are the same; and in which these three variables are pairwise distinct." (This is the axiom of regularity in L_1Set in normalized notation.) Here we could also write down just one copy of each such axiom and generate the rest using Gen, the axiom of specialization, and MP.

The axioms of induction and replacement are shown to be enumerable using the same procedure as in the case of the basis tautologies and the quantifier axioms. We leave the details to the reader.

6 The arithmetical hierarchy

6.1. Using recursion on n, we define the classes Σ_n and Π_n of subsets of $(\mathbf{Z}^+)^m$, $m = 0, 1, 2, \ldots,$ as follows:

(a) $\Sigma_0 = \Pi_0 = \{\text{decidable sets}\}$.
(b) $\Sigma_{n+1} = \{\text{projections of elements of } \Pi_n \text{ having codimension } \geqslant 1\}$.
(c) $\Pi_{n+1} = \{\text{complements of elements of } \Sigma_{n+1} \text{ in their ambient spaces } (\mathbf{Z}^+)^m\}$.

Obviously, Σ_1 consists of all enumerable sets (see Theorem 1.2 of Chapter VI), and Π_1 consists of their complements. The following result justifies calling $\{\Sigma_n, \Pi_n\}$ "the arithmetical hierarchy."

6.2. **Proposition.**

(a) $\forall n \geqslant 0,\ \Sigma_n \cup \Pi_n \subset \Sigma_{n+1} \cap \Pi_{n+1}$.

(b) $\bigcup_{n=0}^{\infty}\Sigma_n = \bigcup_{n=0}^{\infty}\Pi_n = \{$*arithmetical sets*$\}$, *i.e., all sets definable by formulas in* L_1Ar.

(c) *For* $n \geqslant 1$ *the sets in* Σ_n *are precisely those which can be defined by formulas of the following* $\mathfrak{L}_1$ *type (where the quantifiers are taken over variables in* $\mathbf{Z}^+$, *and* E *is a decidable set)*:

$$\exists x_1\, \forall x_2\, \exists x_3 \cdots \forall x_n\, \neg(\langle x_1, \ldots, x_n, x_{n+1}, \ldots, x_m\rangle \in E), \qquad n \text{ even};$$
$$\exists x_1\, \forall x_2\, \exists x_3 \cdots \exists x_n(\langle x_1, \ldots, x_n, x_{n+1}, \ldots, x_m\rangle \in E), \qquad n \text{ odd}.$$

Similarly, for Π_n:

$$\forall x_1\, \exists x_2\, \forall x_3 \cdots \exists x_n(\langle x_1, \ldots, x_n, x_{n+1}, \ldots, x_m\rangle \in E), \qquad n \text{ even};$$
$$\forall x_1\, \exists x_2\, \forall x_3 \cdots \forall x_n\, \neg(\langle x_1, \ldots, x_n, x_{n+1}, \ldots, x_m\rangle \in E), \qquad n \text{ odd}.$$

(d) *The sets in* Σ_n *or* Π_n *are definable by the analogous formulas in* L_1Ar *with the following changes: instead of* $\langle x_1, \ldots, x_m\rangle \in E$ *we have any atomic formula, and the number of quantifiers is* $\geqslant n$, *with exactly* $n-1$ *alternations from* $\exists$ *to* $\forall$ *or* $\forall$ *to* $\exists$.

PROOF.

(a) We use induction on n. For $n = 0$ we have $\Sigma_0 \cup \Pi_0 = \Sigma_1 \cap \Pi_1$, by the definition of decidable sets. If $\Sigma_{n-1} \subset \Sigma_n$, then $\Sigma_n \subset \Sigma_{n+1}$ (since Σ_{n+1} consists of projections of the complements of elements of Σ_n, and Σ_n

consists of projections of the complements of elements of Σ_{n-1}), and also $\Pi_n \subset \Pi_{n+1}$, by the definition of Π. Finally, we have $\Pi_n \subset \Sigma_{n+1}$, from which it trivially follows that $\Sigma_n \subset \Pi_{n+1}$. In fact, if $E \in \Pi_n$, then $E \times \mathbf{Z}^+ \in \Pi_n$ (since taking the product with $\mathbf{Z}^+$ commutes with complements and projections, and takes $\Sigma_0 = \Pi_0$ to itself), and hence $E =$ a projection of $E \times \mathbf{Z}^+ \in \Sigma_{n+1}$.

(b) It follows from (a) that $\bigcup_{n=0}^{\infty}\Sigma_n = \bigcup_{n=0}^{\infty}\Pi_n$. This class of sets is contained in the arithmetical sets, since all enumerable sets are arithmetical, and arithmeticality is preserved when taking projections and complements, which correspond to inserting $\exists$ and $\neg$, respectively.

In order to prove the converse {arithmetical sets} $\subset \bigcup_{n=0}^{\infty}\Sigma_n = \Sigma_\infty$, we first note that all sets definable by atomic formulas are decidable, and the rest of the arithmetical sets are obtained from them by taking projections, complements, unions, and intersections (see §2 of Chapter II). Thus, it suffices to show that Σ_∞ is closed with respect to (finite) unions and intersections. We claim that this is actually true for each Σ_n separately.

We prove this by induction on n. The result has already been proved for Σ_0. If Σ_n is closed with respect to $\cap$, then Π_n is closed with respect to $\cup$. Suppose $E_1, E_2 \in \Sigma_{n+1}$, $E_i =$ a projection of F_i, and $F_i \in \Pi_n$. We can then introduce dummy variables so as to identify the ambient spaces of the F_i and the projection of these spaces onto an ambient space for both the E_i. Then $E_1 \cup E_2 =$ a projection of $F_1 \cup F_2$, so that $E_1 \cup E_2 \in \Sigma_{n+1}$. Thus, Σ_{n+1} is closed with respect to $\cup$.

Similarly, if Σ_n is closed with respect to $\cup$, it follows that Π_n is closed with respect to $\cap$, and an analogous argument shows that Σ_{n+1} is closed with respect to $\cap$. However, here we must embed the products $F_1 \times (\mathbf{Z}^+)^{m_2}$ and $(\mathbf{Z}^+)^{m_1} \times F_2$ for certain m_1 and m_2 in a single space in such a way that, when we identify the two projections, we have $\mathrm{pr}(F_1 \times (\mathbf{Z}^+)^{m_2} \cap (\mathbf{Z}^+)^{m_1} \times F_2) = \mathrm{pr}\, F_1 \cap \mathrm{pr}\, F_2$. In terms of formulas this means that the variables bound by the $\exists$ quantifiers in the formulas corresponding to F_1 and F_2 must be renamed so that they form two disjoint sets.

(c) This assertion is proved by induction on n and a simple examination of the definitions. Here, whenever we take the complement, we must move the corresponding $\neg$ to the right of all the quantifiers by means of the usual commutation rule $\neg\forall = \exists\neg$, $\neg\exists = \forall\neg$. If we have a projection of codimension $m \geqslant 2$, which is defined by a series of quantifiers $\exists x_{i_1} \cdots \exists x_{i_m}$, we must reduce it to a projection of codimension 1 by replacing the set of variables $\langle x_{i_1}, \ldots, x_{i_m}\rangle$ by $\langle t_1^{(m)}(y), \ldots, t_m^{(m)}(y)\rangle$ in E and replacing the series of quantifiers by $\exists y$.

(d) The proof is analogous to that of (c). Here we use the fact that the sets in Σ_0 are Diophantine, and we observe that, in general, $\exists \cdots \exists$ can not be replaced by $\exists$ in this case.

The proposition is proved. □

6.3. Theorem. *For all $n \geqslant 1$*

$$\Sigma_n \setminus \Pi_n \neq \varnothing, \qquad \Pi_n \setminus \Sigma_n \neq \varnothing.$$

PROOF. The assertion that $\Sigma_1 \setminus \Pi_1 \neq \varnothing$ is precisely Theorem 5.8 of Chapter V on the existence of undecidable enumerable sets. We prove the general case by an analogous diagonal process applied to a versal family.

Let $\{E_k\}$ be a versal family of enumerable $(n+1)$-sets over $\mathbf{Z}^+$, and let E be its total space:

$$\langle k, x_0, \ldots, x_n \rangle \in E \Leftrightarrow \langle x_0, \ldots, x_n \rangle \in E_k.$$

To fix ideas, suppose n is even. We set

$$F = \{k \mid \exists x_1 \, \forall x_2 \cdots \forall x_n \, \neg(\langle k, k, x_1, \ldots, x_n \rangle \in E)\} \subset \mathbf{Z}^+.$$

By 6.2(c), we have $F \in \Sigma_n$. Since $\{E_k\}$ is versal, it follows by 6.2(c) that any subset of $\mathbf{Z}^+$ in Π_n can be represented in the form

$$F_{k_0} = \{x_0 \mid \neg(\exists x_1 \, \forall x_2 \cdots \forall x_n \, \neg(\langle k_0, x_0, x_1, \ldots, x_n \rangle \in E))\}$$

for some $k_0 \in \mathbf{Z}^+$. It is clear that k_0 lies either in $F \setminus F_{k_0}$ or in $F_{k_0} \setminus F$. Hence $F \neq F_{k_0}$, and $F \in \Sigma_n \setminus \Pi_n$.

The other cases are handled analogously. □

6.4. *Remarks*

(a) From the point of view of the theorems of Tarski and Gödel, the results in 6.2 and 6.3 show us the tremendous distance from provability to truth: $D \in \Sigma_1$, while T falls not only outside Σ_1, but even outside Σ_∞. In the next section we indicate some mileposts along the way from D to T.

(b) Although not really formally justified by the above considerations, nevertheless it makes sense to classify arithmetic problems, i.e., questions "Is it true that $P \in T$?," according to the number of alternations between $\exists$ and $\forall$ when the closed formula P is written as in 6.2(c).

As we showed in §1 of Chapter I, the Fermat conjecture is expressed by a Π_1-formula, and the Riemann hypothesis is expressed by a Π_3-formula, although there is an assertion of type Π_1 which is equivalent to the RH.

H. Rogers writes that,

> "Almost all statements which (i) have been extensively studied by mathematicians and (ii) are known to be arithmetically expressible can be seen, from a relatively superficial examination, to have quite low level in the Σ_n classification. As has been occasionally remarked, the human mind seems limited in its ability to understand and visualize beyond four or five alternations of quantifier. Indeed, it can be argued that the inventions, subtheories, and central lemmas of various parts of mathematics are devices for assisting the mind in dealing with one or two additional alternations of quantifier."

7 Productivity of arithmetical truth

7.1. In this section we discuss a final feature of Gödel's theorem: the possibility, starting from any enumerable set of truths of arithmetic which we already know, effectively to enlarge this set by adding new truths. To see this more clearly, we examine the original version of the proof, in which the diagonal method is explicit, rather than hidden in the construction of an undecidable enumerable set. It is convenient to describe this version by comparing it with the proof of Tarski's theorem.

7.2. Suppose we are given a language of arithmetic (L_1Ar, SAr, or an extension of one of them). Further suppose that we have chosen a fixed numbering of its alphabet, which determines a fixed numbering N of the formulas. (It is essential to note that the construction which follows is not invariant if we replace our numbering by an equivalent one.)

Both the Tarski and the Gödel arguments are based on the following "self-reference lemma:"

7.3. **Lemma.** *Given any formula $P(x)$ in the language which has one free variable, we can effectively construct a closed formula Q_P which says, "my number does not belong to the set defined by P." In other words, Q_P is true if and only if $P(\bar{N}(Q_P))$ is false, where $\bar{N}(Q_P)$ is the term-name for $N(Q_P)$.*

PROOF. This lemma was proved for SAr in §11 of Chapter II. In L_1Ar we construct the formula Q_P as follows.

If $R(x)$ is a formula with one free variable, we call the formula $R(\bar{N}(R(x)))$ its *diagonalization*. Let diag : $\mathbf{Z}^+ \to \mathbf{Z}^+$ be the partial function

the N-number of a formula with one free variable

$\mapsto$ the N-number of its diagonalization.

It is easy to show, using the results and methods in §4, that diag is computable. Thus, its graph is definable by a formula in L_1Ar which can be explicitly constructed. We denote this formula by "$y = \text{diag}\, x$," construct the formula $R(x) : \exists y(\text{"}y = \text{diag}\, x\text{"} \wedge P(y))$, and finally set:

$$Q_P : \quad \neg R(\bar{N}(\neg R(x))) = \text{the diagonalization of } \neg R(x).$$

By the definitions, we then have:

$$\begin{aligned} Q_P \text{ is true} &\Leftrightarrow \text{the number of } \neg R(x) \text{ does not satisfy } R(x) \\ &\Leftrightarrow \text{the number of the diagonalization of } \neg R(x) \\ &\qquad \text{does not satisfy } P(x) \\ &\Leftrightarrow \text{the number of } Q_P \text{ does not satisfy } P(x). \end{aligned}$$

The lemma is proved. □

We note that it requires a large amount of technical work to verify that "$y = \text{diag}\, x$" is definable in L_1Ar, which is why we used SAr instead in Chapter II.

7.4. The arguments of Tarski and Gödel now take the following parallel form:
Tarski:

(a) Suppose that truth is definable by a formula P.
(b) Then there is a formula Q_P which says "I am not true."
(c) The formula Q_P cannot be false (because of its semantics).
(d) The formula Q_P cannot be true (because of its semantics).
(e) Therefore, truth is not definable.

Gödel:

(a) Provability is definable by a formula P.
(b) There is a formula Q_P which says "I am not provable."
(c) The formula Q_P cannot be false (because of its semantics, since otherwise it would be provable, and hence true).
(d) Therefore, Q_P is true.
(e) Therefore, Q_P is not provable (because of its semantics).

We note that, in the above paraphrasing of Gödel's argument, part (c) explicitly uses the stipulation that only true formulas are provable. When Gödel's paper appeared in 1931, specialists were very busy looking for finitistic proofs that the axioms of arithmetic are consistent, so that stipulating that $D \subset T$ would have run counter to the spirit of the times. Therefore, in Gödel's own original wording the argument looks somewhat different. This distinction is traditionally explained in great detail in all textbooks on logic. However, we shall be satisfied with remarking that, if $D \not\subset T$, then $D \neq T$, and the Incompleteness theorem is trivially true. But in that case we would be in such bad shape that we would no longer care about completeness or incompleteness.

The main point we are interested in is the following: given any fixed conception of provability which leads to an enumerable (or even to an arithmetical) set D of provable true formulas, we can effectively construct a new formula which is true but not provable. We now define more precisely what we mean by "effectively."

7.5. **Definition.** A set $F \subset \mathbf{Z}^+$ is said to be *productive* relative to a versal family $\{E_k\}$ of 1-sets, if there exists a partial recursive function f such that: for all $k \in \mathbf{Z}^+$ with $E_k \subset F$, we have $k \in D(f)$ and $f(k) \in F \setminus E_k$.

7.6. **Proposition.** *Under the conditions in* 7.2, *the set of numbers of true formulas is productive relative to the versal family* $\{E_k\}$ *constructed in* §8 *of Chapter* VI.

PROOF. To fix ideas, we shall work with the language L_1Ar. We first construct an enumerable family $\{P_k(x_1)\}$ of formulas with one free variable x_1 such that P_k defines E_k. To do this, we define a sequence of terms $\bar{f}[k]$ in L_1Ar as in 8.1(a) of §8, Chapter VI, by setting

$$\bar{f}[4k] = \bar{k} = +\big(\cdots +(\bar{1}, \bar{1})\cdots\big), \qquad k \text{ times};$$

$$\bar{f}[4k+1] = x_{k+1} = \text{the } (k+1)\text{st variable in } L_1Ar;$$

$$\bar{f}[4k+2] = +\big(\bar{f}[t_1(k)], \bar{f}[t_2(k)]\big);$$

$$\bar{f}[4k+3] = \cdot\big(\bar{f}[t_1(k)], \bar{f}[t_2(k)]\big).$$

We then write

$$P_k = \exists x_2\Big(\exists x_3 \cdots \Big(\exists x_k\big(\bar{f}[t_1(k)] = \bar{f}[t_2(k)]\big)\Big)\cdots\Big).$$

It is easy to see, using the methods in §4, that the function $k \mapsto N(P_k)$ is recursive. We next fix a translation of "$y = \text{diag } x$" and set

$$R_k = \exists x_{k+1}\big((\text{"}x_{k+1} = \text{diag } x_1\text{"}) \wedge (P_k(x_1))\big),$$

$$Q_{P_k} = \neg\Big(R_k\big(\bar{N}(\neg(R_k))\big)\Big),$$

and finally

$$f(k) = N(Q_{P_k}).$$

This function is computable because $N(P_k)$ is computable. By Lemma 7.3, it satisfies the condition 7.5 with T in place of F. □

7.7. The concept of productivity gives us the following approach to the problem of exhausting T: we begin with the set D_0 of formulas which are provable in the Peano axiom system Ax_0, we define D_0 by a formula P_0; we set $Ax_1 = Ax_0 \cup \{Q_{P_0}\}$; and we similarly construct D_1, P_1, and $Ax_2 = Ax_1 \cup \{Q_{P_1}\}$, and so on. It follows from Gödel's theorem that, as long as we do all this "uniformly effectively," we cannot obtain all of T even after transfinitely many steps. However, S. Feferman has shown that, if we are willing to dispense with effectiveness, we can obtain all of TAr in this way. We conclude this section by formulating Feferman's result, which gives unexpected and philosophically interesting information about TAr. We omit the proof and the technical details (see Feferman's original article "Transfinite recursive progressions of axiomatic theories," *J. Symb. Logic* 27, No. 3 (1962), 259–316).

7.8. *Principles of extension*. In the first place, in order to exhaust TAr it is not enough to add Gödel's formula to Ax_i at every step. There are many other ways of constructing intuitively true formulas which in various ways formalize "having faith in the axioms Ax_i."

Feferman, in particular, uses the following construction. Suppose that we have already constructed the axiom system Ax_α (where α is an ordinal), and that the set of numbers of formulas deducible from Ax_α is defined by the formula D_α. For any formula $P(x)$ with one free variable, we construct a formula B_α^P which has the intuitive meaning: "if $P(\bar{n})$ is provable (from Ax_α) for all term-names $\bar{n}$ of natural numbers, then $\forall x\, P(x)$ is true." These formulas B_α^P must lie in T, and we can set

$$\mathrm{Ax}_{\alpha+1} = \mathrm{Ax}_\alpha \cup \{B_\alpha^P | \text{all } P\};$$

$$\mathrm{Ax}_\beta = \bigcup_{\alpha<\beta} \mathrm{Ax}_\alpha, \quad \text{if } \beta \text{ is a limit ordinal.}$$

Here is a method for giving B_α^P explicitly. The function $n \mapsto N(P(\bar{n}))$ is computable as a function of n and $N(P)$. We define its graph by a formula $M(x, y, z)$, so that, for $l, m, n \in \mathbf{Z}^+$

$$M(\bar{l}, \bar{m}, \bar{n}) \text{ is true} \Leftrightarrow \begin{cases} l \text{ is the number of a formula } P \text{ with one free} \\ \text{variable } x, \text{ and } m \text{ is the number of } P(\bar{n}). \end{cases}$$

We then set:

$$B_\alpha^P = \forall y\, \forall z \big(M(\bar{N}(P), y, z) \Rightarrow D_\alpha(y)\big) \Rightarrow \forall x\, P(x).$$

7.9. *The problem of choosing D_α.* This is the subtlest part of the proof. Here it is crucial to show that D_β exists when β is a limit ordinal.

Feferman shows how the D_α can be constructed for a suitable countable sequence of ordinals with limit γ not exceeding $\omega_0^{\omega_0^{\omega_0}}$ so that the following result will be true.

7.10. **Theorem.** *All true formulas in* $\mathrm{L_1Ar}$ *are deducible from* $\bigcup_{\alpha<\gamma} \mathrm{Ax}_\alpha$.

Thus, suppose we have accepted the Peano axioms. Then, in order to attain the total truth in arithmetic, we must perform a transfinite sequence of acts of faith in our not having been led astray by the previous acts of faith.

8 On the length of proofs

8.1. The title of this section is taken from a short paper written by Gödel in 1936. His article consists of a precise formulation and proof of the following qualitative assertions.

Suppose we are given a formal language L together with some conception of deducibility of a formula P from a (variable) set of formulas $\mathcal{A}$. Suppose, in addition, that we are actually given a function which estimates the "complexity of deduction" of a formula P from the set $\mathcal{A}$. (In languages of $\mathfrak{L}_1$, this "complexity" could be the *minimal size of a deduction of P from $\mathcal{A}$*, i.e., the number of signs of a fixed finite protoalphabet needed for such a deduction; note that the use of the word "complexity" here has nothing to do with the Kolmogorov complexity in §9 of Chapter VI.) We further assume that L contains a certain fragment of the logic of $\mathfrak{L}_1$, that L and $\mathfrak{A}$ are rich enough for the incompleteness principles to take effect, and that the "complexity of deduction" satisfies certain natural axioms. We then have the following facts:

(a) *There exist formulas deducible from $\mathfrak{A}$ whose deduction is arbitrarily more complex than the formula itself.*

Observation shows that this somewhat vaguely defined class includes, if not the most important, at least the most "prized" mathematical facts.

(b) *If we add any independent formula A to the axioms $\mathfrak{A}$, then we can find formulas deducible from $\mathfrak{A}$ whose deduction from $\mathfrak{A} \cup \{A\}$ is arbitrarily less complex than from $\mathfrak{A}$ (the principle of cutting down proofs).*

Compare with the great strength of "analytic" methods in comparison with "elementary" methods in number theory.

The following more precise presentation of these ideas is based on a short article by Ehrenfeucht and Mycielski in *Bull. Amer. Math. Soc.* 17, No. 3 (1971), 366–367.

8.2. We consider the following set of data.

(a) A *countable alphabet A* with a fixed numbering $N : A \to \mathbf{Z}^+$.
(b) A subset $F \subset S(A)$ whose elements are called *formulas*.
(c) A partial function $\mathfrak{D} : \mathfrak{P}(F) \to \mathfrak{P}(F)$ which to certain subsets $\mathfrak{A} \subset F$ corresponds sets $\mathfrak{D}(\mathfrak{A})$ of formulas "*deducible from $\mathfrak{A}$*." We shall often write $\mathfrak{A} \vdash P$ instead of $P \in \mathfrak{D}(\mathfrak{A})$.
(d) The *complexity of deduction*: this is a function $\mathrm{Cd}_{\mathfrak{A}}(P)$ which is defined for pairs $\mathfrak{A} \subset F$, $P \in \mathfrak{D}(\mathfrak{A})$, and takes values in $\mathbf{Z}^+$. It is convenient to take $\mathrm{Cd}_{\mathfrak{A}}(P) = \infty$ if $P \notin \mathfrak{D}(\mathfrak{A})$.

We impose the following conditions on this data:

8.3. (a) *A contains* $\neg$, $\to$, (, *and*).

(b) *If P and $Q \in F$, then $\neg(P)$ and $(P) \to (Q) \in F$.* As usual, we shall write $P \to Q$ instead of $(P) \to (Q)$, and so on.

(c_1) $\mathfrak{A} \subset \mathfrak{D}(\mathfrak{A})$; *if $\mathfrak{A} \subset \mathfrak{A}'$ and $\mathfrak{D}$ is defined at $\mathfrak{A}$, then $\mathfrak{D}$ is defined at $\mathfrak{A}'$ and $\mathfrak{D}(\mathfrak{A}) \subset \mathfrak{D}(\mathfrak{A}')$.*

(c_2) *If $\mathfrak{A} \cup \{P\} \vdash Q$, then $\mathfrak{A} \vdash P \to Q$.*

(c_3) $\mathfrak{A} \vdash P \to (\neg P \to Q)$ for any $P, Q \in F$.

(d_0) *If $\mathfrak{A} \subset \mathfrak{A}'$, then* $\mathrm{Cd}_{\mathfrak{A}'}(P) \leqslant \mathrm{Cd}_{\mathfrak{A}}(P)$.

(d_1) *The set* $\{\langle P, n \rangle | \mathrm{Cd}_{\mathfrak{A}}(P) \leqslant n\} \subset S(A) \times \mathbf{Z}^+$ *is decidable.*

Condition (d_1) does not actually have to hold for all $\mathfrak{A} \subset F$, but we shall only consider those $\mathfrak{A}$ for which it is true. In the case when $\mathrm{Cd}_{\mathfrak{A}}(P)$ is the size of the shortest $\mathfrak{L}_1$-deduction of P from $\mathfrak{A}$ in a finite protoalphabet, and $\mathfrak{A}$ is a decidable set of axioms, (d_1) holds for the following reason. We can write down all the texts in A having size $\leqslant n$—there are a finite number of them—and then verify for each one in turn whether or not it is a deduction of P from $\mathfrak{A}$.

(d$_2$) *There exists a general recursive function $f(x, y, z)$ which is nondecreasing in x, such that*

$$\mathrm{Cd}_{\mathfrak{A}\cup\{P\}}(Q) \leqslant f(\mathrm{Cd}_{\mathfrak{A}}(P\to Q), N(P), N(Q))$$

for all $Q \in \mathfrak{D}(\mathfrak{A})$.

Both sides of this inequality are finite because of the previous conditions: since $\mathfrak{A}\vdash Q$, it follows by (c$_1$) that $\mathfrak{A}\cup\{P\}\vdash Q$, and then by (c$_2$) that $\mathfrak{A}\vdash P\to Q$. We have an estimate of the type in (d$_2$) in languages of $\mathfrak{L}_1$, because, starting with any deduction of $P\to Q$ from $\mathfrak{A}$, we can obtain a deduction of Q from $\mathfrak{A}\cup\{P\}$ by simply adding P and Q (by modus ponens). This increases the size of the deduction of $P\to Q$ by the sizes of P and Q.

(d$_3$) *There exists a general recursive function $g(x, y)$ such that*

$$\mathrm{Cd}_{\mathfrak{A}}(P\to(\neg P\to Q)) \leqslant g(N(P), N(Q)).$$

In languages of $\mathfrak{L}_1$, the formula $P\to(\neg P\to Q)$ is a logical axiom, and, if $\mathfrak{A}$ contains this axiom, then the deduction has length 1 and size equal to the size of the formula itself. Of course, the size of this formula can be represented in the form $g(N(P), N(Q))$.

We now formulate Gödel's theorem on "cutting down proofs." We suppose that the conditions and conventions in 8.2–8.3 are fulfilled.

8.4. Theorem.

(a) *Suppose that $\mathfrak{A}\subset F$ and $\mathfrak{D}(\mathfrak{A})$ is undecidable. Then for any general recursive function l there exist infinitely many formulas $P\in\mathfrak{D}(\mathfrak{A})$ such that*

$$\mathrm{Cd}_{\mathfrak{A}}(P) > l(N(P)).$$

(b) *Suppose that $\mathfrak{A}' = \mathfrak{A}\cup\{A\}$ and the formula A has the property that $\mathfrak{D}(\mathfrak{A}\cup\{\neg A\})$ is undecidable. Then for any general recursive function r there exist infinitely many formulas $P\in\mathfrak{D}(\mathfrak{A})$ such that*

$$\mathrm{Cd}_{\mathfrak{A}}(P) > r(\mathrm{Cd}_{\mathfrak{A}'}(P)).$$

PROOF.

(a) If the first assertion were false, then for a suitable l and for all $P\in\mathfrak{D}(\mathfrak{A})$ we would have $\mathrm{Cd}_{\mathfrak{A}}(P)\leqslant l(N(P))$. But then the set

$$\mathfrak{D}(\mathfrak{A}) = \{P|\mathrm{Cd}_{\mathfrak{A}}(P)\leqslant l(N(P))\}\subset S(A)$$

would be decidable by (d$_1$), since it is obtained by applying a bounded universal quantifier (in n) to the decidable set in (d$_1$). This contradicts the assumption.

(b) Let $P \in \mathfrak{D}(\mathfrak{A} \cup \{\neg A\})$. By ($d_2$) we have

$$\mathrm{Cd}_{\mathfrak{A} \cup \{\neg A\}}(P) \leqslant f(\mathrm{Cd}_{\mathfrak{A}}(\neg A \to P), N(\neg A), N(P)).$$

If we now suppose that the second assertion of the theorem were false, then for a suitable nondecreasing general recursive function r we would obtain:

$$\mathrm{Cd}_{\mathfrak{A}}(\neg A \to P) \leqslant r(\mathrm{Cd}_{\mathfrak{A}'}(\neg A \to P)),$$

or, by (d_2) and (d_3):

$$\begin{aligned}\mathrm{Cd}_{\mathfrak{A}}(\neg A \to P) &\leqslant r \circ f(\mathrm{Cd}_{\mathfrak{A}}(A \to (\neg A \to P)), N(A), N(P)) \\ &\leqslant r \circ f(g(N(A), N(P)), N(A), N(P)).\end{aligned}$$

Substituting this in the above inequality for $\mathrm{Cd}_{\mathfrak{A} \cup \{\neg A\}}(P)$, for fixed A we obtain an estimate of the form

$$\mathrm{Cd}_{\mathfrak{A} \cup \{\neg A\}}(P) \leqslant l(N(P)),$$

where l is general recursive and $P \in \mathfrak{D}(\mathfrak{A} \cup \{\neg A\})$. But this contradicts the assumption that $\mathfrak{D}(\mathfrak{A} \cup \{\neg A\})$ is undecidable by the first assertion of the theorem. □

CHAPTER VIII

Recursive groups

1 Basic result and its corollaries

1.1. We consider a countable "group alphabet"

$$A = \{a_1, a_2, \ldots ; a_1^{-1}, a_2^{-1}, \ldots\}.$$

The expressions in the alphabet A, including the empty expression $\varnothing$, are traditionally called *words*. The word $a_i \cdots a_i$ ($m \geqslant 1$ times) will be written a_i^m; the word $a_i^{-1} \cdots a_i^{-1}$ ($m \geqslant 1$ times) will be written a_i^{-m}; and we agree to take $a_i^0 = \varnothing$. We call a word $a_{i_1}^{m_1} \cdots a_{i_r}^{m_r}$ *reduced* if either it is empty or there are no subwords of the form $a_i^{-1}a_i$ or $a_ia_i^{-1}$ when it is written in expanded form.

The operation of "joining and reducing" (by "reducing" we mean crossing out all subwords of the form $a_ia_i^{-1}$ or $a_i^{-1}a_i$) defines a group structure with unit $\varnothing$ (which we sometimes denote by 1) on the set of reduced words. This is a free group F with a countable set of generators $\{a_1, \ldots, a_n, \ldots\}$. We can also consider nonreduced words as elements in F: we identify such a word with the word obtained by reducing it.

We have a canonical numbering on A : $N(a_i) = 2i$, $N(a_i^{-1}) = 2i - 1$. All properties related to the computability of operations and the enumerability of subsets in A and $S(A)$ will be considered relative to any numbering of A equivalent to N and any numbering of $S(A)$ compatible with N (see the definitions in §1 of Chapter VII). We shall continually be making use of the following facts.

1.2. **Lemma.**

(a) *The set F of reduced words is decidable.*

(b) *The group operations in F are computable.*

(c) *A subgroup $G \subset F$ in enumerable in $S(A)$ if and only if it has an enumerable set of generators.*

(d) *A normal subgroup $H \subset G$ in an enumerable subgroup $G \subset F$ is enumerable if and only if it is generated as a normal subgroup by an enumerable set.*

(e) *A homomorphism $F \to F$ is recursive if and only if the induced map $\{a_1, \ldots, a_n, \ldots\} \to F$ is recursive.*

The proof is a good exercise in using the techniques of Chapter VII, and we leave it to the reader. It is convenient to begin by showing that the operation of reducing is computable; the rest goes through more or less automatically.

1.3. **Definition.** A group is called *recursive* if it is isomorphic to a quotient group of the form G/H, where $G \subset F$ is an enumerable subgroup and $H \subset G$ is an enumerable normal subgroup.

Here we could limit ourselves to subgroups $G \subset F$ which are generated by an enumerable subset of the standard generators $\{a_1, \ldots, a_n, \ldots\}$.

1.4. REMARKS AND EXAMPLES.

(a) Recursive groups have at most countably many elements.
(b) *Finitely presented* (f.p.) groups, i.e., those which have a finite number of generators and relations, are recursive. In particular, finite groups and *finitely generated* (f.g.) abelian groups are recursive.
(c) A subgroup H of an f.p. group G is not necessarily f.p. (or even f.g.). But, *if it is finitely generated, then it is recursive.*

In fact, let $\{h_1, \ldots, h_m\}$ be generators of H. We add generators $\{h_{m+1}, \ldots, h_n\}$ of the group G which are connected by a finite number of relations, and we define a homomorphism $\phi : F \to G$ by setting $\phi(a_i) = h_i$ if $i \leqslant n$ and $\phi(a_i) = 1$ if $i > n$. The kernel E of ϕ is generated by a finite number of relations between the $a_1, \ldots, a_n$ and by the set $\{a_{n+1}, a_{n+2}, \ldots\}$. Hence E is enumerable by Lemma 1.2(d). The subgroup $\overline{H} \subset F$ generated by $a_1, \ldots, a_m$ is also enumerable, by Lemma 1.2(c). Therefore the set $\overline{H} \cap E$ is enumerable. But ϕ induces an isomorphism $\overline{H}/\overline{H} \cap E \rightrightarrows H$. Consequently, H is recursive. □

The basic aim of this chapter is to prove the following remarkable theorem of Higman, which gives the converse of the simple assertion 1.4(c). (G. Higman, Subgroups of finitely presented groups, *Proc. Royal Soc.*, Ser. A, vol. 262 (1961), 455–475.)

1.5. **Theorem.**

(a) *Any recursive group* G/H *(in the notation of* 1.3) *can be embedded in a suitable* f.p. *group* F/N.

(b) *This embedding can be made effective, i.e., it can be induced by a suitable recursive map* $G \to F$.

Here are some applications of this theorem.

1.6. **Corollary** (Universal finitely presented groups). *There exists an* f.p. *group* U *such that any* f.p. *group* G *can be embedded in* U *(and hence, any recursive group can be embedded in* U*).*

In fact, any f.p. group is isomorphic to the quotient of F by a normal subgroup which is generated by a finite set of reduced words in F and by all a_i with $i \geqslant n$ for some n. We let $I \subset S(S(A)) \times \mathbf{Z}^+$ be the decidable set of pairs $\langle$a finite sequence of reduced words, $n\rangle$, and we let N_i (for $i \in I$) denote the corresponding normal subgroup. We construct the "doubly infinite" group alphabet $\{a_{jk}, a_{jk}^{-1} | j, k \geqslant 1\}$, we identify I with $\mathbf{Z}^+$ by choosing a recursive numbering of I, and we define the group U_0 which has generators $\{a_{jk}\}$ and relations "N_j, written in the alphabet $\{a_{j1}, a_{j2}, \ldots\}$." It is clear that U_0 is recursive. It will also be clear from the results in the next section that U_0 is the free product of all the groups F/N_j, so that any f.p. group can be embedded in U_0. Thus, any f.p. group U in which we can embed U_0, using Higman's theorem, is universal. □

In M. K. Valijev's article Examples of universal finitely presented groups, *Dok. AN SSSR*, 1973, vol. 211, No. 2, a universal group U is constructed which has 14 generators and 42 relations, and it is mentioned that such a group can be constructed with only 2 generators and 27 relations.

1.7. F.p. *groups with algorithmically undecidable word problem.*

Let G be the group with four generators a, b, c, d, and with the relations

$$b^{-m}ab^m = d^{-m}cd^m, \quad \text{for all } m \in E,$$

where $E \subset \mathbf{Z}^+$ is an undecidable enumerable set. It easily follows from the results in §2 that the equation

$$b^{-x}ab^x = d^{-x}cd^x$$

only holds in G if $x \in E$. (In fact, the elements $b^{-m}ab^m$ for $m \geqslant 1$ generate a free subgroup of G, so that G contains the free product of the subgroups generated by $\{b^{-x}ab^x | x \geqslant 1\}$ and by $\{d^{-x}cd^x | x \geqslant 1\}$ with amalgamation $\{b^{-x}ab^x = d^{-x}cd^x | x \in E\}$.) Hence, the question of whether or not the equation $b^{-x}ab^{x} = d^{-x}cd^{x}$ holds is undecidable (as a mass problem indexed by x), and, if we embed G effectively in a f.p. group, we may conclude that the word problem is unsolvable in this f.p. group.

The existence of such groups was first established by P. S. Novikov and W. Boone.

1.8. *"Natural" recursive groups*. We find many examples of recursive groups which are not *a priori* finitely presented in algebraic geometry over algebraic number fields. We shall limit ourselves to one typical example.

Let $\mathcal{O}_n(\mathbf{Q})$ be the orthogonal group of automorphisms of an n-dimensional linear space L (over the rational numbers $\mathbf{Q}$) together with a quadratic form f. Let b be the corresponding bilinear form. The symmetry $\tau_x \in \mathcal{O}_n(\mathbf{Q})$ is defined for any vector $x \in L$ with $f(x) \neq 0$:

$$\tau_x(y) = y - \frac{b(x, y)}{f(x)}\, x$$

for all $y \in L$. The involutions $\tau_x \in \mathcal{O}_n(\mathbf{Q})$ give us an enumerable system of generators of $\mathcal{O}_n(\mathbf{Q})$, and all the relations are generated by the enumerable (indeed, decidable) system of relations

$$\tau_x^2 = 1, \qquad (\tau_x\tau_y\tau_z)^2 = 1, \quad \text{for all coplanar } \{x, y, z\}$$

(S. Becken).

The numbering of $L \cong \mathbf{Q}^n$ implicit here is taken to be compatible with any numbering of $\mathbf{Q}$ which is compatible with the standard numbering of $\mathbf{Z}^+$ and in which the field operations are computable.

1.9. Higman's theorem is related to the theorem that enumerable sets are Diophantine (Chapter VI), although it was first proved earlier than the latter result. Perhaps both facts are special cases of some general assertion about recursive algebraic structures.

In any case, the theorem on the Diophantine nature of enumerable sets can be used to simplify considerably the recursion-theoretic part of Higman's proof. This was shown by Valijev, whose construction will be given in §§5–6 (cf. *Algebra i Logika*, vol. 7, No. 3 (1968)). §§2–4 will be devoted to the group theoretic preliminaries; here we shall follow Higman.

2 Free products and HNN-extensions

2.1 Suppose we are given a family of groups (G_i), $i \in I$, and a family of group homomorphisms $\alpha_i : A \to G_i$. We consider the class of families (H, β_i) of homomorphisms $\beta_i : G_i \to H$ such that $\beta_i \circ \alpha_i : A \to H$ does not depend on $i \in I$. This class contains a *universal family* $\phi_i : G_i \to *_A G_k$ which is unique up to isomorphism: any other family (H, β_i) uniquely determines and is uniquely determined by the homomorphism $\gamma : *_A G_k \to H$ for which $\beta_i = \gamma \circ \phi_i$.

In what follows we shall only need the case when all the α_i are embeddings. In this case $*_A G_k$ is called the *free product of the groups* G_i *with amalgamated subgroups* $\alpha_i(A) \subset G_i$. We shall generally denote the structure maps $G_i \to *_A G_k$ by ϕ_i, perhaps with additional indices. We let ϕ denote the structure homomorphism $\phi_i \circ \alpha_i : A \to *_A G_k$, which does not depend on i. If $A = \{1\}$, we write simply $* G_i$ instead of $*_A G_i$; if the set of indices is $\{1, \ldots, n\}$, we write $G_1 * \cdots * G_n$, and so on. We shall continually be making use of the following structure lemma.

Let $\alpha_i : A \to G_i$ be embeddings, and let $S_i \subset G_i$ be subsets such that

$$G_i \setminus \alpha_i(A) = \bigcup_{s \in S_i} \alpha_i(A)s, \quad \text{and}$$

$$\alpha_i(A)s_1 \neq \alpha_i(A)s_2, \quad \text{for } s_1 \neq s_2 \in S_i.$$

2.2. **Proposition.** *Any element in the group $*_A G_i$ can be uniquely represented in the form*

$$\phi(a)\phi_{i_1}(s_1) \cdots \phi_{i_n}(s_n),$$

where $a \in A$, $s_k \in S_{i_k}$, $i_j \neq i_{j+1}$ for all j, and $n \geqslant 0$ depends on the element.

We shall call this the canonical expansion of an element.

For the proof of this fact and for further details, see, for example, Serre's lecture notes *Arbres, amalgames et SL_2*.

2.3. **Corollaries**

(a) *Under the conditions in 2.2, the structure homomorphisms ϕ and ϕ_i are embeddings.*

This allows us to identify A and G_i with subgroups of $*_A G_i$ using ϕ and ϕ_i. We shall do this in the statements that follow. However, in the several-step constructions in the later subsections, one and the same group will be embedded in another group in many different ways using various compositions of the structure maps, and it will be necessary to keep careful track of these embeddings.

(b) $G_i \cap G_j = A$ *(in $*_A G_i$) for $i \neq j$.*

In other words, $\phi_i(G_i) \cap \phi_j(G_j) = \phi(A)$. We can use Proposition 2.2 to prove $\subset$: otherwise we would have $\phi_i(s_i) = \phi_j(s_j)$, which would contradict the uniqueness.

(c) *Suppose we are given a family of embeddings $\beta_i : H_i \to G_i$ and a subgroup $B \subset A$ such that $\beta_i(H_i) \cap \alpha_i(A) = \alpha_i(B)$ for all i. Then the composition*

$$B \overset{\beta_i^{-1} \circ \alpha_i}{\Longrightarrow} H_i \overset{\phi_i \circ \beta_i}{\Longrightarrow} *_A G_i$$

does not depend on i, and therefore gives a canonical map $*_B H_i \to *_A G_i$. *This map is an embedding. In particular, the subgroup of* $*_A G_i$ *generated by* $\phi_i \circ \beta_i(H_i)$ *is isomorphic to* $*_B H_i$.

In fact, the canonical expansion in 2.2 of an element in $*_B H_i$ goes to the canonical expansion of the image of this element in $*_A G_i$.

(d) *With the same notation, we have*

$$\left(*_B H_i\right) \cap A = B \textit{ in } *_A G_i;$$

$$\left(*_B H_i\right) \cap G_j = H_j \textit{ in } *_A G_i.$$

2.4. *Generators and relations.* Let M be a set, and let R be a subset of the free group F_M which is freely generated by M. We let $|M : R|$ denote the quotient group $F_M/\bar{R}$, where $\bar{R}$ is the smallest normal subgroup of F_M containing R. This is what we mean by defining a group by generators (M) and relations (R).

We shall take the following liberties with notation:

(a) If M has a nonempty intersection with a group that has already been defined, then all relations coming from the relations in the earlier group are assumed to be included in R, even if they are not explicitly written out. We might completely omit any reference to R if there are no other relations besides those coming from the earlier group. For example, if E and $F \subset G$ are two subgroups, then $|E \cup F|$ is the subgroup they generate in G, and so on.

(b) Instead of writing, say, $a_1 a_2^{-1}$ is in R, we may write $a_1 = a_2$.

EXAMPLE. If the $\alpha_i : A \to G_i$ are embeddings, then $*_A G_i$ is defined by the following generators and relations:

$$\left| \bigcup_{i \in I} G_i : \alpha_i(a) = \alpha_j(a) \text{ for all } a \in A,\ i, j \in I \right|.$$

We now introduce a construction which will be fundamental for everything that follows (G. Higman, B. Neumann, H. Neumann).

Suppose we are given two embeddings of groups $\alpha, \beta : A \to G$.

2.5. **Definition.** The HNN-extension of the group G (relative to A, α, β) is the group

$$K = |G \cup \{t\} : t^{-1}\alpha(a)t = \beta(a) \text{ for all } a \in A|.$$

2.6. **Proposition.** *The following homomorphisms are embeddings*:

(a) $G \to K : g \mapsto$ *the class of g modulo the relations in K.*

(b) $G *_A t^{-1}Gt \to K$, *where the free product is taken relative to the embeddings* $a \mapsto \beta(a)$ *and* $a \mapsto t^{-1}\alpha(a)t$.

Proof. In the group $G * \{u^n\}$, the subgroup U generated by G and $u^{-1}\alpha(A)u$ is isomorphic to $G * u^{-1}\alpha(A)u$. In fact, the canonical expansion of an element in $G * u^{-1}\alpha(A)u$ has the form $g_1u^{-1}\alpha(a_1)ug_2 \cdots g_nu^{-1}\alpha(a_n)u$, where $g_1 \in G, g_2, \ldots, g_n \in G \setminus \{1\}, a_1, \ldots, a_{n-1} \in A \setminus \{1\}, a_n \in A$, and so this expansion also has the canonical form in $G * \{u^n\}$.

We construct the subgroup $V = G * v\beta(A)v^{-1} \subset G * \{v^n\}$ similarly.

We identify the group $W = G * w^{-1}Aw$ with U and V by means of the isomorphisms which are the identity on G and take $w^{-1}aw$ to $u^{-1}\alpha(a)u$ and $v\beta(a)v^{-1}$, respectively.

We now consider the group $(G * \{u^n\}) *_W (G * \{v^n\})$. The group $G \subset W$ is canonically embedded in it, and for all $a \in A$ the element $t = uv$ satisfies the relation

$$t^{-1}\alpha(a)t = \beta(a),$$

because we have made the identification $u^{-1}\alpha(a)u = v\beta(a)v^{-1}$. In addition, it is clear from Proposition 2.2 that in $(G * \{u^n\}) *_W (G * \{v^n\})$ the groups $u^{-1}Gu$ and vGv^{-1} generate a free product with amalgamation A embedded by means of the maps $a \mapsto u^{-1}\alpha(a)u$ and $a \mapsto v\beta(a)v^{-1}$, respectively. Hence, if we conjugate by v, we see that G and $t^{-1}Gt$ also generate a free product, as described in the statement of 2.6.

Therefore, the subgroup

$$K' = |G \cup \{t = u^v\}| \subset (G * \{u^n\}) *_W (G * \{v^n\})$$

is a homomorphic image of K, and assertions (a) and (b) hold for K'. Moreover, the canonical map $K \to K'$ is an isomorphism. To see this it suffices to note that there exists an isomorphism

$$K * \{v^n\} \xrightarrow{\sim} (G * \{u^n\}) *_W (G * \{v^n\})$$

which takes $t \in K$ to uv. In particular, t has infinite order in K. The proposition is proved. □

We shall need to refine and generalize this result in two directions. In the first place, we want to consider iterated HNN-extensions; in the second place, we are interested in the connection between HNN-extensions of a group and a subgroup. We now bring together all the facts we need into a single statement.

Suppose that we are given an entire family of pairs of embeddings $\alpha_i, \beta_i : A_i \to G$ $(i \in I)$ and a subgroup $H \subset G$ with the property that $\alpha_i^{-1}(\alpha_i(A_i) \cap H) = \beta_i^{-1}(\beta_i(A_i) \cap H) = B_i \subset A_i$ are subgroups. Under these conditions we have

2.7 **Proposition.** *Let*

$$K_G = |G \cup \{t_i | i \in I\} : t_i^{-1}\alpha_i(a)t_i = \beta_i(a) \text{ for all } i \in I, a \in A_i|;$$

$$K_H = |H \cup \{t_i' | i \in I\} : t_i'^{-1}\alpha_i(b)t_i' = \beta_i(b) \text{ for all } i \in I, b \in B_i|.$$

Then

(a) *the* $\{t_i\}$ *freely generate a free subgroup in* K_G;
(b) *the natural maps* $G \to K_G$ *and* $K_H \to K_G$ (*the latter given by* $t_i' \mapsto t_i$) *are embeddings. In addition,* $K_H \cap G = H$ *in* K_G.

PROOF.
(a) If the relations in K_G implied a nontrivial relation between the t_i, this relation would be preserved in the quotient of K_G by the smallest normal divisor containing G. But in this quotient the relations $t_i^{-1}\alpha_i(a)t_i = \beta_i(a)$ become trivial ($1 = 1$), and no restrictions are imposed on the images of the t_i. This proves (a).

(b) We first consider the case when I consists of one element. In the notation used in the proof of Proposition 2.6, we consider K_G as a subgroup of $(G * \{u^n\}) *_W (G * \{v^n\})$. By Proposition 2.2, in $G * \{u^n\}$ we have

$$H * \{u^n\} \cap G * u^{-1}\alpha(A)u = H * u^{-1}\alpha(B)u,$$

and, similarly, in $G * \{v^n\}$ we have

$$H * \{v^n\} \cap G * v\beta(A)v^{-1} = H * v\beta(B)v^{-1}.$$

The above identifications of U and V with W identify these intersections with the subgroup

$$W_0 = H * w^{-1}Bw \subset G * w^{-1}Aw = W.$$

By Corollary 2.3(c), we have a canonical embedding

$$(H * \{u^n\}) *_{W_0} (H * \{v^n\}) \to (G * \{u^n\}) *_W (G * \{v^n\}).$$

But, as at the end of the proof of 2.6, the group on the left is $K_H * \{v^n\}$ and the group on the right is $K_G * \{v^n\}$, so we obtain an embedding $K_H \to K_G$.

Furthermore (the intersection is taken in $(G * \{u^n\}) *_W (G * \{v^n\})$):

$$(H * \{u^n\}) *_{W_0} (H * \{v^n\}) \cap G * u^{-1}\alpha(A)u = H * u^{-1}\alpha(B)u,$$

so that, if we now intersect with G, we obtain H. It follows *a fortiori* that $K_H \cap G = H$.

We prove (b) for finite I by an easy induction on n, and then for infinite I by passing to the inductive limit (which here is a union). We leave the details to the reader. □

3 Embeddings in groups with two generators

In this section we prove a result which will be used later and which shows vividly in a simple situation how the number of generators can be decreased using embeddings.

3.1. **Proposition.**

(a) *Any countable or finite group G can be embedded in a group with two generators.*

(b) *If G is recursive, then there is such an embedding which is recursive.*

PROOF.

(a) The group $\mathbf{Z} * \mathbf{Z} = \{b^n\} * \{v^n\}$ has a free subgroup of countable rank, for example,

$$S = \left|\{b^{-i}vb^i | i \geqslant 0\}\right|.$$

It immediately follows from Proposition 2.2 that there are no relations between the generators $b^{-i}vb^i$.

Thus, if G is a free countable group, it embeds in $\mathbf{Z} * \mathbf{Z}$. If G is not, we could try to represent G in the form F/N, where F is countable and free, then embed F in $\mathbf{Z} * \mathbf{Z}$ and consider the induced homomorphism $F/N \to \mathbf{Z} * \mathbf{Z}/N'$, where N' is the normal subgroup in $\mathbf{Z} * \mathbf{Z}$ generated by N. Unfortunately, $N' \cap F$ may be strictly larger than N, so that this homomorphism does not have to be an embedding. The following construction shows how to deal with this problem.

Let $\{g_1, g_2, g_3, \ldots\}$ be a countable system of generators of G, where $g_i \neq 1$. We successively construct the following extensions of G:

(1) $G * \{u^n\}$;

(2) the HNN-extension of $G * \{u^n\}$

$$\left|G * \{u^n\} \cup \{t_i | t_i^{-1}ut_i = ug_i, i = 1, 2, \ldots\}\right|,$$

(note that u and the ug_i generate infinite cyclic subgroups in $G * \{u^n\}$);

(3) the free product P of this HNN-extension and the group $\{b^n\} * \{v^n\}$ with subgroups $|\{t_1, t_2, \ldots\}|$ and $|\{b^{-i}vb^i | i \geqslant 1\}|$ amalgamated by means of the isomorphism

$$t_i = b^{-i}vb^i, \qquad i \geqslant 1.$$

(4) P has the two rank 2 free subgroups $|\{b, v\}|$ and $|\{u, b\}|$. There are no relations between u and b because there can be no relations in the quotient by the smallest normal subgroup containing G, t_i, and v.

Finally, we construct the following HNN-extension of P:

$$Q = |P \cup \{a\} : a^{-1}ba = u, a^{-1}va = b|.$$

To complete the proof, it remains to verify that *Q is generated by the elements a and b.*

In fact, Q has the obvious system of generators $\{g_i, t_i (i \geqslant 1); u, v, a, b\}$. The relations $g_i = u^{-1}t_i^{-1}ut_i$ allow us to eliminate the g_i; the relations $t_i = b^{-i}vb^i$ allow us to eliminate the t_i; and the relations $u = a^{-1}ba$ and $v = aba^{-1}$ allow us to eliminate u and v. This proves the first part of the proposition. The following analysis of the construction establishes part (b).

If we express g_i in terms of a and b in Q using the above relations, we find that $g_i = e_i$ modulo the relations in Q, where

$$e_i = a^{-1}b^{-1}ab^{-i}ab^{-1}a^{-1}b^ia^{-1}bab^{-i}aba^{-1}b^i.$$

Hence, the subgroup $E = |\{e_i | i \geqslant 1\}|$ in the group $\{a^n\} * \{b^n\}$ has the following remarkable property: any normal subgroup $N \subset E$ generates a normal subgroup N' in $\{a^n\} * \{b^n\}$ such that $E \cap N' = N$ (compare with the remark at the beginning of the proof).

In particular, if $\{g_i\}$ is an enumerable system of generators of G which is connected by an enumerable set of relations, it follows that the map $g_i \mapsto e_i$ (mod the relations) induces a recursive embedding of G in the recursive group E/N', since N' is enumerable whenever N is. □

4 Benign subgroups

4.1. **Definition-Lemma.** *Let G be a finitely presented group, and let $H \subset G$ be a subgroup. H is called* benign *if the following equivalent conditions are fulfilled*:

(a) *There exists a finitely presented group K, a finitely generated subgroup $L \subset K$, and an embedding $G \subset K$ such that $G \cap L = H$.*

(b) *The* HNN-*extension*

$$K_G = |G \cup \{t\} : t^{-1}ht = h, \textit{for all } h \in H|$$

can be embedded in a finitely presented group.

(c) *$G *_H G$ can be embedded in a finitely presented group.*

Proof of the equivalence

(a)⇒(b). Suppose that $G \subset K$ and L satisfy (a). Then it follows by 2.6 that K_G-is embedded in the HNN-extension

$$|K \cup \{t\} : t^{-1}lt = l, \text{ for all } l \in L|.$$

This group is finitely presented: we add t to the generators of K, and add the relations $t^{-1}l_it = l_i$, for a finite system of generators $\{l_i\}$ of L, to the relations between the generators of K.

(b)⇒(c). The group $G *_H G$ is embedded in K_G by 2.6(b), and K_G can be embedded in an f.p. group because we have assumed condition (b).

(c)⇒(a). Suppose that $G *_H G \subset M$, where M is finitely presented. We set $K = M$, we set $L =$ the image of G under the composite embedding $\phi_2 : G \to G *_H G \to M$, and we embed G in K by means of $\phi_1 : G \to G *_H G \to M$. Since $\phi_1(G) \cap \phi_2(G) = H$, we have $G \cap L = H$ in K, as required. □

The basic goal of this section is to reduce Higman's theorem 1.5 to proving that all enumerable subgroups in $\mathbf{Z} * \mathbf{Z}$ are benign. For this purpose and for later uses we shall need the following lemma.

4.2. Lemma. *Let R be a benign subgroup of an* f.g. *free group F, and let $\bar{R}$ be the normal subgroup it generates. Then $F/\bar{R}$ can be embedded in an* f.p. *group.*

PROOF. Let i be an embedding of $F *_R F$ in an f.p. group K (see 4.1(c)), and let $\phi_1, \phi_2 : F \to F *_R F$ be the structure maps. We consider two embeddings of F in $K \times F/\bar{R}$:

$$\alpha : f \mapsto \langle i \circ \phi_1(f), f\bar{R} \rangle;$$

$$\beta : f \mapsto \langle i \circ \phi_2(f), 1 \rangle.$$

They obviously coincide on the subgroup $R \subset F$. Hence they are induced by a homomorphism

$$\gamma : F \underset{R}{*} F \to K \times F/\bar{R},$$

which has a trivial kernel, since the composition of γ with the projection onto K coincides with i.

We construct an HNN-extension which takes $i \times \{1\} : F *_R F \to K \times F/\bar{R}$ to γ:

$$L = \Big| K \times F/\bar{R} \cup \{t\} : t^{-1} \langle i \circ \phi_1(f), 1 \rangle t = \big\langle i \circ \phi_1(f), f\bar{R} \big\rangle,$$
$$t^{-1} \langle i \circ \phi_2(f), 1 \rangle t = \langle i \circ \phi_2(f), 1 \rangle \text{ for all } f \in F \Big|.$$

L obviously contains $F/\bar{R}$. We show that L is finitely presented.

Generators of L : $\{t\} \cup$ finite system of generators of $K \cup$ finite system of generators of F. This system is finite.

Relations in L :

(a) {the relations between the generators of K}.

(b) {the commutation relations between the generators of K and the generators of F}.

After imposing these relations, we may consider that we are working in $K \times F$.

(c) $t^{-1} \langle i \circ \phi_1(f), 1 \rangle t = \langle i \circ \phi_1(f), f \rangle$,

$t^{-1} \langle i \circ \phi_2(f), 1 \rangle t = \langle i \circ \phi_2(f), 1 \rangle$,

where f runs through the system of generators of F.

(d) The relations in R between the generators of F.

We can take the system of relations $R_0 = \text{(a)} \cup \text{(b)} \cup \text{(c)}$ to be finite. We need only verify that the relations in (d) follow from R_0.

Let $R' \subset F$ be the normal subgroup generated by R_0, i.e., the kernel of the natural homomorphism $F \to |K \cup F \cup \{t\} : R_0|$. We want to show that $R' = \bar{R}$. The inclusion $R' \subset \bar{R}$ is obvious. We verify the converse.

If $f \in F$, we set $f' = f \bmod R'$ and $f_{1,2} = i \circ \phi_{1,2}(f) \in K$. It then follows from the relations (b) and (c) that in $K \times F/R'$ we have

$$t^{-1} \langle f_1, 1 \rangle t = \langle f_1, 1 \rangle \langle 1, f' \rangle \quad \text{and} \quad t^{-1} \langle f_2, 1 \rangle t = \langle f_2, 1 \rangle.$$

On the other hand, if $f \in R$, then, since $F *_R F$ is embedded in K, it follows from the relations (a) that $f_1 = f_2$. Hence $f' = 1$, so that $R \subset R'$. □

This lemma gives us the following reduction step.

4.3. **Proposition.** *If all enumerable subgroups in* $\mathbf{Z} * \mathbf{Z}$ *are benign, then Higman's theorem is true.*

PROOF. Let G be the free group generated by an enumerable set of free generators $\{g_i\}$, $i = 1, 2, 3, \ldots,$ and let $N \subset G$ be an enumerable normal subgroup. We shall show how to embed G/N into an f.p. group.

We first consider the embedding $G \to \{a^n\} * \{b^n\}$ given by $g_i \mapsto e_i$, where the e_i are as defined at the end of §3. Let the image of N under this embedding generate the normal subgroup $N' \subset \{a^n\} * \{b^n\}$. By the remark at the end of §3, G/N embeds in $\{a^n\} * \{b^n\}/N'$. But N' is enumerable by Lemma 1.2(d), since it is generated by the image of an enumerable set under a recursive map. Therefore, N' is a benign normal subgroup. Lemma 4.2 then shows that $\{a^n\} * \{b^n\}/N'$ can be embedded in an f.p. group. □

We conclude this section by establishing several basic properties of benign subgroups.

4.4. **Lemma.** *Let* $E, F \subset G$ *be benign subgroups of* G. *Then*:

(a) $E \cap F$ *is a benign subgroup*;
(b) $|E \cup F|$ ("*the sum of* E *and* F *in* G") *is a benign subgroup.*

PROOF. Let $\phi_1, \phi_2 : G \to G *_E G$ and $\phi_1', \phi_2' : G \to G *_F G$ be the structure homomorphisms. Let M_1 and M_2 be f.p. groups such that $G *_E G \subset M_1$ and $G *_F G \subset M_2$. We identify $\phi_1(G) \subset M_1$ and $\phi_1'(G) \subset M_2$ with G, and construct the group $M_1 *_G M_2$. This group is finitely presented (since it suffices to add to the relations in M_1 and M_2 the relations $\phi_1(g_i) = \phi_1'(g_i)$ for a finite system of generators of G). Let $\phi_1'', \phi_2'' : M_1, M_2 \to M_1 *_G M_2$ be the structure embeddings. We set $K = M_1 *_G M_2$ and $L = \phi_1'' \circ \phi_2(G)$, and we embed G in K by means of $\phi_2'' \circ \phi_2'$.

We claim that $G \cap L = E \cap F$ (as a subgroup of G in K). In fact, $\phi_1''(M_1) \cap \phi_2''(M_2) = G$ with its canonical embedding in $M_1 *_G M_2$. If we only take $\phi_2(G)$ in M_1 and $\phi_2'(G)$ in M_2, then intersecting with the amalgamation G gives E and F, respectively, and intersecting $\phi_2(G)$ with $\phi_2'(G)$ gives $E \cap F$.

(b) The subgroups $\phi_1(|E \cup F|)$ and $\phi_2(G)$ have the same intersection with the amalgamation in $G *_E G$, since they actually contain it. Hence, by 2.3(d), we have $|\phi_1(|E \cup F|) \cup \phi_2(G)| \cap \phi_1(G) = |E \cup F|$ in $G *_E G$, i.e., since E is the amalgamation,

$$|\phi_1(F) \cup \phi_2(G)| \cap \phi_1(G) = |E \cup F|.$$

Similarly, we have

$$|\phi_1'(E) \cup \phi_2'(G)| \cap \phi_1'(G) = |E \cup F|$$

in $G *_F G$. The notation is compatible with the fact that these two intersections are identified in the amalgamation of the product $M_1 *_G M_2$, which is constructed as in part (a).

Applying 2.3(d) to this product, we find that

$$|\phi_1''(|\phi_1(F) \cup \phi_2(G)|) \cup \phi_2''(|\phi_1'(E) \cup \phi_2'(G)|)| \cap G = |E \cup F|.$$

But the group $|\phi_1'' \circ \phi_2(G) \cup \phi_2'' \circ \phi_2'(G)| \cap G$ obviously contains the right-hand side and is contained in the left hand side of this equality, so that it also coincides with $|E \cup F|$.

Finally, $|\phi_1'' \circ \phi_2(G) \cup \phi_2'' \circ \phi_2'(G)|$ is a finitely generated subgroup of the finitely presented group $M_1 *_G M_2$. The proof is complete. □

4.5. **Lemma.** *Let G and H be* f.g. *subgroups of* f.p. *groups. Then any homomorphism from G to H takes benign subgroups of G to benign subgroups of H.*

PROOF.

(a) If $A \subset G$ is benign, then $A \times \{1\} \subset G \times H$ is also benign, since, given an embedding of (G, A) in (K, L) as in 4.1(a), we can construct the obvious embedding of $(G \times H)$ in $(K \times M, L \times \{1\})$, where M is the f.p. group containing H, which also satisfies the conditions in 4.1(a). Conversely, if $A \times \{1\} \subset G \times H$ is benign, then from an embedding of $(G \times H, A \times \{1\})$ in (K, L) as in 4.1(a) we construct the corresponding embedding of (G, A) in $(K, L \cap G \times \{1\})$.

(b) Now let $\phi : G \to H$ be any homomorphism, let F be its graph, and let $A \subset G$ be a benign subgroup. Then in $G \times H$ we have:

$$\{1\} \times \phi(A) = |(|A \times \{1\} \cup \{1\} \times H| \cap F) \cup G \times \{1\}| \cap \{1\} \times H.$$

It is clear from the assumptions regarding G and H that F is a benign subgroup in $G \times H$. By part (a), the other subgroups on the right in the formula are also benign. By Lemma 4.4, $\{1\} \times \phi(A)$ is a benign subgroup. Hence, $\phi(A)$ is also benign. □

5 Bounded systems of generators

5.1. Let $G' = |\{a_1, \ldots, a_n\}|$, $n \geqslant 1$, be the group freely generated by the a_i. We call a subset $R' \subset G'$ *bounded* if there exists an $r > 1$ such that any element in R' can be represented in the form $a_{i_1}^{x_1} \cdots a_{i_r}^{x_r}$, $x_i \in \mathbf{Z}$. In this section we prove the following special case of the hypothesis of Proposition 4.3:

5.2. **Proposition.** *If the subgroup $H' \subset G'$ is generated by a bounded enumerable subset $R' \subset G'$, then it is benign.*

Corollary. *The same is true if G' is an* f.g. *subgroup of an* f.p. *group* (using Lemma 4.5).

In the next section we show how the general case follows from this special case.

The proof of 5.2 consists of a series of reduction steps.

5.3. *First reduction.* In the free group $G = |\{a_1, b_1, c_1; \cdots ; a_{rn}, b_{rn}, c_{rn}\}|$ we shall consider a set of "layered" words of the form

$$R = \{a_1^{x_1} b_1 c_1^{x_1} \cdots a_{rn}^{x_{rn}} b_{rn} c_{rn}^{x_{rn}}\}$$

and the subgroup $H \subset G$ it generates. We shall later show that, *if R is enumerable, then H is benign.* This is a special case of 5.2 to which the general case reduces using the following technique.

Suppose we are given G' and R' as in 5.1. For each element $g' = a_{i_1}^{x_1} \cdots a_{i_r}^{x_r} \in R'$ we construct an element $g \in G$ as follows. We represent g' in the form

$$\prod_{i=1}^{n} a_i^{x_{1,i}} \prod_{i=1}^{n} a_i^{x_{2,i}} \cdots \prod_{i=1}^{n} a_i^{x_{r,i}},$$

where

$$x_{k,i} = \begin{cases} x_k, & \text{for } i = i_k, \\ 0, & \text{for } i \neq i_k. \end{cases}$$

We then set

$$g = \left(\prod_{i=1}^{n} a_i^{x_{1,i}} b_i c_i^{x_{1,i}} \right) \left(\prod_{i=1}^{n} a_{n+i}^{x_{2,i}} b_{n+i} c_{n+i}^{x_{2,i}} \right) \cdots \left(\prod_{i=1}^{n} a_{(r-1)n+i}^{x_{r,i}} b_{(r-1)n+i} c_{(r-1)n+i}^{x_{r,i}} \right).$$

If R' is enumerable, then the set R of all elements g obtained from all the $g' \in R'$ is enumerable.

We consider the surjective homomorphism $\phi : G \to G'$ given by $\phi(a_{nj+i}) = a_i$ $(1 \leqslant i \leqslant n, 0 \leqslant j \leqslant r-1)$, $\phi(b_i) = \phi(c_i) = 1$ for all $i = 1, \ldots, rn$. Clearly $\phi(R) = R'$, and hence $\phi(H) = H'$. It then follows from Lemma 4.5 that, if R is benign in G, then R' is benign in G'. □

5.4. *Using the theorem that all enumerable sets are Diophantine.*

From this point on, we fix a pair (G, enumerable R), as in 5.3. We shall write $l \geqslant 1$ in place of rn. We define the set $E \subset \mathbf{Z}^{l+1}$ by the condition

$$R = \{a_0^{x_0} b_0 c_0^{x_0} \cdots a_l^{x_l} b_l c_l^{x_l} | \langle x_0, \ldots, x_l \rangle \in E\}.$$

It is not hard to see that R is enumerable if and only if E is enumerable.

We now show that E *can be represented as the projection onto the first* $l+1$ *coordinates of a set*

$$\bigcap_{s=1}^{N} E_s \subset \mathbf{Z}^{l+1} \times \mathbf{Z}^{m-l}, \qquad m \geqslant l+2,$$

where each of the E_s *is defined by an equation of one of the following forms*:

$$\begin{aligned} x_i &= c, & c &\in \mathbf{Z};\\ x_i &= x_j, & 0 &\leqslant i,j \leqslant m;\\ x_k &= x_j + x_i, & l+1 &\leqslant k<j<i \leqslant m;\\ x_k &= x_j \cdot x_i, & l+1 &\leqslant k<j<i \leqslant m. \end{aligned}$$

In fact, let $\varepsilon_0, \ldots, \varepsilon_l \in \{1, -1\}$, and let $\bar{\varepsilon} = \langle \varepsilon_0, \ldots, \varepsilon_l \rangle$. We consider the enumerable sets

$$E^{\bar{\varepsilon}} = \left\{ \langle x_0, \ldots, x_l \rangle \in (\mathbf{Z}^+ \cup \{0\})^{l+1} | \langle \varepsilon_0 x_0, \ldots, \varepsilon_l x_l \rangle \in E \right\}.$$

By the fundamental theorem in Chapter VI, there exist polynomials $P^{\bar{\varepsilon}}$ with integral coefficients such that

$E^{\bar{\varepsilon}} =$ the projection of the 0-level of $P^{\bar{\varepsilon}}$ in $(\mathbf{Z}^+ \cup \{0\})^{l+1} \times (\mathbf{Z}^+)^{n-l}$ onto the first $l+1$ coordinates $\langle x_0, \ldots, x_l \rangle$.

Here we can take n large enough so that the sets of variables which actually occur in $P^{\bar{\varepsilon}'}$ and in $P^{\bar{\varepsilon}''}$ and which "drop out" in the projection do not intersect if $\bar{\varepsilon}' \neq \bar{\varepsilon}''$. If we add the $(n+1)2^{l+3}$ new variables $y_{ij\bar{\varepsilon}}$ $(0 \leqslant i \leqslant n, j = 1, 2, 3, 4)$ to the variables which drop out in the projection, we find that E can be represented as the projection onto the first $l+1$ coordinates of the 0-level of the following polynomial, where the 0-level is now in $\mathbf{Z}^{l+1} \times \mathbf{Z}^{n+(n+1)2^{l+3}-l}$:

$$Q = \prod_{\bar{\varepsilon}} \left[\left(P^{\bar{\varepsilon}}(\varepsilon_0 x_0, \ldots, \varepsilon_l x_l, x_{l+1}, \ldots, x_n) \right)^2 + \sum_{i=0}^{l} \left(\varepsilon_i x_i - \sum_{j=1}^{4} y_{ij\bar{\varepsilon}}^2 \right)^2 + \sum_{i=l+1}^{n} \left(x_i - 1 - \sum_{j=1}^{4} y_{ij\bar{\varepsilon}}^2 \right)^2 \right].$$

Finally, in order to represent the set $Q=0$ as a projection of an intersection $\cap_{s=1}^{N} E_s$ of the required type, we introduce additional variables as follows. Let $x_0, \ldots, x_t$ be the variables which occur in Q. Instead of $Q=0$ we write $Q_1 = Q_2$, where Q_1 is the sum of the monomials in Q with positive coefficients, and Q_2 is the sum of the monomials with negative coefficients. Then

0-level of $Q =$ a projection of $(x_{t+1} = Q_1) \cap (x_{t+2} = Q_2) \cap (x_{t+1} = x_{t+2})$.

If Q_1 and Q_2 are constants or variables, this gives us the desired representation. Otherwise, we write, say, Q_1 in the form $Q_1' + Q_1''$ or $Q_1' \cdot Q_1''$ and, after introducing two more variables, we have, for example,

$$(x_{t+1} = Q_1' + Q_1'') = \text{a projection of } (x_{t+3} = Q_1') \cap (x_{t+4} = Q_1'') \cap (x_{t+1} = x_{t+3} + x_{t+4}).$$

We complete the proof by induction on the sum of the absolute values of the coefficients and on the degree of Q. □

5.5. *Second reduction*. We now assume that, along with the pair (G, R) described in 5.3, we have fixed a representation of E in the form $\bigcap_{s=1}^{N} E_s$ as in 5.4. In this subsection we show that *the subgroup $H \subset G$ generated by R is benign if all of the following subgroups $\bar{H}_s \subset \bar{G}$, $s = 1, \ldots, N$, are benign*:

$$\bar{G} = \left|\{a_0, b_0, c_0; \cdots; a_m, b_m, c_m; \bar{a}_1, \bar{b}_1, \bar{c}_1, \ldots, \bar{a}_l, \bar{b}_l, \bar{c}_l\}\right|;$$

$$\bar{H}_s = \left|\left\{\left(\prod_{i=l+1}^{m} a_i^{x_i} b_i c_i^{x_i}\right)^{-1}\left(\prod_{i=1}^{l} \bar{a}_i^{x_i} \bar{b}_i \bar{c}_i^{x_i}\right)^{-1} \prod_{i=0}^{m} a_i^{x_i} b_i c_i^{x_i};\ \langle x_0, \ldots, x_m\rangle \in E_s\right\}\right|.$$

To show this, we first set

$$a(x_0, \ldots, x_m) = \left(\prod_{i=l+1}^{m} a_i^{x_i} b_i c_i^{x_i}\right)^{-1}\left(\prod_{i=1}^{l} \bar{a}_i^{x_i} \bar{b}_i \bar{c}_i^{x_i}\right)^{-1} \prod_{i=0}^{m} a_i^{x_i} b_i c_i^{x_i}. \quad (1)$$

The set of words $\{a(x_0, \ldots, x_m); \langle x_0, \ldots, x_m\rangle \in \mathbf{Z}^{m+1}\}$ is free, since, when we join two such words (or when we join such a word with the inverse of another such word), any cancellation cannot involve the "middle part" of each word, which consists of the symbols $\bar{a}_i, \bar{b}_i, \bar{c}_i$.

It hence follows that

$$\bigcap_{s=1}^{N} \bar{H}_s = \left|\left\{a(x_0, \ldots, x_m), \langle x_0, \ldots, x_m\rangle \in \bigcap_{i=1}^{N} E_s\right\}\right|,$$

and the subgroup $\bar{H} = \bigcap_{s=1}^{N} \bar{H}_s \subset \bar{G}$ is benign if all of the $\bar{H}_s$ are benign. Finally, we have

$$\left|\bar{H} \cup \{a_{l+1}, b_{l+1}, c_{l+1}, \ldots, a_m, b_m, c_m; \bar{a}_1, \bar{b}_1, \bar{c}_1, \ldots, \bar{a}_l, \bar{b}_l, \bar{c}_l\}\right|$$

$$= \left|\left\{\prod_{i=0}^{l} a_i^{x_i} b_i c_i^{x_i}, \langle x_0, \ldots, x_l\rangle \in E = \text{projection of } \bigcap_{s=1}^{N} E_s\right\} \cup \{a_{l+1}, b_{l+1}, c_{l+1}, \ldots, \bar{a}_l, \bar{b}_l, \bar{c}_l\}\right|,$$

so that

$$H = \left|\overline{H} \cup \{a_{l+1}, b_{l+1}, \ldots, \bar{b}_l, \bar{c}_l\}\right| \cap |\{a_0, \ldots, b_l, c_l\}|.$$

Therefore, H is benign whenever $\overline{H}$ is benign. □

5.6. *Construction of the group* K. We use the criterion 4.1(a) to verify that the $\overline{H}_s \subset \overline{G}$ are benign subgroups. That is, we explicitly construct a finitely presented group $K \supset \overline{G}$ and finitely generated subgroups $L_s \subset K$ such that $L_s \cap \overline{G} = \overline{H}_s$ for all $s = 1, \ldots, N$. We construct K as a multiple HNN-extension of $\overline{G}$.

(a) *The first* HNN-*extension*. We set

$$K_0 = \left|\overline{G} \cup \{t_0, \ldots, t_m\} : R_0\right|,$$

where R_0 is the set of relations

$$\Big\{ t_i^{-1} b_i t_i = a_i b_i c_i \quad \text{and} \quad t_i^{-1} \bar{b}_i t_i = \bar{a}_i \bar{b}_i \bar{c}_i, \text{ for } i = 0, \ldots, m;$$

$$\text{the } t_i \text{ commute with all the other generators of } \overline{G} \Big\}. \quad (2)$$

(b) *The second* HNN-*extension*. We set

$$K = \left|K_0 \cup \{t_{ijk};\ l+1 \leqslant k < i,\ k < j,\ i \neq j;\ i, j, k \leqslant m\} : R\right|,$$

where R is the set of relations

$$\Big\{ t_{ijk}^{-1} b_i t_{ijk} = a_i b_i c_i,\ t_{ijk}^{-1} c_j t_{ijk} = t_k c_j;$$

$$\text{the } t_{ijk} \text{ commute with the } t_k \text{ and with the other generators of } \overline{G} \Big\}. \quad (3)$$

Unlike in 5.6(a), here it is not completely obvious that K is an HNN-extension of K_0. To check this it suffices to show that the map ϕ_{ijk} (i, j, k fixed, $i \neq k$, $j \neq k$) from the set {generators of $\overline{G}$} $\cup$ $\{t_k\}$ to itself which takes

$$b_i \mapsto a_i b_i c_i, \qquad c_j \mapsto t_k c_j, \qquad t_k \mapsto t_k,$$

and leaves the other generators of $\overline{G}$ fixed, extends to an automorphism of the subgroup $|\overline{G} \cup \{t_k\}| \subset K_0$. We have:

$$\left|\overline{G} \cup \{t_k\}\right| = \left|\overline{G} * \{t_k^n\} : t_k^{-1} b_i t_k = b_i,\ t_k^{-1} c_j t_k = c_j, \ldots\right|,$$

where the $\cdots$ stands for relations which do not involve b_i and c_j, and so are taken to themselves under ϕ_{ijk}. On the other hand, the two relations which are written out are taken to relations which follow from the defining relations in K_0: the first goes to

$$t_k^{-1} a_i b_i c_i t_k = a_i b_i c_i,$$

and the second goes to

$$t_k^{-1} t_k c_j t_k = t_k c_j.$$

It remains to use the stipulation that $i \neq k$ and $j \neq k$.

It is clear from the definition of K that K is finitely presented. It follows from the properties of HNN-extensions that $\overline{G} \subset K$.

5.7. *Construction of the subgroups* $L_s \subset K$. The form of L_s will depend on the equation defining the set E_s (see 5.4). We define a large number of groups which will include all the L_s:

$$L_i^c = |\{a(\underbrace{0 \cdots 0}_{i} c0 \cdots 0), t_r(r \neq i)\}|,$$

$$L_{ij}^= = |\{a(0 \cdots 0), t_i t_j, t_r(r \neq i, j)\}|,$$

$$L_{ijk}^+ = |\{a(0 \cdots 0), t_i t_k, t_j t_k, t_r(r \neq i, j, k)\}|,$$

$$L_{ijk}^\times = |\{a(0 \cdots 0), t_{ijk}, t_{jik}, t_r(r \neq i, j, k)\}|,$$

and analogously, in the notation of 5.5,

$$\overline{H}_i^c = |\{a(x_0, \ldots, x_m), x_i = c\}|,$$

$$\overline{H}_{ij}^= = |\{a(x_0, \ldots, x_m), x_i = x_j\}|,$$

$$H_{ijk}^+ = |\{a(x_0, \ldots, x_m), x_k = x_j + x_i\}|,$$

$$H_{ijk}^\times = |\{a(x_0, \ldots, x_m), x_k = x_j \cdot x_i\}|.$$

The L_s are clearly finitely generated. It remains to perform one final series of verifications:

5.8. $\overline{H}_i^c = \overline{G} \cap L_i^c$, $\overline{H}_{ij}^= = \overline{G} \cap L_{ij}^=$, *and so on.*

First of all, it follows from (1), (2), and (3) that

$$t_i^{-1} a(x_0, \ldots, x_m) t_i = a(x_0, \ldots, x_{i-1}, x_i + 1, x_{i+1}, \ldots, x_m), \tag{4}$$

$$t_{ijk}^{-1} a(x_0, \ldots, x_m) t_{ijk} = a(y_0, \ldots, y_m), \tag{5}$$

where $y_i = x_i + 1$, $y_k = x_k + x_j$, and $y_s = x_s$ for $s \neq i, k$. (To verify (5) recall that, since $k \geqslant l + 1$, it follows that t_k commutes with the middle part of the word $a(x_0, \ldots, x_m)$, which consists of $\bar{a}_i, \bar{b}_i, \bar{c}_i, i \leqslant l$.)

It hence follows that

$$L_i^c = |\overline{H}_i^c \cup \{t_r | r \neq i\}|,$$

$$L_{ij}^= = |\overline{H}_{ij}^= \cup \{t_i t_j, t_r | r \neq i, j\}|,$$

$$L_{ijk}^+ = |\overline{H}_{ijk}^+ \cup \{t_i t_k, t_j t_k, t_r | r \neq i, j, k\}|,$$

$$L_{ijk}^\times = |\overline{H}_{ijk}^\times \cup \{t_{ijk}, t_{jik}, t_r | r \neq i, j, k\}|.$$

In fact, the inclusions $\subset$ are obvious. Next, if we begin with $a(x_0, \ldots, x_m)$ and conjugate by t_r, it follows by (4) that we can vary the rth coordinate arbitrarily. This immediately gives the inclusion $L_i^c \supset H_i^c$, and hence the first required equality. The second equality is obtained analogously.

The third equality: conjugating by $t_i t_k$ increases the ith and kth coordinates by 1, and conjugating by $t_j t_k$ increases the jth and kth coordinates by 1, so that we can obtain any vector with $x_k = x_j + x_i$ starting from a vector with zeros in these places.

The fourth equality: conjugating by t_{ijk} increases x_i by 1 and increases x_k by x_j, and conjugating by t_{jik} increases x_j by 1 and x_k by x_i. Hence, we can obtain any vector with $x_k = x_j \cdot x_i$ starting from the zero vector.

This new characterization of the groups L_s shows that $L_s \cap \bar{G} \supset H_s$ for all s. It remains to prove the converse.

To do this, we note that, using (4) and (5), we can represent any element in L_s in the form Th, where $T \in |\{t_i, t_{ijk}\}|$ (here the set of admissible indices i and ijk depends on s) and $h \in H_s$. This follows by the same argument as above. But, by Proposition 2.7(a), all the $\{t_i, t_{ijk}\}$ generate a free subgroup which has a trivial intersection with $\bar{G}$ (see the proof of 2.7(a)). Consequently, if $Th \in \bar{G}$, it follows that $T = 1$ and $h \in H_s$, which completes the proof. □

6 End of the proof

6.1. In this section we finish the verification of Proposition 4.3, and hence the proof of Higman's theorem.

Let $G = |\{a, b\}|$, and let $H \subset G$ be an enumerable subgroup. We shall show that H is benign. The first step is to reduce the problem to proving that a certain special subgroup

$$H' \subset G' \cong \mathop{*}_{1}^{7} \mathbf{Z},$$

which does not depend on H, is benign. To define H', we first introduce the following recursive enumeration $\gamma : \mathbf{Z}^+ \to G$ (which covers each $g \in G$ infinitely many times):

$$\gamma(2^{m_0} 3^{m_1} \cdots p_r^{m_r} \cdots) = \prod_{i=0}^{\infty} a^{m_{4i} - m_{4i+1}} b^{m_{4i+2} - m_{4i+3}}.$$

We then set

$$G' = |\{a, b, t, v, c, d, e\}|;$$

$$\tau : S(\{a, b, a^{-1}, b^{-1}\}) \to G' : \prod_{i \geqslant 0} a^{m_{2i}} b^{m_{2i+1}}$$
$$\mapsto t \prod_{i \geqslant 0} (v^{-i} a v^i)^{m_{2i}} (v^{-i} b v^i)^{m_{2i+1}};$$

$$H' = |\{\tau(g) c^n d e^n \mid g \in S(\{a, b, a^{-1}, b^{-1}\}), n \in \mathbf{Z}^+, g = \gamma(n)\}| \subset G'.$$

The formula for τ defines τ on words which are not necessarily reduced, and reducing a word can change its image under τ. Also note that a generator $\tau(g) c^n d e^n$ of H' is uniquely determined by n.

6.2. **Lemma.** *If $H' \subset G'$ is a benign subgroup, then any enumerable subgroup $H \subset G$ is benign.*

PROOF.
(a) We set

$$H'' = |\{\tau(h)c^n de^n | \text{image of } h \in H, n \in \mathbf{Z}^+, h = \gamma(n)\}| \subset H'.$$

Then

$$H'' = H' \cap |\{a, b, t, v, c^n de^n | n \in \gamma^{-1}(H)\}|.$$

In fact, the inclusion $\subset$ is obvious. The converse follows because the set of images of the elements $c^n de^n$, $n \geqslant 1$, in the quotient of G' by the kernel generated by a, b, t, and v, is free. Hence, in any reduced word in the generators $\tau(g)c^n de^n$, the sequence of n's can be uniquely recovered from the word, and, if all the n's lie in $\gamma^{-1}(H)$, it follows that the word lies in H''.

Thus, H'' is the intersection of H' with the subgroup generated by a bounded enumerable set of generators (since $\gamma^{-1}(H)$ is enumerable whenever H is). Consequently, H'' is benign if H' is benign.

(b) We set

$$\overline{H} = |\{\tau(h) | h \in H\}| \subset G'.$$

It is easy to see that

$$|\overline{H} \cup \{c, d, e\}| = |H'' \cup \{c, d, e\}|.$$

Hence,

$$\overline{H} = |H'' \cup |\{c, d, e\}|\, | \cap |\{a, b, v, t\}|.$$

By Lemma 4.4, $\overline{H}$ is benign if H'' is benign.

(c) Finally, we consider the homomorphism $\phi : G' \to G$ which takes a to a, b to b, and t, v, c, d, and e to 1. Obviously, $\phi(\overline{H}) = H$. By Lemma 4.5, H is benign if $\overline{H}$ is benign. □

6.3. We now prove that the subgroup $H' \subset G'$ is benign. To do this, we construct a commutative diagram of group embeddings

$$\begin{array}{ccccc} G' & \longrightarrow & K' & \longrightarrow & K \\ \uparrow & & \uparrow & & \uparrow \\ H' & \longrightarrow & L' & \xrightarrow{\sim} & L \end{array}$$

with the following properties:

(a) K is defined by a finite set of generators and a bounded enumerable set of relations; L is generated by a bounded enumerable set of words in the generators of K.
(b) $L' \xrightarrow{\sim} L$ is an isomorphism.
(c) $H' = G' \cap L'$ in K'.

It will then follow that H' is benign. In fact, let $K = F/\bar{R}$, where F is the free group generated by a finite system of generators of K, R_0 is a bounded enumerable set of relations between these generators, and $\bar{R}$ is the normal subgroup generated by these relations. It follows from Proposition 5.2 that R_0 generates a benign subgroup R in F, and then Lemma 4.2 implies that $K = F/\bar{R}$ can be embedded in an f.p. group M. When we embed K in M, the bounded enumerable set of generators of L remains a bounded enumerable set in M (relative to the generators of M), and hence $L \subset M$ is benign by the corollary to Proposition 5.2. Therefore, by (b) and (c) we have: the subgroup $H' = G' \cap L$ is benign as a subgroup of M whenever G and L are benign. Hence, there is an embedding of (M, H') in $(\bar{M}, \bar{H})$ such that $\bar{M}$ is finitely presented, $\bar{H}$ is finitely generated, and $H' = \bar{H} \cap M$. This embedding induces an embedding of the pair (G', H') in $(\bar{M}, \bar{H})$ with the same properties. Consequently, H' is also benign in G'.

It remains to construct the diagram of embeddings with properties (a), (b), and (c).

6.4. *The group* K'. This will be a multiple HNN-extension of G' which, as in Proposition 2.7, we define using four countable sequences of non-trivial isomorphisms of the subgroup $|\{t, c, d, e, v^{-i}av^i, v^{-i}bv^i | i \geqslant 0\}| \subset G'$ with G'. Since the elements listed here freely generate this subgroup, it is sufficient to indicate where our isomorphisms take these elements. These isomorphisms will be induced in K' by conjugation by four sequences of generators x_i, $\bar{x}_i$, y_i, and $\bar{y}_i$, $i \geqslant 0$ (instead of the t_i, $i \in I$, in §2). The following table gives the action of these generators. We use the notation: $a_i = v^{-i}av^i$, $b_i = v^{-i}bv^i$, $p_j =$ the jth prime number. The element in the table, say, in the c-row and the $\bar{x}_i$-column, is $\bar{x}_i^{-1}cx_i$.

	x_i	$\bar{x}_i$	y_i	$\bar{y}_i$
t	ta_i	ta_i^{-1}	tb_i	tb_i^{-1}
c	$c^{p_{4i}}$	$c^{p_{4i+1}}$	$c^{p_{4i+2}}$	$c^{p_{4i+3}}$
d	d	d	d	d
e	$e^{p_{4i}}$	$e^{p_{4i+1}}$	$e^{p_{4i+2}}$	$e^{p_{4i+3}}$
a_j	$\begin{cases} a_i^{-1}a_ja_i, & j \leqslant i \\ a_j, & j \geqslant i \end{cases}$	$\begin{cases} a_ia_ja_i^{-1}, & j < i \\ a_j, & j \geqslant i \end{cases}$	$\begin{cases} b_i^{-1}a_jb_i, & j < i \\ a_j, & j \geqslant i \end{cases}$	$\begin{cases} b_ia_jb_i^{-1}, & j < i \\ a_j, & j \geqslant i \end{cases}$
b_j	$\begin{cases} a_i^{-1}b_ja_i, & j < i \\ b_j, & j \geqslant i \end{cases}$	$\begin{cases} a_ib_ja_i^{-1}, & j < i \\ b_j, & j \geqslant i \end{cases}$	$\begin{cases} b_i^{-1}b_jb_i, & j < i \\ b_j, & j \geqslant i \end{cases}$	$\begin{cases} b_ib_jb_i^{-1}, & j < i \\ b_j, & j \geqslant i \end{cases}$.

We finally set:

$$K' = |G' \cup \{x_i, \bar{x}_i, y_i, \bar{y}_i | i \geqslant 0\}: \text{the relations in the table}|,$$

and we take $G' \to K'$ to be the natural embedding.

6.5. *The group L'*. We set

$$L' = |\{tcde, x_i, \bar{x}_i, y_i, \bar{y}_i | i \geqslant 0\}| \subset K',$$

and we take $L' \to K'$ to be the natural embedding. In subsection 6.7 we shall verify that H' is embedded in L' (as a subgroup of K', in view of the commutativity of the diagram).

6.6. *The groups K and L*. We set

$$K = |G' \cup \{u_1, u_2, u_3, u_4, v_1, v_2, v_3, v_4\} : R|,$$

where the relations R and the embedding $K' \to K$ are both defined by the conditions

$R =$ the image of the relations in the table after making the substitutions

$$x_i \mapsto u_1^{-i} v_1 u_1^i, \qquad \bar{x}_i \mapsto u_2^{-i} v_2 u_2^i, \qquad y_i \mapsto u_3^{-i} v_3 u_3^i, \qquad \bar{y}_i \mapsto u_4^{-i} v_4 u_4^i;$$

$K' \to K$ is the homomorphism which is the identity on G' and acts by these substitutions on the other generators.

The homomorphism $K' \to K$ is an embedding. In fact, the elements $u_j^{-i} v_j u_j^i$ are free in $|\{u_j, v_j\}|$, so that K can be considered as the free product of K' and $|\{u_j, v_j | 1 \leqslant j \leqslant 4\}|$ with the amalgamation given by the above substitutions (here we take into account Proposition 2.7(a)).

Finally, we set

$$L = \text{the image of } L' \text{ under the embedding } K' \to K.$$

6.7. The diagram has now been constructed. It follows immediately from the definitions that it satisfies 6.4(a) and (b). It remains to show that $H' = G' \cap L'$ in K'.

(a) We set $[n] = \tau(g)c^n de^n$ for $n \in \mathbf{Z}^+$ and $g = \gamma(n)$ in the notation of 6.1. We recall that H' is generated by all the $[n]$ in G', and hence in K' as well.

The table of relations in K' was composed in such a way so that the following relations would be fulfilled:

$$x_i^{-1}[n]x_i = [p_{4i}n], \qquad \bar{x}_i^{-1}[n]\bar{x}_i = [p_{4i+1}n],$$
$$y_i^{-1}[n]y_i = [p_{4i+2}^{n}], \qquad \bar{y}_i^{-1}[n]\bar{y}_i = [p_{4i+3}n].$$

For example, we verify the first relation. Let $n = \Pi p_j^{m_j}$. Then, according to the definitions,

$$\gamma(n) = \prod_j a^{m_{4j} - m_{4j+1}} b^{m_{4j+2} - m_{4j+3}},$$

$$[n] = t \prod_j a_j^{m_{4j} - m_{4j+1}} b_j^{m_{4j+2} - m_{4j+3}} c^n de^n,$$

so that, by the first column of the table in 6.4,

$$x_i^{-1}[n]x_i = ta_i a_i^{-1} \prod_{j<i} (\cdots);\ a_i \prod_{j\geqslant i} (\cdots);\ c^{p_{4i}n} d e^{p_{4i}n} = [p_{4i}n].$$

If we further take into account that $[1] = tcde \in L'$, we may conclude from these conjugation formulas that $[n] \in L'$ for all n, and that $H' \subset L'$, as promised in 6.5. Moreover, $|H' \cup \{x_i, \bar{x}_i, y_i, \bar{y}_i | i \geqslant 0\}| = L$, since the inclusion $\subset$ has been verified, and the inclusion $\supset$ is obvious.

(b) We now show that in K' we have

$$|H' \cup \{x_i, \bar{x}_i, y_i, \bar{y}_i | i \geqslant 0\}| \cap G' = H'.$$

Since K' is an HNN-extension of G', it suffices to show that we are in the situation of Proposition 2.7 (as described in the paragraph preceding the proposition, at the end of 2.6), and then to apply 2.7(b).

We verify these conditions, for example, for the first series of isomorphisms of the subgroup of G', as described at the beginning of 6.4. This series corresponds to conjugating by x_i in K'. The conditions take the following form in our case:

$$x_i^{-1}\big[H' \cap |\{t, c, d, e;\ a_j, b_j | j \geqslant 0\}|\big]x_i$$
$$= H' \cap x_i^{-1}|\{t, c, d, e;\ a_j, b_j | j \geqslant 0\}|x_i;$$

i.e., if we use the definition of H' and the table,

$$x_i^{-1}H'x_i = H' \cap |\{t, c^{p_{4i}}, d, e^{p_{4i}};\ a_j, b_j | j \geqslant 0\}|.$$

Since $x_i^{-1}[n]x_i = [p_{4i}n]$, the inclusion $\subset$ is obvious. Conversely, suppose we are given an element in H' which is written as a reduced word in the $[n]$: $\Pi_{j \geqslant 0}[n_j]^{\varepsilon_j}$, $\varepsilon_j = \pm 1$. We consider the corresponding reduced word g in G'. We show that if all the powers of c and d which occur in g are divisible by p_{4i}, then all the n_j with nonzero ε_j in the above product are divisible by p_{4i}, i.e., $[n_j] \in x_i^{-1}H'x_i$.

In fact, let $\bar{g}$ = the image of g in $|\{c, d, e\}|$ under the homomorphism which takes t, a_j, and b_j to 1. Since $[\bar{n}] = c^n d e^n$, it follows that all the $[\bar{n}]$ are free, and that $\bar{g}$ uniquely determines the sequence $\{\varepsilon_j n_j\}$. It is not hard to see that the formulas which express $\varepsilon_j n_j$ in terms of the powers of c and e which occur in the reduced word $\bar{g}$ are linear with integer coefficients (more precisely, they are a disjunction of linear formulas accompanied by inequality conditions). Therefore, if all these powers are divisible by p_{4i}, then so is n_j.

This completes the proof. □

Index

Graduate Texts in Mathematics

Soft and hard cover editions are available for each volume up to vol. 14, hard cover only from Vol. 15

1 TAKEUTI/ZARING. Introduction to Axiomatic Set Theory. vii, 250 pages. 1971.
2 OXTOBY. Measure and Category. viii, 95 pages. 1971.
3 SCHAEFFER. Topological Vector Spaces. xi, 294 pages. 1971.
4 HILTON/STAMMBACH. A Course in Homological Algebra. ix, 338 pages. 1971. (Hard cover edition only)
5 MACLANE. Categories for the Working Mathematician. ix, 262 pages. 1972.
6 HUGHES/PIPER. Projective Planes. xii, 291 pages. 1973.
7 SERRE. A Course in Arithmetic. x, 115 pages. 1973.
8 TAKEUTI/ZARING. Axiomatic Set Theory. viii, 238 pages. 1973.
9 HUMPHREYS. Introduction to Lie Algebras and Representation Theory. xiv, 169 pages. 1972.
10 COHEN. A Course in Simple Homotopy Theory. xii, 114 pages. 1973.
11 CONWAY. Functions of One Complex Variable. 2nd corrected reprint. xiii, 313 pages. 1975. (Hard cover edition only.)
12 BEALS. Advanced Mathematical Analysis. xi, 230 pages. 1973.
13 ANDERSON/FULLER. Rings and Categories of Modules. ix, 339 pages. 1974.
14 GOLUBITSKY/GUILLEMIN. Stable Mappings and Their Singularities. x, 211 pages. 1974.
15 BERBERIAN. Lectures in Functional Analysis and Operator Theory. x, 356 pages. 1974.
16 WINTER. The Structure of Fields. xiii, 205 pages. 1974.
17 ROSENBLATT. Random Processes. 2nd ed. x, 228 pages. 1974.
18 HALMOS. Measure Theory. xi, 304 pages. 1974.
19 HALMOS. A Hilbert Space Problem Book. xvii, 365 pages. 1974.
20 HUSEMOLLER. Fibre Bundles. 2nd ed. xvi, 344 pages. 1975.
21 HUMPHREYS. Linear Algebraic Groups. xiv. 272 pages. 1975.
22 BARNES/MACK. An Algebraic Introduction to Mathematical Logic. x, 137 pages. 1975.
23 GREUB. Linear Algebra. 4th ed. xvii, 451 pages. 1975.
24 HOLMES. Geometric Functional Analysis and Its Applications. x, 246 pages. 1975.
25 HEWITT/STROMBERG. Real and Abstract Analysis. 3rd printing. viii, 476 pages. 1975.
26 MANES. Algebraic Theories. x, 356 pages. 1976.

27 KELLEY. General Topology. xiv, 298 pages. 1975.
28 ZARISKI/SAMUEL. Commutative Algebra I. xi, 329 pages. 1975.
29 ZARISKI/SAMUEL. Commutative Algebra II. x, 414 pages. 1976.
30 JACOBSON. Lectures in Abstract Alegbra I: Basic Concepts. xii, 205 pages. 1976.
31 JACOBSON. Lectures in Abstract Algebra II: Linear Algebra. xii, 280 pages. 1975.
32 JACOBSON. Lectures in Abstract Algebra III: Theory of Fields and Galois Theory. ix, 324 pages. 1976.
33 HIRSCH. Differential Topology. x, 222 pages. 1976.
34 SPITZER. Principles of Random Walk. 2nd ed. xiii, 408 pages. 1976.
35 WERMER. Banach Algebras and Several Complex Variables. 2nd ed. xiv, 162 pages. 1976.
36 KELLEY/NAMIOKA. Linear Topological Spaces. xv, 256 pages. 1976.
37 MONK. Mathematical Logic. x, 531 pages. 1976.
38 GRAUERT/FRITZSCHE. Several Complex Variables. viii, 207 pages. 1976
39 ARVESON. An Invitation to C^*-Algebras. x, 106 pages. 1976.
40 KEMENY/SNELL/KNAPP. Denumerable Markov Chains. 2nd ed. xii, 484 pages. 1976.
41 APOSTOL. Modular Functions and Dirichlet Series in Number Theory. x, 198 pages. 1976.
42 SERRE. Linear Representations of Finite Groups. 176 pages. 1977.
43 GILLMAN/JERISON. Rings of Continuous Functions. xiii, 300 pages. 1976.
44 KENDIG. Elementary Algebraic Geometry. viii, 309 pages. 1977.
45 LOEVE. Probability Theory. 4th ed. Vol. 1. xvii, 425 pages. 1977.
46 LOEVE. Probability Theory. 4th ed. Vol. 2. approx. 350 pages. 1977.
47 MOISE. Geometric Topology in Dimensions 2 and 3. x, 262 pages. 1977.
48 SACHS/WU. General Relativity for Mathematicians. xii, 291 pages. 1977.
49 GRUENBERG/WEIR. Linear Geometry. 2nd ed. x, 198 pages. 1977.
50 EDWARDS. Fermat's Last Theorem. xv, 410 pages. 1977.
51 KLINGENBERG. A Course in Differential Geometry. approx. 192 pages. 1977.
52 HARTSHORNE. Algebraic Geometry. approx. 500 pages. 1977.
53 MANIN. A Course in Mathematical Logic. xiii, 286 pages. 1977.
54 GRAVER/WATKINS. Combinatorics with Emphasis on the Theory of Graphs. approx. 350 pages. 1977.
55 BROWN/PEARCY. Introduction to Operator Theory. Vol. 1: Elements of Functional Analysis. approx. 500 pages. 1977.
56 MASSEY. Algebraic Topology: An Introduction. xxi, 261 pages. 1977.
57 CROWELL/FOX. Introduction to Knot Theory. x, 182 pages. 1977.
58 KOBLITZ. p-adic Numbers, p-adic Analysis, and Zeta-Functions. x, 122 pages. 1977.

GPSR Compliance
The European Union's (EU) General Product Safety Regulation (GPSR) is a set of rules that requires consumer products to be safe and our obligations to ensure this.

If you have any concerns about our products, you can contact us on

ProductSafety@springernature.com

In case Publisher is established outside the EU, the EU authorized representative is:

Springer Nature Customer Service Center GmbH
Europaplatz 3
69115 Heidelberg, Germany

www.ingramcontent.com/pod-product-compliance
Ingram Content Group UK Ltd.
Pitfield, Milton Keynes, MK11 3LW, UK
UKHW020107200726
13856UKWH00002B/418
9781475743869